Magnetic Monopoles

NATO Advanced Science Institutes Series

A series of edited volumes comprising multifaceted studies of contemporary scientific issues by some of the best scientific minds in the world, assembled in cooperation with NATO Scientific Affairs Division.

This series is published by an international board of publishers in conjunction with NATO Scientific Affairs Division

A	**Life Sciences**	Plenum Publishing Corporation
B	**Physics**	New York and London
C	**Mathematical and Physical Sciences**	D. Reidel Publishing Company Dordrecht, Boston, and London
D	**Behavioral and Social Sciences**	Martinus Nijhoff Publishers The Hague, Boston, and London
E	**Applied Sciences**	
F	**Computer and Systems Sciences**	Springer Verlag Heidelberg, Berlin, and New York
G	**Ecological Sciences**	

Magnetic Monopoles

Edited by

Richard A. Carrigan, Jr.

Fermi National Accelerator Laboratory
Batavia, Illinois

and

W. Peter Trower

Virginia Polytechnic Institute and State University
Blacksburg, Virginia

Plenum Press
New York and London
Published in cooperation with NATO Scientific Affairs Division

Proceedings of a NATO Advanced Study Institute on
Magnetic Monopoles,
held October 14–17, 1982
in Wingspread, Wisconsin

Library of Congress Cataloging in Publication Data

NATO Advanced Study Institute on Magnetic Monopoles (1982: Wingspread, Wis.)
    Magnetic monopoles.

    (NATO advanced science institute series. Series B, Physics; v. 102)
    "Proceedings of a NATO Advanced Study Institute on Magnetic Monopoles, held October 14–17, 1982, in Wingspread, Wisconsin"—T.p. verso.
    Includes bibliographical references and index.
    1. Magnetic monopoles—Congresses. 2. Astrophysics—Congresses. I. Carrigan, R. A. II. Trower, W. Peter. III. Title. IV. Series.
QC760.4.M33N37 1983   538′.3   83—11120
ISBN 0-306-41399-X

CREDITS

The Wingspread Magnetic Monopole Workshop, organized with the invaluable assistance of Robin Craven, was supported by the North Atlantic Treaty Organization, and the U.S. Department of Energy, and the U.S. National Science Foundation. The timely support of the Johnson Foundation particularly in providing the wonderful ambiance of Wingspread was most important.

Special thanks go to the International Advisory Committee for their advice and assistance in preparing the conference program and to Sue Grommes for her good work in preparing this book.

CO-CHAIRMEN

Dick Carrigan (Fermilab, USA)

Peter Trower (Virginia Tech, USA)

ADVISORS

Giorgio Giacomelli (University of Bologna, Italy)

David Schramm (University of Chicago, USA)

Qaisar Shafi (International Centre for Theoretical Physics, Italy)

Frank Wilczek (University of California at Santa Barbara, USA)

PREFACE

In 1269 Petrus Peregrinus observed lines of force around a lodestone and noted that they were concentrated at two points which he designated as the north and south poles of the magnet. Subsequent observation has confirmed that all magnetic objects have paired regions of opposite polarity, that is, all magnets are dipoles.

It is easy to conceive of an isolated pole, which J.J. Thomson did in 1904 when he set his famous problem of the motion of an electron in the field of a magnetic charge. In 1931 P.A.M. Dirac solved this problem quantum mechanically and showed that the existence of a single magnet pole anywhere in the universe could explain the mystery of charge quantization.

By late 1981, theoretical interest in monopoles had reached the point where a meeting was organized at the International Centre for Theoretical Physics in Trieste. Many mathematical properties of monopoles were discussed at length but there was only a solitary account describing experiments. This imbalance did not so much reflect the meeting's venue as it indicated the relative theoretical and experimental effort at that point.

1982 was the year where the concrete problems of monopoles were confronted. Motivated only in part by Cabrera's startling candidate event, the field began to bloom with experimental and observational activities. The Wingspread Magnetic Monopole Workshop was organized to bring together interested scientists from particle theory and experiment and other disciplines for a status review of monopoles. This volume has been produced as a clearing house for these monopole activities. Hopefully, it also provides a guide as to how to proceed. A theoretical and experimental overview of the magnetic monopole is presented (Section I). The particle models and cosmological consequences (Section II) and the place of monopoles in the universe (Section III) are presented. The current and proposed experiments involving induction (Section IV) and ionization (Section V) are discussed, despite the yet inconclusive struggle to predict energy loss for slow monopoles.

While many physicists remain skeptical about monopoles, a growing number are speculating that they may have been made in the first blaze of creation. If found, monopoles would provide a profound clue as to the origin and nature of the universe.

Richard A. Carrigan, Jr.   W. Peter Trower
Batavia, Illinois   Blacksburg, Virginia

CONTENTS

MONOPOLES AND GAUGE THEORIES

Alfred S. Goldhaber

Institute for Theoretical Physics
State University of New York at Stony Brook
Stony Brook, Long Island, New York 11794

INTRODUCTION

Since its first expression by Pierre of Maricourt, in a letter written during the siege of Lucera, Italy in 1269,[1] the notion of a magnetic pole has been seen as simple and useful. With the passage of time, and especially in recent years, the idea of isolated poles increasingly has been perceived as beautiful and rich in implications for the fundamental structure of physics, perhaps even of the Universe.

In 1820 Oersted's[2] discovery that electric current deflects a compass needle led Ampere[3] to the picture of magnets as nothing but solenoids, a picture widely accepted during the 19th century, but most strongly confirmed by Dirac's theory of the electron,[4] which explained its magnetic dipole moment as due entirely to a "little Amperian current."[5]

Ironically and wonderfully, it was also Dirac[6] who in 1931 gave a foundation to the possibility of isolated poles, noting a necessary connection between this possibility and the observation that electric charge appears in integral multiples of a smallest unit. Since that time numerous theoretical investigations have confirmed the consistency with quantum physics of Dirac monopoles, although their existence has yet to be confirmed by observation.

In 1974 't Hooft[7] and Polyakov[8] discovered that monopoles arise as stable solutions to the classical equations of motion for spontaneously broken non-Abelian gauge field theories. Ramifications of their result are still being uncovered, but perhaps the most dramatic so far is the proposal[9] that there may be very heavy "primordial" monopoles present in the Universe. This suggestion has revolutionized experimental approaches to monopole hunting, since it implies that the poles are slow, ionizing little if at all, as well as being hard to manipulate or trap in the laboratory. Furthermore, no existing or foreseeable accelerator could produce such massive objects.

QUANTIZATION OF CHARGE

In the following pages I shall attempt to give an informal but nonetheless accurate view of fundamental physics with monopoles. The starting point is surely Dirac's quantization condition, which has been derived in many ways. While the condition is understandable theoretically as a basic consequence of quantum mechanics, it may be worth noting that known experimental results are sufficient to require quantization.

Consider a monopole passing very slowly through a superconducting loop (see Fig. 1). Since the flux through the loop may not change (either by Faraday's law or by the Meissner effect), the net change in flux due to passage of the pole must be balanced exactly by a change in current in the loop. This means that the

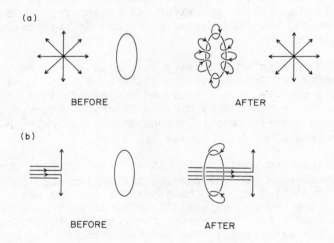

(a)

BEFORE   AFTER

(b)

BEFORE   AFTER

Fig. 1.   (a) Monopole passes through superconducting loop leaving an integral number of flux quanta threading the loop. This implies quantization of the pole strength; (b) Solenoid or magnetic needle penetrates the loop. Net flux through does not change; hence, there is no quantization condition.

change in current-generated flux between pre-passage and post-passage must be equal and opposite to the total flux from the pole. It is found by experiment[10,11] that magnetic flux through a superconducting loop is quantized in units of hc/Q, where Q is the minimum charge of a current carrier in the superconductor.

This gives the equation

$$4\pi g = Nhc/Q \tag{1}$$

where N is an integer. This would be exactly Dirac's condition except that Q for known superconductors is twice $\underline{e}$, the charge of the electron. However, in principle we may envision a superconductor with carriers of charge $\underline{e}$, supposing which we arrive at

$$eg/\hbar c = N/2, \tag{2}$$

exactly Dirac's result.[6]

It is interesting to ask what happens if we try to "fool" the loop by putting through one end of a long magnet, so that afterwards the loop girdles the magnet at a central point far from either end. Surely the flux from one pole of the magnet need not be quantized! Indeed, everything is consistent, because the change in loop current flux is equal and opposite to the flux within the magnet, so that the total (locally measurable) flux through the loop has not changed at all.

DYONS AND FRACTIONAL CHARGE

The quantization condition must also apply to cases involving particles with both electric and magnetic charge, called dyons.[12] For a pair of such particles the condition takes the form[13]

$$s = q_1 g_2 - g_1 q_2 = N\hbar/2. \tag{3}$$

Imagine a world with ordinary charges $q = e$, and also dyons. Then (3) will hold for all pairs if and only if the magnetic charge of each dyon obeys (2), and the electric charge of each dyon is equal modulo $\underline{e}$ to a constant times its magnetic charge. It is simple algebra to check that in this case the contributions to s of the irrational electric charges cancel exactly.

It is worth studying the dyon electric charge more carefully, specializing to the case of minimum strength Dirac poles. Then all north poles may bear an electric charge of $q = e\theta/\pi$ mod e, provided all south poles bear a charge $q = -e\theta/\pi$ mod e. The reason for the fancy notation is that Witten[14] found such a dyon charge in a

particular model where θ is an already discussed parameter
characterizing the vacuum state of the theory, with all physical
phenomena depending only on θ mod 2π.

Certainly that is true of the dyon charge q. If θ takes an
arbitrary value the dyons will violate CP symmetry, since electric
charge is odd under CP while pole strength is even. Witten's
result now seems to be a theorem, derivable from general
arguments[15] and also verified in more complex and realistic
models.[16]

One very interesting case is that for θ = π/2, in which CP may
be conserved so long as a north dyon possesses two degenerate
forms, one with q = +e/2, the other with q = -e/2. Such a
possibility of fractional charge was suggested by Jackiw and
Rebbi,[17] who found a one-dimensional model with degenerate states
bearing fermionic charge or fermion "number" ±1/2. Since then the
phenomenon in one dimension has been proposed as applicable to
charge and electron density localization around topological defects
in certain polymers.[18] However, in this case there are two possible
spin states of the electron, and so there are four levels, two with
spin zero and charge ±e, and two with charge zero and spin ±1/2
(see Fig. 2). I believe that this is an example of a general
phenomenon, that in real physical systems possessing charge
conjugation symmetry, the half-integral fermion number of Jackiw
and Rebbi is literally concealed by a duplicity of Nature, leaving
as its trace an anomalous correlation of fermion number with other
quantum numbers. A detailed study on this point is in preparation.

THE GOLDEN TRIANGLE

Charge Quantization and Gauge Theories

When Dirac introduced the notion of monopoles, he observed
that one of the most attractive aspects is the consistency
requirement that electric charge comes in multiples of a smallest
unit, if even one monopole exists. This would explain an
extraordinary fact of Nature, a fact which otherwise seems quite
enigmatic.

There is another way of explaining charge quantization which
seems very different. If and only if charge is quantized, when the
wave function of a charged particle is multiplied by the phase
factor $e^{i\zeta Q}$ (where Q is the charge in units of the smallest
possible charge), the factor will be equal to unity whenever ζ is
an integral multiple of 2π. We say that charge is the generator of
a compact group, since the phase factors may all be parameterized
by a finite range of ζ values, e.g., between 0 and 2π.[19] If
different particles had incommensurate charges, then all real

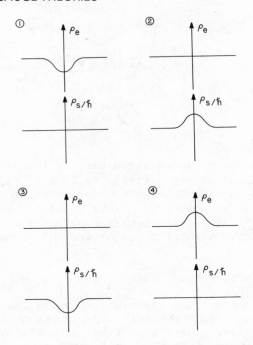

Fig. 2.   In the Su et al. model of polyacetylene, a soliton or
          defect possess four nearly degenerate states.[18] While each
          has integral electron number, the connection between that
          number and the total spin localized about the soliton is
          unusual. Odd electron number goes with spin zero, even,
          with spin one-half.

values of ζ would give distinct arrays of phase factors, and charge
would generate a noncompact group. To explain charge quantization
then, we have to explain why charge generates a compact group.

     An analogy is helpful here. Consider a particle moving in
uniform magnetic and electric fields which point parallel to the z
axis (see Fig. 3). What are the possible values of angular
momentum about the axis? Of course, we know the answer is integral
or half-integral multiples of $\hbar$, but why? The reason is that J_z is
one of three generators of the rotation group, and these generators
have quantized eigenvalues, as may be derived from their
commutation relations. Even though the electric field breaks full
rotational symmetry down to axial symmetry, J_z remains quantized.
More formally, J_z generates a subgroup of a compact group. Such a
subgroup is automatically compact itself. Thus, a sufficient
condition for electric charge to be quantized is that it be one of
the generators of a compact group, even if some or all of the other

Fig. 3. A charged particle moving under the influence of parallel
         electric and magnetic fields follows a trajectory in which
         the component of angular momentum along the field
         direction is conserved. This component is quantized even
         though the two perpendicular components and the magnitude
         of the total angular momentum are not conserved.

generators are not conserved, like J_x and J_y in our example.

     Non-Abelian gauge field theories, of which the prototype was
introduced by Yang and Mills,[20] are characterized by compact
symmetry groups, and therefore lead naturally to quantized
"charges." In these theories, all observable quantities are
unchanged by any transformation in the symmetry group, even when
the   transformation is an arbitrary function of position in
space-time. Consequently, there are too many conserved, quantized
charges.   If the symmetry is broken as in our analogy, then the
leftover conserved charges will still be quantized.   The Higgs
mechanism (in which the symmetry is "spontaneously" broken because
a scalar field which is not invariant under the group has a nonzero
magnitude)   works   in   precisely   this   way.[21]   Therefore, a
spontaneously broken non-Abelian gauge theory implies quantization
of charge.[22]

Monopole and Gauge Theories

     As mentioned earlier, such theories also provide a natural
mechanism for the appearance of magnetic monopoles. What about the
converse? Do monopoles imply gauge theories? While the link in
this direction may not be a logical deduction, it seems at least as
compelling as the motivation which charge quantization gives for
considering either monopoles or gauge theories.

     Imagine a scattering at arbitrarily high energy of an electron
on a monopole.   In the lowest angular momentum state, the
projection of the electron spin in the radial direction is constant
($\pm 1/2$ depending on the sign of the product eg). As a result, if no
other degrees of freedom are allowed, the scattering must lead

either to bounce-back of the electron or spin-flip, or both. This implies infinitely strong forces, which would require infinite mass for the monopole. Of course, we know that there is at least one other degree of freedom, namely, radiation of photons. It is conceivable that such radiation could inevitably carry the charge-pole system out of the minimum angular momentum state, allowing preservation of the helicity as well as the direction of relative motion. However, if this were not the case, then the extra degrees of freedom would have to be ascribed to the monopole.

There is a stronger reason to require an intrinsic structure for a monopole. If the pole were localizable to within a Compton wavelength λ_c, then the associated energy in the static magnetic field outside a radius equal to λ_c would be greater than the mass by a factor $g^2/2$.[23] Thus, the pole must be spread out over a distance r larger than λ_c by at least this factor, implying that an electron wave packet with energy comparable to the monopole mass could plunge into the interior of the pole. Again, to preserve helicity, it must exchange some other quantum number with the pole. Even if helicity is ignored, the conservation of angular momentum along the velocity direction requires that the projectile exchanges either or both charge and angular momentum with the pole.

For such an extended object, the minimum energy required for an angular momentum excitation is of order $(mr^2)^{-1} \approx 1/rg^2 \approx \alpha/r \approx \alpha^2 m$, while the minimum energy for dyon excitation (i.e., adding a quantum of electric charge) is of order $e^2/r \approx \alpha^2 m$ also. The fact that there are levels with different charge or angular momentum, nearly degenerate on the scale of the monopole mass, suggests that high-energy scattering would be described best by a scheme in which the first approximation is to specify the orientation of the monopole in angle variables conjugate to angular momentum and to electric charge.

As in the Blair model,[24] where the scattering amplitude for a nucleon striking a deformed nucleus is computed as a function of deformation axis direction, so the interaction of a charge with the pole might be described by a classical field with no singularities. There may be more than one type of field variable, but at least there must be a vector potential if the long-range charge-pole interaction is to be described correctly.

The only way such a potential can be nonsingular is to be a non-Abelian gauge field, since in ordinary Maxwell theory, the condition $\nabla \cdot \nabla \times \vec{A} = 0$ precludes a spread-out magnetic pole density. The desire to find a nonsingular interaction between pole and charge led me to what is now recognized as the simplest kind of nonsingular vector potential, arising in SU(2) gauge theory. I showed that this interaction is equivalent under a gauge transformation to the usual singular Dirac vector potential.[25] Wu and Yang[26] found the same nonsingular potential independently, but did not identify it as a magnetic monopole because they were considering unbroken SU(2). Since we observe only one kind of long-range field, the full gauge symmetry must be broken so that no other kinds of massless "photon" are allowed to propagate freely. Thus, the existence of monopoles would invite consideration of spontaneously broken non-Abelian gauge theories, at least for economical description of interactions between a pole and massive, pointlike electric charges. This reasoning does not indicate that the fields in question necessarily are fundamental, but surely they are useful. By taking such fields as given, 't Hooft[7] and Polyakov[8] discovered monopole solutions and launched a new wave of interest in these objects.

Because the issue of the finite magnetic energy of the monopole is so critical, it may be worthwhile to approach the energy requirement from a slightly different point of view. In conventional Abelian theory, the energy is $E_m = \int d^3r \vec{B}^2/8\pi$. To keep this finite for a field whose long-range character is that expected for a magnetic monopole, we must alter the definition of $\vec{B}$ to remove the singularity at small distances, or we must alter the definition of energy density, or both.

The latter type of alteration must be carried out in a way which preserves the structure of electromagnetic interactions of charged particles. One way to achieve this is to assume that the geometry of space is altered in the vicinity of the monopole, so that it becomes a black hole carrying a nonzero magnetic charge.

This might seem like an escape from introducing non-Abelian gauge fields, but there is a price to pay. If a black hole is formed, magnetic flux lines may be pulled in, but, since in Abelian theory these flux lines have no beginning or end, save on a singularity, no net magnetic flux may be pulled in. Therefore, formation of such "monoholes" would have to be by intrinsically quantum mechanical transitions, always producing the objects in pairs.

What would be the minimum energy of a monohole? In units of the Planck mass, the contribution of the gravitational field to the energy is approximately $r_S/2$, where r_S is the Schwarzschild radius, while the magnetic Coulomb energy is about $g^2/2r_S$. The total energy is minimized for $r_S = gr_P$, $M = gm_P$, where P stands for Planck. Thus the object is geometrically big enough so that the classical field configuration should be a good first approximation. David Gross and Malcolm Perry at Princeton have found a monopole of this size and mass as a solution of Kaluza-Klein theory, which is Einstein gravity and Maxwell electrodynamics supplemented by specific couplings to an extra scalar field. While not a black hole, it shares the feature that the magnetic field has a singularity even though the energy remains finite. Consequently, the singularity may be thought of as a coordinate effect which reflects a nontrivial topology of the field configuration. Such "geometrical" monopoles deserve further study, but if they are left aside then only the non-Abelian type are consistent with the finite-energy requirement.

We have found that among the trio of concepts, charge quantization, spontaneously broken non-Abelian gauge theories, and magnetic monopoles, each suggests the other two. While some of the links appear more heuristic than others, all of the connections are so close that one is tempted to view the trio as three aspects of a single phenomenon (see Fig. 4).

Charge quantization has been verified in stringent experimental tests. Broken gauge theories have led to great insight about the weak interactions and their relation to electromagnetism. These results lend encouragement to pursuit of more grandiose gauge theories, which almost surely must accommodate magnetic poles. Unless some additional fundamental effect bars their appearance in isolation, it becomes a detailed question about the evolution of our Universe whether monopoles are present today. That they form part of the fabric of fundamental microphysics is quite likely, whatever the answer to the observational question.

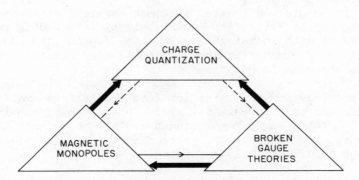

Fig. 4.   The Golden Triangle. Each directed line indicates the strength of logical connection between two vertices. Charge quantization follows either from monopoles or spontaneously broken non-Abelian gauge theories, and therefore suggests that either might be true. The existence of monopoles as stable classical field configurations has been demonstrated for broken gauge theories. Energy considerations make such theories natural if not inevitable correlates of monopoles.

OUTLOOK

In studying these proceedings, as well as any later results on monopoles, it is helpful to keep in mind an array of questions.

(1) Would the existence of magnetic poles be consistent with our understanding of microphysics? As indicated above, the answer at the moment is surely positive, but, regardless of precise model assumptions, the masses of such objects would almost surely be far greater than those of previously observed particles, since the pole radius could be no larger than the Compton wavelength of the heaviest vector meson whose presence is required for its composition, and that meson would itself be far heavier than the $W^{\pm}$ or Z bosons. Thus, a conservative lower bound on the mass would be about 10^5 GeV, and the oftmentioned estimate of 10^{16} GeV is quite credible. Over the years, there have been many questions raised about the consistency of theories including monopoles, but up to now, no question has been answered in the negative, while many apparent paradoxes have been resolved favorably.

Let me mention only two. Quantum theory requires that magnetic fields be described by vector potentials, but such fields have vanishing divergence, so how could there be monopoles? Dirac[6] observed that the charge-pole interaction could be described by a singular vector potential formally equivalent to that of an infinitely long thin solenoid containing a quantum of magnetic flux. Since the solenoid, or string, would not influence the motion, it could be treated as a mathematical artifact. Indeed, the string may be "exorcised" completely by introducing two vector potentials, each nonsingular in its domain of use.[27] Thus, nonrelativistic quantum theory with poles is consistent provided that pole strength is quantized.

Another paradox is that a spinless charge (presumably a boson) and a spinless pole (presumably also a boson) could combine to form a dyon with total angular momentum $\hbar/2$. This object might seem to be a spin 1/2 boson, in contradiction to the spin-statistics theorem, but in fact it is a fermion![28]

One point of principle which has been addressed here by Preskill,[29] and also recently by Lazarides, Shafi and Trower,[30] is the consistency of minimum-strength Dirac monopoles and free fractional electric charges, superficially violating the Dirac quantization condition. Preskill remarks that if a fractional charge and a pole have an additional long-range interaction which ordinary particles do not feel, then their total interaction may obey the Dirac condition. He assumes that this total interaction is spherically symmetric. However, if there is no extra long-range field to carry this interaction, then Dirac's string must become observable, hence carry a finite energy per unit length, so that an

isolated pole would have infinite energy. This is observed by
Lazarides, et al., to occur in a particular model of broken color
symmetry which accommodates free fractional charges, the "glow"
model of Slansky, Goldman and Shaw.[31] Chapline and I[32] have thought
about what this picture would mean phenomenologically. In view of
the enormous ratio of monopole mass to energy in the string, poles
and antipoles could be separated by macroscopic distances, and
interact with ordinary matter like free Dirac monopoles. It should
be noted that conservation of magnetic charge would keep these
poles from disappearing unless they could annihilate with their
antipoles.

   (2) Would the existence of poles be consistent with our
picture of the evolution of the Universe? This topic has been
discussed recently by Lazarides, et al.,[30] who point out that in
the "new inflationary Universe" described in these proceedings by
Guth,[33] the "reheating" which gives rise to the baryons we see
today would produce negligible monopole density because of a
Boltzmann suppression. However, Guth, Pi and I[34] are investigating
the possibility that monopoles could be produced during the
expansion which precedes the reheating, in numbers which are
neither overwhelmingly high nor negligible. In any case, one
should not be surprised if models which have their error bars in
the exponent could make room for a sprinkling of monopoles.

   (3) Would poles be consistent with our knowledge of the Galaxy
and nearby astronomical objects? These matters have been discussed
here by several speakers. Let me simply state a prejudice that if
poles are found and their properties analyzed experimentally,
astrophysics will fall into harmony with the observations.
Nevertheless, asking just how this could come about is likely to be
a productive enterprise.

   (4) How should matter interact with slow, heavy monopoles?
This question is most relevant for detection methods and has been a
principal topic here. Probably the newest thought was described by
Parke:[35] Passage of a pole through a light atom would induce level
crossings, permitting substantial energy deposition. The initial
ground state rises and crosses descending levels rather than mixing
and being pushed back down, because conservation of angular
momentum prevents the mixing.

   In any case, we may expect the next few years to bring a new
generation of experiments using slow charged particles, along with
extensive theoretical modeling to fit the data obtained. The
result should be substantial improvement in the precision of
existing descriptions of energy loss mechanisms and hence of
desirable experimental approaches to pole hunting.

   (5) Should monopoles act as catalysts for baryon decay? If

there is a single concept which has inspired both excitement and bewilderment, this is surely it. In chronological order of publication, the concept was proposed by Rubakov,[36] Wilczek,[15] and Callan.[37] Callan and Rubakov have each pursued the subject in detail and Callan has discussed it here.[38] Therefore, only some general remarks are in order. Catalysis is really a two-step process. First the monopole must capture a nucleon. The capture cross section may be estimated using knowledge of electromagnetic interactions as well as strong interactions on distance scales corresponding to the size of a nucleon. While there has yet to be published a careful analysis of capture, I believe the estimates can be made reliably and that for a wide range of situations, the cross section is as large or larger than typical nucleon-nucleon total cross sections (see Fig. 5).

The real issue becomes whether a nucleon after capture may be transmuted into leptons. The point I want to make is that such an event could occur at a very low rate and still give a signal in the laboratory almost indistinguishable from catalysis in the strong interaction time of 10^{-23} sec. The reason is that capture in normal matter is relatively slow because of the long mean free path, so that the catalysis rate is limited by the capture rate. Thus, transmutation in times of a microsecond or less could give multiple events in a large detection chamber. This means that the Rubakov-Callan effect could be suppressed as much as seventeen orders of magnitude and still be easy to observe.

The second point I want to emphasize is that, once the assumption is made that coupling of light fermions in the lowest (total J = 0) angular momentum channel to other channels may be neglected, the conclusion of Callan and Rubakov is almost unavoidable. It follows directly from conservation laws built into the model that conservation of baryon number must be violated. The only remaining loophole is associated with conservation of the charge coupled to the Z boson. Since this particle is supposed to have a mass of about 100 GeV, the associated charge need not be conserved for processes on the scale of 100 MeV. I suspect that neglected couplings to other channels will reduce the decay rate, but not to an unobservable level. It also seems possible from Callan's formulation that transmutation, in contrast to free baryon decay, would be characterized by isotropic emission of low energy electron-positron pairs, a unique signal.

A final point: The internal structure of the monopole is not directly exploited in the Callan and Rubakov analyses, so that a Dirac monopole with the same ordinary and color magnetic pole strengths should have precisely the same effect. This is plausible, in view of the enormously small length scale of the internal structure, but it is also mysterious, since the transmutation in no sense can be ascribed to virtual X or Y bosons.

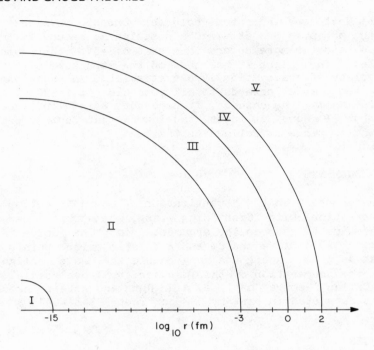

Fig. 5.   A fanciful picture of one quadrant of an SU(5) monopole in
          cross section.   (I).   SU(5) is a good symmetry in this
          region, where all the gauge and Higgs fields contributing
          to pole structure become small; (II). The enormous
          magnetic field of the pole breaks color and weak isospin
          symmetry, so that SU(3)$_{color}$ is reduced to SU(2)$_{color}$ ×
          U(1), where the U(1) is generated by $\Lambda_8$$_{color}$, while (SU(2)
          × U(1))$_{el}$ is reduced to U(1)$_{el}$ × U(1)$_Z$, where the last
          U(1) measures the charge coupling to the Z° boson;
          (III). The mass of the Z° is no longer negligible on this
          length scale, so that the only electroweak conservation
          law remaining accurate is that for ordinary electric
          charge; (IV). Color confinement allows only ordinary
          magnetic field to emerge this far. The vacuum is easily
          polarized by excitation of spherically symmetric
          electron-positron oscillations; (V). In this regime,
          nonrelativistic physics should be an accurate guide to the
          interactions of baryons or leptons with the pole. Even
          here, there remain important unanswered questions.

Rather, in the presence of the pole, low mass scalar fermion pair
excitations with the quantum numbers of a pair of X mesons (or of a
hydrogen atom!) can be formed or destroyed. This may be the
aspect most in need of further illumination.

(6) What does the future hold for monopole physics? I believe that the monopole has attained a position as secure as that of the positron, and far more secure than that of any other undiscovered particle. The light it has shed on the structure of physics, and the subtlety of observation it has stimulated in those who seek it, have given ample rewards to all monopole aficionados. Like that wonderful animal the unicorn, it need not exist in order to be true, but I hope it does, and I know that further prizes await those who continue to follow its trail.

ACKNOWLEDGEMENTS

The approach in the "Quantization of Charge" section arose in a conversation with Chen Ning Yang, whose criticisms also have stimulated me to sharpen the arguments in "The Golden Triangle" section. David Olive and Jogesh Pati also raised points important for that section. Among the many people who have enlightened me during the preparation of this manuscript, I owe special thanks to Curtis Callan, George Chapline, Alan Guth and Robert Shrock. This work was supported in part by NSF contract PHY-81-09110.

REFERENCES

1. Petrus Peregrinus, in "A Source Book in Medieval Science,"
    E. Grant, ed., Harvard University Press, Cambridge (1974)
    p. 368.
2. H. C. Oersted, in "A Source Book in Physics," W. F. Magie,
    ed., Harvard University Press, Cambridge (1965) p. 437.
3. A. M. Ampère, in Magie, op. cit., p. 446.
4. P. A. M. Dirac, Proc. Roy. Soc. A $\underline{117}$, 610; A $\underline{118}$, 351 (1928).
5. F. Bloch, Phys. Rev. $\underline{51}$, 994 (1937).
6. P. A. M. Dirac, Proc. Roy. Soc. A $\underline{133}$, 60 (1931).
7. G. 't Hooft, Nucl. Phys. B $\underline{79}$, 276 (1974).
8. A. M. Polyakov, JETP Lett. $\underline{20}$, 194 (1974).
9. T. W. B. Kibble, J. Phys. A $\underline{9}$, 1387 (1976); Ya. B. Zeldovich
    and M. Yu Khlopov, Phys. Lett. $\underline{79B}$, 239 (1978), J. P.
    Preskill, Phys. Rev. Lett. $\underline{43}$, 1365 (1979); M. B. Einhorn,
    D. L. Stein and D. Toussaint, Phys. Rev. D $\underline{21}$, 3295 (1980).
10. B. S. Deaver and W. H. Fairbank, Phys. Rev. Lett. $\underline{7}$, 43
    (1961).
11. R. Doll and M. Näbauer, ibid. $\underline{7}$, 51 (1961).
12. J. Schwinger, Science $\underline{165}$, 757 (1969).
13. D. Zwanziger, Phys. Rev. $\underline{176}$, 1489 (1968).
14. E. Witten, Phys. Lett. $\underline{86B}$, 283 (1979).
15. F. Wilczek, Phys. Rev. Lett. $\underline{48}$, 1146 (1982).
16. H. Yamagishi, Phys. Rev. D., in press; B. Grossman, Phys. Rev.
    Lett., in press.

17. R. Jackiw and C. Rebbi, Phys. Rev. D 13, 3398 (1976).
18. W. Su, J. Schrieffer and A. Heeger, Phys. Rev. Lett. 42, 1698
      (1979), Phys Rev. B 22, 2099 (1980).
19. C. N. Yang, Phys. Rev. D 1, 2360 (1970).
20. C. N. Yang and R. L. Mills, Phys. Rev. 96, 191 (1954).
21. P. W. Higgs, Phys. Lett. 12, 132 (1964); F. Englert and R.
      Brout, Phys. Rev. Lett. 13, 321 (1969); G. S. Guralnik, C.
      R. Hagen and T. W. B. Kibble, ibid., p. 585.
22. P. Goddard and D. I. Olive, Rep. Prog. Phys. 41, 1357 (1978);
      Nucl. Phys. B191, 511 (1981).
23. M. A. Ruderman and D. Zwanziger, Phys. Rev. Lett. 22, 146
      (1969).
24. J. S. Blair, Phys. Rev. 115, 928 (1959).
25. A. S. Goldhaber, Phys. Rev. 140, B1407 (1965).
26. T. T. Wu and C. N. Yang in "Properties of Matter Under Unusual
      Conditions," H. Mark and S. Fernbach, eds., Interscience,
      New York (1969) p. 349.
27. T. T. Wu and C. N. Yang, Phys. Rev. D 12, 3845 (1975).
28. A. S. Goldhaber, Phys. Rev. Lett. 36, 1122 (1976).
29. J. P. Preskill, these proceedings; S. M. Barr, D. B. Reiss and
      A. Zee, Phys. Rev. Lett. 50, 317 (1983).
30. G. Lazarides, Q. Shafi, and P. Trower, Phys. Rev. Lett. 49 ,
      1756 (1982).
31. R. Slansky, T. Goldman, and G. L. Shaw, Phys. Rev. Lett. 47,
      887 (1981).
32. G. Chapline and A. S. Goldhaber, in preparation.
33. A. H. Guth, these proceedings.
34. A. S. Goldhaber, A. H. Guth and S.-Y. Pi, in preparation.
35. S. J. Parke, these proceedings; S. D. Drell, N. M. Kroll,
      M. T. Mueller, S. J. Parke, and M. A. Ruderman, Phys. Rev.
      Lett. 50, 644 (1983).
36. V. A. Rubakov, JETP Lett. 33, 644 (1981); Nucl. Phys. B203,
      311 (1982), in press.
37. C. G. Callan, Jr., Phys. Rev. D 25, 2141 (1982); D 26, 2058
      (1982), in press.
38. C. G. Callan, Jr., these proceedings.

TESTING GUTS: WHERE DO MONOPOLES FIT?*

John Ellis§

Stanford Linear Accelerator Center
Stanford University
Stanford, California 94305

APPETIZER

It is impossible to resist the temptation to develop an anatomical analogy for the introduction to this talk. You have had the bare bones of monopole theory exposed to you.[1] My brief is now to enter the belly of the beast and discuss whether grand unified monopoles[2] (GUMs) should be regarded as a minor appendix to GUTs. We will indeed find that GUMs could provide crucial tests of GUTs, particularly through their possible propensity[3,4,5] to eat matter as they pass by it.

The skeletal outline of this talk is as follows: the second section describes why the inadequacies of the "standard model" of elementary particles impel some theorists[6] toward embedding the strong, weak and electromagnetic interactions in a simple GUT group, and explains why[7] the grand unification scale and hence the GUM mass are expected to be so large ($\geq 10^{14}$ GeV). The third section goes on to describe some model GUTs, notably minimal SU(5)[6] and supersymmetric (susy) GUTs.[8] We introduce the grand unified analogues of generalized Cabibbo mixing angles,[9] relevant to the prediction of baryon decay modes in different theories as well as to the "Decay" modes catalyzed by GUMs.[10] Phenomenologies of conventional and susy GUTs are contrasted including the potential increase in the grand unification scale[11] as well as possible different baryon decay modes in susy GUTs.[12] It is emphasized that

*Work supported by the Department of Energy, contract DE-AC03-76SF00515.
§On leave of absence from CERN, CH-1211 Geneva 23, Switzerland.

although the central hypothesis of GUTs--namely the existence of a primordial simple group broken down to include an electromagnetic U(1) factor at low energies--necessarily requires the existence of GUMs, nevertheless their masses are uncertain within the range $O(10^{16}$ to $10^{19})$ GeV. The fourth section discusses the phenomenology of GUMs, principally their ability[3,4,5] to catalyze baryon "decays." It is shown that while at distances outside the core of radius $O(1/m_X) < O(10^{-28})$ cm a gauge theory 't Hooft-Polyakov[2] GUM closely resembles a Dirac monopole,[13] nevertheless GUTs specify boundary conditions[14] at the core which cause monopole-fermion scattering to violate fermion number in general and baryon number in particular. The resulting large GUM-baryon $\Delta B \neq 0$ cross sections are then estimated,[10] and some possible experimental signatures[15,10] are mentioned (hierarchy of catalyzed "decay" modes, a possible "chain" of "decays" along the GUM's path, an apparent excess of Fermi motion due to recoil momentum $O(300)$ MeV). The fifth section briefly introduces some of the astrophysical[16] and cosmological[17] constraints on GUMs, which make it difficult to imagine ever seeing a GUM and may impose serious restrictions on GUT model-building via their behavior in the very early universe. We can get useful information about GUTs already from the abundance of GUMs as well as from their $\Delta B \neq 0$ interactions if they are ever seen. Finally, the sixth section summarizes the reasons why GUMs are crucial aspects and tests of GUTs.

WHY GUTS?

The "standard model" of elementary particle physics is very unsatisfactory, possessing as it does a "random" gauge group SU(3)×SU(2)×U(1) with three independent factors having three independent gauge couplings g_3, g_2 and g_1. Furthermore the known left-handed fermions sit in rather "random" looking representations of this group: each generation such as $(u, d, e^-, \nu_e)_L$ transforms as

$$(3,2) + (\bar{3},1) + (\bar{3},1) + (1,2) + (1,1)$$
$$(u,d)_L + \bar{u}_L + \bar{d}_L + (\nu_e,e^-)_L + e_L^+$$

$$(1)$$

Furthermore the U(1) hypercharge Y assignments are rather puzzling: they are all rational numbers so that the electromagnetic charges $Q_{em} = I_3 + Y$ are integer or fractional. Why are none of the hypercharges irrational or transcendental? This is another way of restating the old puzzle of the quantization of electromagnetic charge--why is $|Q_e/Q_p| = 1 + O(10^{-21})$? Even if one accepts as God-given all these fermion representation assignments, the "standard model" still has at least twenty arbitrary parameters, starting of course with the three gauge couplings g_3, g_2 and g_1 mentioned earlier.

A natural philosophy is to search for a simpler non-Abelian gauge group with a single gauge coupling. This was first tried for the weak and electromagnetic interactions alone, leaving the strong interactions to one side and postulating a gauge theory based on $SU(3)_{color} \times G_{weak}$ with Q_{em} a generator of G_{weak}. This is unaesthetic because it still requires two gauge couplings, and furthermore it is difficult to arrange because when Q_{em} is a generator of a non-Abelian group one must have

$$\sum_{representation} Q_{em} = 0 \ . \tag{2}$$

Since quarks and leptons have different colors, the cancellation (2) must be arranged for them separately

$$\sum_{leptons} Q_{em} = 0 = \sum_{quarks} Q_{em} \ . \tag{3}$$

This is not possible with the known generations of quarks and leptons, which each contribute -1 to the left-hand sum and $+1$ to the right-hand sum. One possibility might be to add in additional particles to enforce the cancellation. Another possibility is to note that if one adds together the known quarks and leptons, then the condition (2) is satisfied.

Combining quarks and leptons in this way entails postulating a simple group containing both the strong and the electroweak interactions

$$G \supset SU(3)_{color} \times SU(2) \times U(1)_Y \tag{4}$$

which therefore implies a single gauge coupling g from which the observed g_3, g_2 and g_1 derive. This is the GUT philosophy[6] we shall follow. Of course one may anticipate that there will be constraints on the fermionic $SU(3) \times SU(2) \times U(1)$ representations because they must all be contained in representations of the underlying group G. In particular, the U(1) hypercharges will be constrained and charge quantization will be automatic. Now Q_{em} is a generator of the group, implying

$$\sum_{q+\ell} Q_{em} = 0 \tag{5}$$

and all the fermion charges are related by simple Clebsch-Gordan coefficients.

The main obstacle to the GUT philosophy[6] outlined above is the fact that at present energies, the different gauge couplings are grossly disparate:

$$g_3 \gg g_2, g_1 . \tag{6}$$

Fortunately, this difficulty is resolved[7] by the realization that couplings vary logarithmically as a function of energy (momentum) scale. In particular, if no new physics intervenes, the SU(3) and SU(2) couplings approach each other (see Fig. 1) as

$$\frac{1}{\alpha_3(Q^2)} - \frac{1}{\alpha_2(Q^2)} = -\frac{(11+(N_H/2))}{12\pi} \ln\left(\frac{m_X^2}{Q^2}\right) \tag{7}$$

where N_H is the number of light (mass $\lesssim O(100)$ GeV) Higgs boson doublets which is 1 in the minimal Weinberg-Salam model, and m_X is the energy scale Q at which $g_3 = g_2$ and grand unification becomes possible: $\alpha_{GUT} = \alpha_3 = \alpha_2$ ($\alpha_i \equiv g_i^2/4\pi$). Because of the logarithmic rate of variation (7) the grand unification scale m_X will be exponentially high:

$$\frac{m_X}{\Lambda_{QCD}} = \exp\left\{\frac{O(1)}{\alpha_{em}} + O\left(\ln \alpha_{em}\right) + O(1) + \ldots\right\} \tag{8}$$

where Λ_{QCD} is the strong interaction scale of order 100 MeV to 1 GeV. As we will see in a moment, the grand unification scale must be $\geq 10^{14}$ GeV if baryons are to have lifetimes longer than

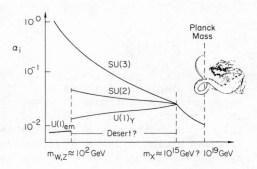

Fig. 1.   An illustration of the GUT philosophy. The SU(3), SU(2) and U(1) couplings come together at an energy scale $m_X = O(10^{15})$ GeV if the desert is unpopulated. This grand unification scale appears to be significantly less than the Planck mass of order 10^{19} GeV at which quantum gravity effects are O(1).

about 10^{30} years as required by experiment. Moreover, m_X must be less than $O(10^{19}$ GeV) if we are to be able to get away without including gravitation in our GUTs. [This is because quantum gravity effects become $O(1)$ at an energy $Q = O(m_p) = O(G_{Newton}^{-1/2})$.] The relation (8) then tells us[18] that the low energy α_{em} must lie in a relatively narrow range

$$\frac{1}{120} > \alpha_{em} > \frac{1}{170} \tag{9}$$

if the GUT philosophy is to make any sense. The observed value of $\alpha_{em} = 1/137$ actually corresponds to $m_X \approx (10^{14}$ to $10^{15})$ GeV, encouraging us to hope that baryons may decay relatively soon, as we will see more quantitatively in a moment. It should however be emphasized that this analysis rests on the absurd and ludicrous assumption that no new physics intervenes between here and 10^{15} GeV (the "Desert Hypothesis"). If this is not valid the grand unification scale may be moved around, and we will see an example soon in the shape of susy GUTs. However, this possible variation in m_X does not vitiate the GUT philosophy of unification in a simple group at very high energy, which carries with it the necessary existence of GUMs with masses[1]

$$m_M = O\left(\frac{m_X}{\alpha_{GUT}}\right) \geq O\ (10^{16})\text{GeV} . \tag{10}$$

There is one empirical test[7,19,20] of the GUT philosophy which is a calculation of the effective weak neutral mixing angle θ_W at an energy scale Q:

$$\sin^2\theta_W(Q^2) = \frac{3}{8}\ \left\{1 - \left[\frac{\alpha_{em}}{4\pi}\ \frac{110}{9}\ \ell n\ \frac{m_X^2}{Q^2}\right]\right\}\ +\ \cdots . \tag{11}$$

In this formula the prefactor 3/8 is the symmetry value obtained from SU(5) Clebsch-Gordan coefficients, while the square bracket is a renormalization factor arising when the GUT symmetry is broken. Including higher order corrections[20] (the $\cdots$ in Eq. (11)) we get

$$\sin^2\theta_W^{eff} = 0.215 \pm 0.002 \tag{12}$$

for the effective value of $\sin^2\theta_W$ measured in experiments at present energies, if Λ_{QCD} = 100 to 200 MeV. The prediction (11) is relatively insensitive to the actual GUT, as long as it obeys the Desert Hypothesis. For comparison, the present experimental value is

$$\sin^2\theta_W^{eff} = 0.216 \pm 0.012 \tag{13}$$

when radiative corrections are included,[21] in encouraging agreement with the prediction (12). The symmetry aspect of the prediction (11) results from the fact that hypercharge and $U(1)_{em}$ are embedded in a GUT. The renormalization correction results from setting $g_3 = g_2 = g_1$ at the same energy scale $Q = m_X$ as illustrated in Fig. 1. Hence the success of the prediction (12) "checks" both the GUT philosophy: $U(1)_{em}$, $Y \subset G$ and the large scale m_X at which it applies.

We now have several reasons for expecting GUMs. We know that quantization of magnetic charge h is to be expected in a theory with magnetic monopoles:

$$\frac{eh}{4\pi} = \frac{n}{2} ; \quad n = 1, 2, \ldots \tag{14}$$

Conversely, one might have expected that the observed quantization of electric charge would be more easily understood in a theory with magnetic monopoles. Indeed, charge quantization emerges automatically when $U(1)_{em}$ is embedded in a simple group. All such theories harbor monopoles, and GUTs are examples of such theories. Therefore we expect to have GUMs.

WHAT GUTS?

In order to specify the properties of GUMs more precisely, we now go on to look at definite GUTs, starting off with the minimal version[6] of the minimal GUT group SU(5). This is broken down to the exact low energy $SU(3)_{color} \times U(1)_{em}$ symmetry as follows:

$$SU(5) \longrightarrow SU(3)_{color} \times SU(2) \times U(1)_Y \longrightarrow SU(3)_{color} \times U(1)_{em}$$

$$10^{15} \text{ GeV} \qquad\qquad 10^2 \text{ GeV}$$

adjoint 24 of Higgs ϕ \qquad adjoint 5 of Higgs H \tag{15}

$$m_{X,Y} \qquad\qquad m_{W,Z}$$

Each fermion generation is assigned (somewhat inelegantly) to a reducible $\bar{5} + 10$ representation of SU(5). For the first generation, neglecting generalized Cabibbo mixing, we have

$$
\bar{5} = \begin{pmatrix} \bar{d}_R \\ \bar{d}_Y \\ \bar{d}_B \\ --- \\ e^- \\ \nu_e \end{pmatrix}_L \left.\begin{matrix} \\ \\ \end{matrix}\right\} SU(3) \quad \left.\begin{matrix} \\ \end{matrix}\right\} SU(2) \qquad 10 = \frac{1}{\sqrt{2}} \begin{pmatrix} 0 & \bar{u}_B & -\bar{u}_Y & u_R & d_R \\ -\bar{u}_B & 0 & \bar{u}_R & u_Y & d_Y \\ \bar{u}_Y & -\bar{u}_R & 0 & u_B & d_B \\ ----&----&----&----&---- \\ -u_R & -u_Y & -u_B & 0 & e^+ \\ -d_R & -d_Y & -d_B & -e^+ & 0 \end{pmatrix}_L \left.\begin{matrix} \\ \\ \end{matrix}\right\} SU(3) \quad \left.\begin{matrix} \\ \end{matrix}\right\} SU(2) \qquad (16)
$$

$$\underbrace{\qquad}_{SU(3)} \quad \underbrace{\qquad}_{SU(2)}$$

where we have indicated explicitly the subspaces on which the strong SU(3) and weak SU(2) subgroups act. We can read off immediately from the $\bar{5}$ representation (16) the traceless diagonal operator corresponding to electromagnetic charge

$$
Q_{em} = diag\left(+\frac{1}{3}, +\frac{1}{3}, +\frac{1}{3}, -1, 0\right) . \qquad (17)
$$

This can be represented as a sum of the weak SU(2) generator T_3 and a traceless hypercharge Y

$$
Q_{em} = \left[T_3 = diag\left(0,0,0, -\frac{1}{2}, +\frac{1}{2}\right)\right] + \left[Y = diag\left(\frac{1}{3}, \frac{1}{3}, \frac{1}{3}, -\frac{1}{2}, -\frac{1}{2}\right)\right].
$$

$$(18)$$

Of relevance both to spontaneous baryon decay and to baryon "decays" catalyzed by GUMs is the structure of generalized Cabibbo mixing in GUTs.[9,19,22] Here we will just quote the results. In minimal SU(5) and related theories one can choose a fermion basis in such a way that:

• there is no mixing between elements of the $\bar{5}$

• there is generalized Cabibbo mixing between elements of the 10 which is the Kobayashi-Maskawa matrix U_{KM} acting on rows and columns 1 to 4 relative to the fifth row and column:

$$
10 = \frac{1}{\sqrt{2}}
\begin{pmatrix}
\bar{u}_{3\times 3} & u_{3\times 1} & d_{3\times 1} \\
-u_{1\times 3} & 0 & e^{+} \\
-d_{1\times 3} & -e^{+} & 0 \quad L
\end{pmatrix}
\qquad (19)
$$

(with U_{KM} bracket over columns, $e^{i\phi}$ bracket over columns 1 to 3)

- there are[9,19] also relative phases $e^{i\phi}$ between rows and columns 1 to 3 and the fourth row and column.

The appearance of the Cabibbo-Kobayashi-Maskawa matrix U_{KM} enables one to make predictions for Cabibbo-favored and -suppressed decay modes. The CP-violating phases $e^{i\phi}$ do not affect decay rates but they may have played a crucial role in Big Bang Baryonsynthesis. They may also give a nonintegral electric charge to the GUT monopoles.[23]

Baryon decay in minimal SU(5) is mediated by the exchange of the super heavy X and Y bosons which couple together the (1,2,3) and (4,5) of the fermion representations (16). The basic interaction is illustrated in Fig. 2(a), and the conventional model for the baryon decay amplitude is illustrated in Fig. 2(b). The amplitude is proportional to $1/m_X^2$ and hence the nucleon lifetime $\propto m_X^4$. Taking $m_X = (1$ to $4) \times 10^{14}$ GeV corresponding to $\Lambda_{QCD} = (100$ to 200) MeV one obtains a baryon lifetime

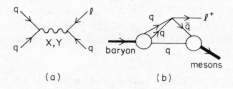

(a)   (b)

Fig. 2.   (a) Lowest order heavy gauge boson exchange diagram giving rise to baryon decay in conventional GUTs, and (b) the most popular way of estimating the baryon decay rate.

$$\tau_B = 10^{29\pm2} \text{ years} \tag{20}$$

Presumably m_X must be greater than 10^{14} GeV if baryons are to live longer than the present experimental limit.[24] The hierarchy of expected decay modes is [9,25]

$$B \rightarrow \quad e\pi, \ e\rho, \ e\omega, \ \ldots \quad \text{(Cabibbo-favored)}$$

$$> \quad \nu\pi, \ \nu\rho, \ \ldots \quad \text{(0(20)\%)}$$
$$\tag{21}$$

$$> \quad \mu K \quad \text{(0(10)\%, phase-space suppressed)}$$

$$\gg \quad eK, \ \mu\pi, \ \ldots \quad \text{(few \%, Cabibbo-suppressed)} \quad .$$

We will see in a moment how these decay patterns differ from those expected[12] in susy GUTs, and from the modes[10] of baryon "decay" catalyzed by GUMs.

Many conventional GUTs closely resemble minimal SU(5) in their predictions for the grand unification scale, $\sin^2\theta_W$, baryon decay modes, et cetera. However, recently much interest has developed in GUTs with N = 1 global supersymmetry (susy GUTs).[8] The motivation for susy GUTs is an attempt to accommodate the required hierarchy of mass scales

$$m_W(m_H) = 0 \ (10^2)\text{GeV} \ll m_X = 0(10^{15})\text{GeV} \underset{?}{\ll} m_P = 0(10^{19})\text{GeV} \ . \tag{22}$$

The difficulty resolved by susy GUTs is that even if the hierarchy (22) is imposed on the Lagrangian at the tree level, it tends to be destroyed by radiative corrections such as the boson loops in Fig. 3(a) which give

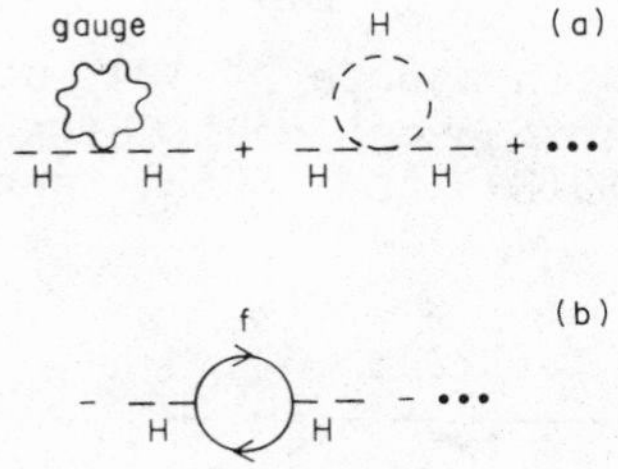

Fig. 3. (a) Boson loop diagrams which contribute positively to δm_H^2, and (b) fermion loop diagrams which contribute negatively and cancel in a supersymmetric theory.

$$\delta m_H^2, \delta m_W^2 \;=\; 0\left(\alpha_{GUT}\right) \; m_B^2 \;>>\; m_H^2, m_W^2 \quad \text{if } m_B^2 \simeq m_X^2 \;.\qquad(23a)$$

The solution proposed by susy is to invoke a cancellation by fermion loops as in Fig. 3(b). Because of their negative sign, if the bosons and fermions have similar couplings, Eq. (23a) gets replaced by

$$\delta m_H^2, \delta m_W^2 \;=\; 0\left(\alpha_{GUT}\right)\left(m_B^2 - m_F^2\right)$$

$$(23b)$$

$$\approx\; m_H^2,\; m_W^2 \quad \text{if } \left|m_B^2 - m_F^2\right| \lesssim 1 \text{ TeV}^2 \;.$$

The difference $\left|m_B^2 - m_F^2\right|$ is a measure of susy breaking for the particles appearing in the loops of Fig. 3. If it is sufficiently small, then the radiative corrections to m_H^2 and m_W^2 are sufficiently small for their values of $0(100 \text{ GeV})^2$ to seem "natural." For the cancellations (23b) to work, there must be bosonic partners for all known fermions (and vice versa) with essentially identical couplings, as seen in the following table:

Spin	Particles		
1	vector boson		
1/2	gaugino	quark, lepton	shiggs
0		squark, slepton	Higgs

No known particles can be supersymmetric partners of each other. Therefore a susy GUT contains at least twice as many particles as a conventional GUT, and in general even more since it requires at least two light Higgs doublets: $N_H \geq 2$.

The new particles with masses $<< 10^{15}$ GeV populate the desert and therefore modify the conventional GUT phenomenology. The rate of approach (7) of the SU(3) and SU(2) gauge couplings becomes significantly slower,[11]

$$\frac{1}{\alpha_3(Q)} - \frac{1}{\alpha_2(Q)} = + \frac{9+(3/2)N_H}{12\pi} \ln \frac{Q^2}{m_X^2}\qquad(24)$$

implying an increase in the grand unification scale

$$m_X \to m_X \times 0(40) \simeq 10^{16} \text{ GeV}\qquad(25)$$

in the most economical susy GUTs. There are more complicated variants[26] with more heavily populated deserts whose grand unification scale may be as large as 10^{19} GeV. Maintaining the successful (13) prediction (12) of $\sin^2\theta_W$ can also be a problem:

$$\sin^2\theta_W^{eff} = 0.236 \pm 0.002 \qquad\qquad (26)$$

in the most economical susy GUTs.[11,12] One might naively have expected the increase in the m_X to increase the baryon lifetime in a susy GUT. However, this is not necessarily the case as there is[27] a new class of diagrams like that in Fig. 4 which can give an interaction amplitude comparable[12] with that from Fig. 2(a), and hence a similar baryon lifetime to the estimate (20). In the simplest susy GUTs, though, the hierarchy of baryon decay modes is different[12] from that (21) in nonconventional GUTs.

$B \rightarrow \quad \bar{\nu}K \quad$ (favored by Cabibbo angles and quark mass factors)

$\quad \gg \bar{\nu}\pi \quad$ (suppressed by quark mass factors)

$\quad \gg \mu^+\pi \quad$ (Cabibbo suppressed) $\qquad\qquad (27)$

$\quad \gg e^+K \quad$ (Cabibbo suppressed)

$\quad \gg e^+\pi \quad$ (suppressed by quark mass factors)

It should be emphasized that while (26) and (27) are the predictions of the most economical susy GUTs, the ability to fix particle masses in such a way that they are not disturbed by radiative corrections (23) means that one can populate the desert in such a way as to vary the predictions of $\sin^2\theta_W$, m_X and the baryon decay modes almost at will. For example, there are susy GUTs where baryons decay predominantly into μ^+K or even into old-fashioned $e^+\pi$.[26]

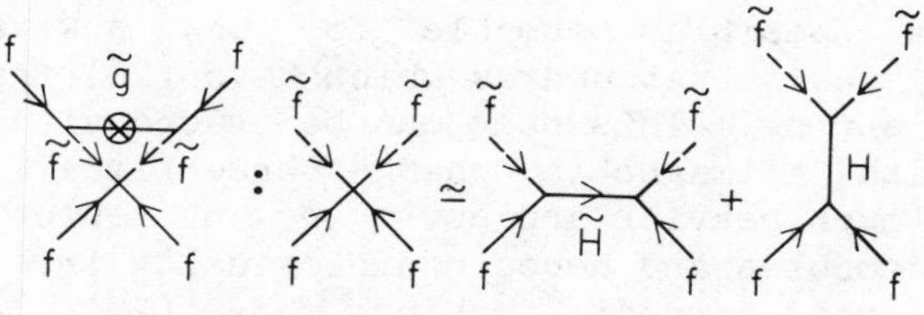

Fig. 4. Lowest order loop diagram contributing to baryon decay in a supersymmetric GUT, showing how the dimension 5 $\tilde{f}\tilde{f}ff$ operator is related to superheavy Higgs H and shiggs $\tilde{H}$ exchange.

The moral of this rapid review of GUTs is that within the general GUT philosophy[6] there are considerable phenomenological ambiguities. While there are "canonical" conventional and susy GUT predictions for m_x (and hence the monopole mass $m_M = m_x/\alpha_{GUT}$) and for the hierarchy of baryon decay modes (to be contrasted with the "decay" modes catalyzed by GUMs) one should be alive to other possibilities. GUM hunters should be prepared for GUM masses anywhere between 10^{16} and 10^{19} GeV, and should be aware that the hierarchy[10] of baryon "decay" modes catalyzed by GUMs may not be specific: they should keep their eyes open for other signatures as well.

GUMS IN GUTS

Previous speakers[1] have shown you that monopoles arise inevitably[2] in theories where a simple non-Abelian group is broken down to give $U(1)_{em}$ at low energies (large distances). Therefore we expect grand unified monopoles (GUMs) in GUTs. At radii much larger than the size of the monopole core (of order m_x^{-1} in GUTs) non-Abelian 't Hooft-Polyakov[2] monopoles look just like Dirac[13] monopoles corresponding to some U(1) subgroup of the exact low-energy $SU(3)_{color} \times U(1)_{em}$ gauge group. The corresponding gauge interactions of the first generation of fermions are

$$= g_3\left(\bar{u}\not{G}u + \bar{d}\not{G}d\right) + e\left(\frac{2}{3}\bar{u}\not{A}u - \frac{1}{3}\bar{d}\not{A}d - \bar{e}\not{A}e\right) \tag{28}$$

where G and A stand for gluon and photon fields respectively. A conventional Dirac monopole would sit in the $U(1)_{em}$ subgroup and have a magnetic charge h obeying the Dirac quantization condition

$$eh = 2\pi \ . \tag{29}$$

This is not a possible monopole for us, however, since the condition (29) means that quarks which have (28) fractional charge can detect its string. This snag can be evaded either by going to a monopole with a magnetic charge three times larger, which is expected to be much heavier and so to be unstable against decay into lighter monopoles and hence cosmologically irrelevant, or else by adding to the magnetic charge h (29) an additional chromomagnetic charge corresponding to some U(1) subgroup of $SU(3)_{color}$. All such options are gauge equivalent to looking at monopoles sitting in the U(1) subgroup generated by the λ_8 of color SU(3), and the minimal GUM has chromomagnetic charge h_3

$$g_3 \ h_3 \ = \ 2\pi \begin{pmatrix} \frac{1}{3} & 0 & 0 \\ 0 & \frac{1}{3} & 0 \\ 0 & 0 & -\frac{2}{3} \end{pmatrix} \qquad (30)$$

as well as the $U(1)_{em}$ charge h (29). It is easy to check that all the first generation left-handed fermions have an integer charge g of the $U(1)$ generated by (λ_8, Q_{em}):

$$g = \begin{cases} +1: \ u_{R_L}, \ u_{Y_L}, \ \bar{d}_{B_L}, \ e_L^+ \\ 0: \ u_{B_L}, \ d_{R_L}, \ d_{Y_L}, \ \bar{u}_{B_L}, \ \bar{d}_{R_L}, \ \bar{d}_{Y_L}, \ \nu_L \\ -1: \ \bar{u}_{R_L}, \ \bar{u}_{Y_L}, \ d_{B_L}, \ e_L^- \end{cases} \qquad (31)$$

and hence cannot see the string of the GUM with the charges (h_3, h) given by (29, 30).

It has been pointed out[14] that fermions may change their nature when scattering off a Dirac monopole in the S-wave. One way to see this is to recall that a particle of electric charge g moving in the field of a monopole of magnetic charge h has an associated angular momentum

$$J = \frac{gh}{4\pi} \ \hat{r} \ . \qquad (32)$$

This acquires a sign change when the fermion passes through the monopole core, because $\hat{r} \to -\hat{r}$. Angular momentum can be conserved if there is a simultaneous change of sign $g \to -g$. In general this requires a change in the flavor of the fermion (cf., the charge assignments (31) of conventional fermions). There is an ambiguity[14] in how one pairs up fermions into doublets (f,f') with equal and opposite electric charges relative to the monopole of interest, which then determines how flavor is violated in scattering events. This ambiguity cannot be resolved in the context of old-fashioned Dirac monopole theory. It can only be resolved by specifying fermion boundary conditions at the core of the monopole, which can be done[10] in the context of a GUT. We recall that our monopole which looks Dirac-like at large distances is expected to have a regular core specified by our choice of GUT. Our (λ_8, Q_{em}) U(1) group must be embedded in some SU(2) subgroup of our GUT. This means that there must be a non-Abelian gauge generator coupling our doublets (f, f'), and hence these transitions must have definite color and electromagnetic charge corresponding to the SU(3)×U(1) transformation properties of a massive gauge boson. The only consistent doublet assignments for the first generation fermion, (31), are

$$\begin{pmatrix} u_R \\ \\ \bar{u}_Y \end{pmatrix}_L, \begin{pmatrix} u_Y \\ \\ \bar{u}_R \end{pmatrix}_L, \begin{pmatrix} e^+ \\ \\ d_B \end{pmatrix}_L, \begin{pmatrix} \bar{d}_B \\ \\ e^- \end{pmatrix}_L . \tag{33}$$

We must allow for "Cabibbo" mixing permutations of these doublets when one takes into consideration multiple generations, a point we return to[10] in a moment. Using the doublets (33) and their friends involving heavier fermions one can construct effective interactions involving even numbers of fermions:

$$(ff'), (ff'f''f'''), (6f), (8f), \ldots . \tag{34}$$

The interesting interactions have $\Delta Q_{em} = 0$, so that they do not involve monopole $\leftrightarrow$ dyon transitions but can be catalyzed by GUMs alone. In general these $\Delta Q_{em} = 0$ interactions will have ΔB, $\Delta L \neq 0$.

So far we have not invoked any specific GUT: let us now see what happens if we embed our monopole in SU(5). One possible SU(2) subgroup of SU(5) that we can exploit for the non-Abelian core of our GUM is

$$I \equiv \frac{1}{2} \begin{pmatrix} 0 & 0 & 0 & 0 & 0 \\ 0 & 0 & 0 & 0 & 0 \\ 0 & 0 & & & 0 \\ 0 & 0 & \underline{\tau} & & 0 \\ 0 & 0 & 0 & 0 & 0 \end{pmatrix} \tag{35}$$

and alternatives involve replacing the blue (third) color by either yellow or red. We now see that the possible doublets of fermions are determined by the generalized "Cabibbo" mixing analysis[9,22] discussed in the third section (see Eq. (19)). The facts that there is no mixing of fermions in the $\bar{5}$ representations and no relative mixing between the third and fourth rows of the 10 representations, but only phase factors $e^{i\phi}$, mean that the prospective doublets (33) are essentially correct apart from phase factors, with a similar structure for heavier generations. For particles weighing less than 1 GeV:

$$
\begin{pmatrix} u_R \\ \bar{u}_Y e^{i\phi} \end{pmatrix}_L , \begin{pmatrix} u_Y \\ -\bar{u}_R e^{i\phi} \end{pmatrix}_L , \begin{pmatrix} \bar{d}_B \\ e^- \end{pmatrix}_L , \begin{pmatrix} e^+ \\ -d_B \end{pmatrix}_L , \begin{pmatrix} \bar{s}_B \\ \mu^- \end{pmatrix}_L , \begin{pmatrix} \mu^+ \\ s_B \end{pmatrix}_L . \tag{36}
$$

These yield the following effective interactions[3,10,15] which may be particularly relevant to low-energy GUM collisions with baryons:

2f: $\bar{d}_B d_B$, $\bar{u}_Y u_Y$, $\bar{u}_R u_R$, $e^+ e^-$

4f: $u_Y d_B u_R e^+$, $u_R e^- u_Y d_B$, $u_Y s_B u_R \mu^+$, $u_R \mu^- u_Y s_B$ $\qquad$ (37)

6f: $u\ d\ u\ e^- \mu^+ \mu^-$, $u\ e^- u\ s\ e^+ \mu^-$.

We see that the 4 and 6 fermion interactions in (37) have $\Delta B = \pm 1$ (while conserving $B - L$) and therefore expect them to lead to $\Delta B \neq 0$ GUM-Baryon cross sections. Condensates like (37) are expected to exist in any region of space where the GUM looks like an apparent Dirac monopole with the magnetic and chromomagnetic charges (29, 30). We expect the chromomagnetic field to extend as far as the confinement radius of order 1 fermi. Therefore the $\Delta B \neq 0$ cross section may have the magnitude of a conventional strong-interaction cross-section. One of the 4 or 6 fermion interactions (37) could take place whenever the GUM overlaps with a baryon. Since the duration of the overlap during a collision is given by a $1/\beta$ flux factor at low velocities β, we parametrize the cross section in the form

$$
\sigma_{GUM\ B \neq 0} = \frac{1}{\beta} \left(\frac{\sigma_0}{1\ GeV^2} \right) \tag{38}
$$

at low velocities, where σ_0 is a dimensionless reduced cross-section factor. It is then a problem in strong interaction phenomenology to estimate σ_0. A priori one might imagine[3,5] that it could be $O(1)$, but quite honestly we do not know at the present time how big it might be. Since the forms of the $\Delta B \neq 0$ interactions (37) are similar to those due to X, Y boson exchange in conventional GUTs, one way[10] of estimating σ_0 (see Fig. 5) is by analogy with the conventional calculation of the spontaneous baryon decay rate $\Gamma(B \rightarrow e^+X)$ mediated by X and Y boson exchange (cf., Fig. 2). One then estimates

$$\sigma_{\text{GUM } B\neq0} \approx \frac{1}{\beta} \frac{\hat{\Gamma}(B\rightarrow e^+X)}{1 \text{ GeV}^3} \,, \tag{39}$$

where $\hat{\Gamma}$ is obtained from Γ by the replacement

$$\frac{g^2}{m^2_{X,Y}} \rightarrow \left(\frac{1}{4\pi^2}\right)^2 \left(\frac{1}{1 \text{ GeV}^2}\right). \tag{40}$$

The factors of $(1/1 \text{ GeV}^2)$ in Eqs. (39) and (40) come from dimensional analysis. The factors of $1/4\pi^2$ in Eq. (40) come from Rubakov's analysis[3] of fermion condensates around a monopole. Keeping track of all the factors of $1/2\pi$ that we can identify, we guess[10] wildly that

$$\sigma_0 = O(10^{-4}) \times O(10^{\pm2}) \text{ ?} \tag{41}$$

In this case an astrophysically plausible slow-moving GUM with $\beta = O(10^{-3})$ might have a B-violating cross section

$$\sigma_{\text{GUM } B\neq0} = O(10^{-28}) \text{ cm}^2 \text{ ??} \tag{42}$$

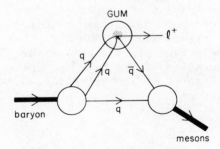

Fig. 5. Sketch of mechanism for baryon "decay" catalyzed by a GUM. Two quarks coming withing one fermi of the GUM core may be sucked into it and change their flavors in a similar way to that in Fig. 2.

While keeping this possibility in mind, we will try to keep σ_0 as a free parameter in the subsequent analysis.

One can deduce[10] from Eq. (37) a likely hierarchy of baryon "decay" modes catalyzed by GUMs. This is

$$B \longrightarrow e^+ \text{ pions} \qquad \text{("Cabibbo"-favored)}$$

$$\gg \mu^+ K \qquad \text{(suppressed by phase space, quark mass factors?)}$$

$$\gg e^+\mu^+\mu^-(\pi), \mu^+e^+e^-K \quad \text{(suppressed by condensate factors)} \qquad (43)$$

$$B \not\longrightarrow \bar{\nu}K \qquad (\nu \text{ has no magnetic charge})$$

$$\not\longrightarrow e^+K, \mu^+\pi \qquad \text{("Cabibbo" disallowed)}$$

This is to be compared with the conventional GUT hierarchy (21) and the simplest susy GUT predictions (27). There are some differences which may serve as signatures for baryon "decays" catalyzed by GUMs. Other possible experimental signatures include the possibility of a three-momentum transfer to the "decay" products. There is no reason why the three-momentum transfer should be zero and we might expect it to be of conventional strong interaction magnitude

$$|\Delta q| = 0(300) \text{ MeV ?} \qquad (44)$$

This would act in the same way as conventional Fermi motion for a decaying nucleon in a heavy nucleus, causing the baryon "decay" products not to come out back-to-back. It might be difficult to conclude that there was an excess of Fermi momentum of order (44), except possibly if one were looking for baryon "decays" in very light nuclei such as hydrogen. One also expects a net energy transfer

$$|\Delta E| = 0(\beta) \text{ GeV} \qquad (45)$$

to the baryon "decay" products, which is undetectable for slow monopoles. A potentially interesting possibility[15] is the observability of multiple baryon "decays" occurring in a chain across a detector. One expects a mean free path between catalyzed events of

$$\lambda = \frac{43m}{\rho} \; \frac{\beta}{\sigma_0} \qquad (46)$$

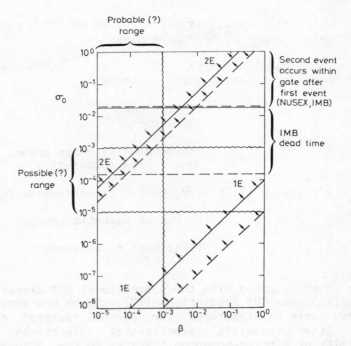

Fig. 6. Indication of the capabilities of baryon decay experiments
to search for B "decays" catalyzed by GUMs.[10] The probable
(?) range of β and the possible (?) range of the ΔB ≠ 0
cross-section strength σ_0 (38) are indicated. Experiments
can observe a catalyzed "decay" in the region marked 1E if
GUMs have the Cabrera flux.[29] In the region 2E two events
can occur within the same apparatus. The solid lines
refer to the NUSEX experiment, the dashed lines to the IMB
experiment.

where ρ is the matter density in gm/cc. This corresponds to a mean
free time between events of

$$\tau = \frac{\lambda}{\beta} = \frac{0.14}{\rho\sigma_0} \text{ microseconds} \tag{47}$$

which would be O(1) milliseconds if σ_0 = O(10^{-4}). Baryon decay
detector designers should bear this point in mind. Several of the
early detectors[28] had electronics dead times of about a millisecond
after each baryon decay candidate occurred, while it was being put
on tape. For this reason they would not have been sensitive to
"decays" in coincidence within the time difference (47). The NUSEX
and IMB experiments are presently considering modifications of
their electronics so as to avoid this dead time. Figure 6

exhibits[10] the sensitivity of possible experimental searches for Baryon-number violating GUM interactions using baryon decay experiments. We see that they can see GUM-catalyzed "decays" if the GUM flux is within a few orders of magnitude of Cabrera's limit,[29] and that they also have a fair chance of seeing double "decays."

ASTROPHYSICAL AND COSMOLOGICAL LIMITS ON GUMS

Since these are the subjects of two full sessions at this Workshop, I will not do much more than just remind you what issues[16,17] are involved. First of all, there are unimpeachable constraints[30,31] on the density and hence flux of GUMs following from upper limits on the mass density in the Universe [relevant if the local GUM velocities β are larger than $O(10^{-3})$ in which case monopoles are not confined to galaxies], or from the missing mass in the galaxy [relevant if $O(10^{-4}) < \beta < O(10^{-3})$ in which case monopoles are not bound in the solar system but are confined to galaxies]. Then there are more arguable constraints on the flux of monopoles due to the persistence of the galactic magnetic field.[32,33] Normally one believes that monopoles act as a drain on the galactic field energy as it accelerates them,[32] but there is a controversial minority viewpoint[33] that monopole plasma oscillations might actually be responsible for generating the galactic magnetic field, in which case the GUM flux could (should) be considerably larger. Another suggestion[34] for making large fluxes of GUMs more tolerable is that their density may be locally enhanced due to our proximity to a local source, most probably the Sun. This might work if $\beta < O(10^{-4})$, but it is not clear how a solar cloud of monopoles could have formed, nor how much local enhancement in the flux could be attained. Finally there are distinctly less reliable constraints on the GUM flux which are conditional on their having large $\Delta B \neq 0$ cross sections. An important constraint comes from neutron stars.[39,10,36] GUMs could get stuck in them and "eat" their baryons, producing energy which is eventually thermalized and radiated as X-ray or ultraviolet light. Upper limits[37] on the flux of X rays either from known point-source neutron stars or from a diffuse background of older unresolved neutron stars give quite stringent limits on the product of GUM flux F and dimensionless reduced cross section σ_0. The most conservative constraint from the X-ray background is[10]

$$F\sigma_0 < 6.6 \times 10^{-15} \beta^2 \text{ cm}^{-2} \text{ s}^{-1} \text{ sr}^{-1} . \tag{48}$$

This constraint has been derived assuming that we and the neutron star are both bathed in a continuing galactic flux, and neglects the possibility[34] that there may be local flux enhancements around stars when $\beta < O(10^{-4})$. In this case the neutron star limit may be somewhat relaxed.[10] The various astrophysical constraints are of

varying relative importance for different GUM masses. Figure 7
shows versions[10] of the different astrophysical constraints for
masses of 10^{16} GeV (Fig. 7a) and of 10^{19} GeV (Fig. 7b). Also shown
is the present constraint[10,15] from baryon decay experiments

$$F\sigma_0 < 2 \times 10^{-12} \, \beta \; cm^{-2} \; s^{-1} \; sr^{-1} \tag{49}$$

and the maximum possible sensitivity of future baryon decay
experiments. While the astrophysical constraints make it difficult
to imagine seeing $\Delta B \neq 0$ interactions catalyzed by GUMs, it is not
impossible at least if $\beta < O(10^{-4})$ and the GUM mass is close to the
Planck mass. As also shown in Fig. 7, it may also be possible to
detect $O(100)$ MeV neutrinos originating from $\Delta B \neq 0$ processes in
the Sun.[3,34]

For completeness, I will also mention the problem[17] of the
cosmological production and abundance of GUMs, emphasizing that the
possibility of an observable GUM flux is not excluded.
Conventionally[38,39] one expects $O(1)$ GUM in every horizon volume at

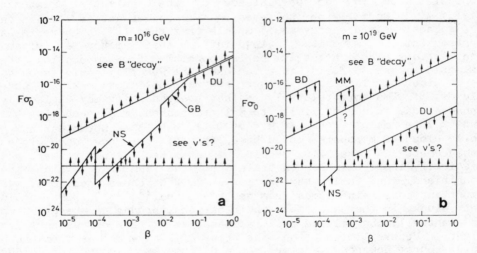

Fig. 7.   Plots of the astrophysically allowed regions for the GUM
flux F (measured in cm^{-2} s^{-1} sr^{-1}) for different GUM
velocities and masses (a) 10^{16} GeV and (b) 10^{19} GeV.[10]
Catalyzed B "decays" and $O(100)$ MeV neutrinos from the Sum
(if GUMs stop inside it) are detectable above the
indicated lines corresponding to an apparent "lifetime" of
10^{33} years and $F_\nu = 1$ cm^{-2} s^{-1} sr^{-1}. The astrophysical
bounds come from neutron stars (NS), the galactic magnetic
field (GB), the density of the Universe (DU) and the
missing mass in the galaxy (MM). Also shown on (b) is the
upper limit on $F\sigma_0$ coming from present-day baryon decay
experiments (BD).

the epoch when GUMs were formed at a critical temperature $T_c \gtrsim 0(10^{15})$ GeV. This gives many more GUMs than are allowed by the present mass density of the Universe, and to compound the problem it seems unlikely[30] that monopole-antimonopole annihilation could have reduced the original GUM abundance to an acceptable level today. An attractive way out of this impasse is the new inflationary cosmology,[39] according to which the entire visible Universe sits within one bubble of correlated Higgs fields and the closest topological knot corresponding to a GUM is (much) more than 10^{10} light years away. (Un)fortunately the new inflationary Universe has difficulties with the fine-tuning of parameters[40] and the magnitude of fluctuations.[41] These can be alleviated or resolved if the underlying theory is supersymmetric[40] and if the magnitude V of the Higgs vacuum expectation value driving the inflation is close to the Planck mass.[42] This is fine for solving the purely cosmological problems which inflation is supposed to cure, but the GUM abundance problem persists if $V >> m_X$ ("primordial inflation"). It may be possible to resolve the GUM problem if primordial inflation is combined with one of the other "solutions" to the GUM problem discussed a while ago. These are "supercosmology" in which the GUT phase transition is delayed to $T_c = 0(10^{10})$ GeV resulting in larger horizon volumes and hence perhaps fewer monopoles,[43] or a second-order GUT phase transition in which the monopole abundance may be thermodynamically suppressed.[44] In either of these cases the abundance of GUMs may avoid being unobservably small.[42] Therefore GUM hunters should not allow themselves to be discouraged by inflationary cosmologists! However, it is clear that getting an acceptably small abundance of GUMs, or still better an observably large flux, imposes nontrivial cosmological constraints on the nature of the GUT. This is yet another way in which monopoles can be used to test and discriminate between different GUTs.

DESSERT

We have seen that gauge theories do not expect monopoles[2] with low masses ($\leq 0(10)$ TeV), but that one does expect grand unified monopoles (GUMs) with masses $\geq 0(10^{16})$ GeV. GUMs are unavoidable consequences of the GUT philosophy[6] of embedding $U(1)_{em}$ in a simple group along with all the other interactions. The mass of the monopole tells us about the grand unification scale: $m_X = \alpha_{GUT} m_M$. There is the exciting possibility[3,4,9] that GUMs may catalyze $\Delta B \neq 0$ interaction with a strong interaction cross section. If so, observation of the modes of catalyzed "decay" could tell us about the grand unified generalized Cabibbo mixing.[9] The production or lack of production of GUMs in the early Universe tells us about the desired behavior of GUTs at temperatures $0(m_X)$. GUMs could therefore provide us with crucial tests of GUTs. Figure 8 shows the ideal, ultimate GUM detector. It contains a more-or-less

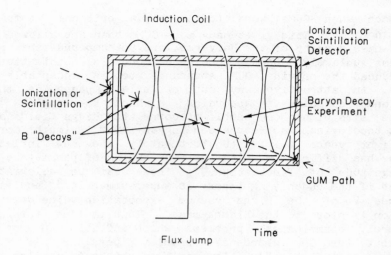

Fig. 8.   Sketch of the ideal, ultimate GUM detector.   It includes
          an induction coil and an ionization or scintillation
          detector mounted around a conventional baryon   decay
          detector.   Also shown is a dream GUM event.

conventional baryon decay detector, hopefully made of   light
material so that "anomalous Fermi motion" can be detected. Around
this is an ionization or scintillation detector,   hopefully
sensitive to β as low as 10^{-4}. Around this is an induction coil
looking for flux jumps due to GUMs passing through. Also shown in
Fig. 8 is a dream event in which the coil flux jumps, the
ionization detector fires and there is a chain of catalyzed baryon
"decays."   Such an event would certainly probe the viscera of our
GUTs.

ACKNOWLEDGMENTS

        It is a pleasure to thank F. A. Bais, K. A. Olive   and
D. V. Nanopoulos for an enjoyable collaboration on the topics
discussed here, to thank participants in the   Workshop   for
discussions, and the organizers for the opportunity to meet in such
congenial circumstances.

REFERENCES

1. A. S. Goldhaber and S. Coleman, talks at this Workshop. See
    also: S. Coleman, "The Magnetic Monopole Fifty Years
    Later," Harvard University preprint HUTP-82/A032 (1982),
    lectures presented at Erice, et cetera, during 1981.

2. G. 't Hooft, Nucl. Phys. B 79, 276 (1974; A. M. Polyakov, Zh.
    Eksp. Teor. Fiz. Pis'ma Red. 20, 430 (1974) [JETP Lett. 20,
    194 (1974)].

3. V. A. Rubakov, Zh. Eksp. Teor. Fiz. Pis'ma Red. 33, 658 (1981)
    [JETP Lett. 33, 644 (1982)]; Nucl. Phys. B 203, 311 (1982);
    U.S.S.R. Academy of Sciences Institute for Nuclear Research
    preprint P-0211 (1982).

4. F. A. Wilczek, Phys. Rev. Lett. 48, 1146 (1982).

5. C. G. Callan, Phys. Rev. D 25, 2141 (1982), D 26, 2058 (1982)
    and Princeton University preprint, "Monopole Catalysis of
    Baryon Decay," (1982); talk at this Workshop.

6. H. Georgi and S. L. Glashow, Phys. Rev. Lett. 32, 438 (1974).
    For a different approach to grand unification, see J. C.
    Pati and A. Salam, Phys. Rev. Lett. 31, 661 (1973); Phys.
    Rev. D 8, 1240 (1973) and D 10, 275 (1974).

7. H. Georgi, H. R. Quinn and S. Weinberg, Phys. Rev. Lett. 33,
    451 (1974).

8. S. Dimopoulos and H. Georgi, Nucl. Phys. B 193, 150 (1981); N.
    Sakai, Z. Phys. C 11, 153 (1982).

9. J. Ellis, M. K. Gaillard and D. V. Nanopoulos, Phys. Lett.
    88B, 320 (1980).

10. F. A. Bais, J. Ellis, D. V. Nanopoulos and K. A. Olive, CERN
    priprint TH-3383 (1982).

11. S. Dimopoulos, S. Raby and F. A. Wilczek, Phys. Rev. D 24,
    1681 (1981); L. E. Ibáñez and G. G. Ross, Phys. Lett. 105B,
    439 (1981); M. B. Einhorn and D. R. T. Jones, Nucl. Phys. B
    196, 475 (1982).

12. S. Dimopoulos, S. Raby and F. A. Wilczek, Phys. Lett. 112B,
    133 (1982); J. Ellis, D. V. Nanopoulos and S. Rudaz, Nucl.
    Phys. B 202, 43 (1982).

13. P.A.M. Dirac, Proc. Roy. Soc. A 133, 60 (1931); Phys. Rev. 74,
    817 (1948).

14. A. S. Goldhaber, Phys. Rev. D 16, 1815 (1977); Y. Kazama, C.
    N. Yang and A. S. Goldhaber, Phys. Rev. D 15, 2287 (1977).

15. J. Ellis, D. V. Nanopoulos and K. A. Olive, Phys. Lett. 116B,
    127 (1982).

16. M. S. Turner, talk at this Workshop.

17. G. Lazarides and A. H. Guth, talks at this Workshop.

18. J. Ellis and D. V. Nanopoulos, Nature 292, 436 (1981).

19. A. J. Buras, J. Ellis, M. K. Gaillard and D. V. Nanopoulos,
    Nucl. Phys. B 139, 66 (1978).

20. W. J. Marciano and A. Sirlin, Phys. Rev. Lett. 46, 163 (1981);
    C. H. Llewellyn Smith, G. G. Ross and J. F. Wheater, Nucl.
    Phys. B 177, 263 (1981).

21. W. J. Marciano and A. Sirlin, Phys. Rev. D 22, 2695 (1980); C.
    H. Llewellyn Smith and J. F. Wheater, Phys. Lett. 105B, 486
    (1981).

22. R. N. Mohapatra, Phys. Rev. Lett. 43, 893 (1982).

23. C. G. Callan, talk at this Workshop.

24. M. R. Krishnaswamy, et al., Phys. Lett. 106B, 339 (1981) and 115B, 349 (1982); M. L. Cherry, et al., Phys. Rev. Lett. 47, 1507 (1981).
25. G. Kane and G. Karl, Phys. Rev. D 22, 1808 (1980).
26. A. Masiero, D. V. Nanopoulos, K. Tamvakis and T. Yanagida, Phys. Lett. 115B, 298 (1982) and Y. Igarashi, J. Kubo and S. Sakakibara; Phys. Lett. 116B, 349 (1982); S. Dimopoulos and S. Raby, Los Alamos preprint LA-UR-82-1282 (1982).
27. S. Weinberg, Phys. Rev. D 25, 287 (1982); N. Sakai and T. Yanagida, Nucl. Phys. B 197, 533 (1982).
28. NUSEX: G. Battistoni et al., proposal for an experiment on nucleon decay with a fine grain calorimeter (1979); IMB: M. Goldhaber et al., proposal for a nucleon decay detector (1979).
29. B. Cabrera, Phys. Rev. Lett. 48, 1378 (1982); and talk at this Workshop.
30. J. P. Preskill, Phys. Rev. Lett. 43, 1365 (1979); see also Y. B. Zel'dovich and M. Yu. Khlopov, Phys. Lett. 79B, 239 (1978).
31. G. Lazarides, Q. Shafi and T. F. Walsh, Phys. Lett. 100B, 20 (1981).
32. E. N. Parker, Ap. J. 139, 951 (1964); M. S. Turner, E. N. Parker and T. J. Bogdan, Phys. Rev. D 26, 1296 (1982).
33. E. E. Salpeter, S. L. Shapiro and I. Wasserman, Phys. Rev. Lett. 49, 1114 (1982); I. Wasserman, talk at this Workshop.
34. S. Dimopoulos, S. L. Glashow, E. M. Purcell and F. A. Wilczek, Nature 298, 824 (1982).
35. E. W. Kolb, S. A. Colgate and J. A. Harvey, Phys. Rev. Lett. 49, 1373 (1982).
36. S. Dimopoulos, J. P. Preskill and F. A. Wilczek, Phys. Lett. 119B, 320 (1982).
37. J. Silk, Ann. Rev. Astron. Astrophys. 11, 269 (1973).
38. T.W.B. Kibble, J. Phys. A 9, 1387 (1976).
39. A. D. Linde, Phys. Lett. 108B, 389 (1982); A. Albrecht and P. J. Steinhardt, Phys. Rev. Lett. 48, 1220 (1982).
40. J. Ellis, D. V. Nanopoulos, K. A. Olive and K. Tamvakis, Phys. Lett. 118B, 335 (1982).
41. A. H. Guth and S.-Y. Pi, Phys. Rev. Lett. 49, 1110 (1982). S. W. Hawking, Phys. Lett. 115B, 295 (1982); A. A. Starobinskii, J. Bardeen, P. J. Steinhardt and M. S. Turner, private communications (1982).
42. J. Ellis, D. V. Nanopoulos, K. A. Olive and K. Tamvakis, CERN preprint TH-3404, (1982) and Phys. Lett. 120B, 331 (1983).
43. M. Srednicki, Nucl. Phys. B 202, 327 (1982); D. V. Nanopoulos and K. Tamvakis, Phys. Lett. 110B, 449 (1982); D. V. Nanopoulos, K. A. Olive and K. Tamvakis, Phys. Lett. 115B, 15 (1982).
44. F. A. Bais and S. Rudaz, Nucl. Phys. B 170, [FS1], 507 (1980).

EXPERIMENTAL STATUS OF MONOPOLES

G. Giacomelli

Istituto di Fisica dell'Università di Bologna
Istituto Nazionale di Fisica Nucleare
Sezione di Bologna

INTRODUCTION

In this talk I shall try to give a broad overview on the experimental status of magnetic monopole searches. Most of what I will say will be discussed again, in much more detail, in the next talks. Therefore I shall try to be general, without discussing any detail.

The field of magnetic monopoles has grown considerably in the last few years. It is now a fascinating subject, involving many fields of physics, from particle physics to astrophysics, from the extremely small to cosmology.[1] In fact, it now looks almost like an interdisciplinary field where are involved mathematical physicists, particle physicists, astrophysicists, applied physicists, and so on.

Magnetic monopoles were introduced by P.A.M. Dirac in 1931 in order to explain the quantization of the electric charge.[2] He established the basic relation between the elementary electric charge e and the hypothesized magnetic charge g:

$$g = ng_0 = \frac{\hbar c}{2e} n = \frac{137}{2} e n \qquad (1)$$

where g_0 is the smallest magnetic charge and n is an integer, which in the original proposal could assume the values $n = \pm 1, \pm 2, \pm 3 \ldots$ In this formulation there was no prediction for the monopole mass. More recently the Grand Unified Theories of strong and electroweak

41

interactions predicted the existence of magnetic monopoles with extremely large masses.[1] At this workshop, we have heard of different types of magnetic monopoles. For the purpose of this talk and for a discussion of the experimental searches, I shall classify the magnetic monopoles in the following way:

(i) "Classical" monopoles, defined as the original Dirac monopoles. For them, we hypothesize a relatively small mass, so that they could be produced at existing high-energy accelerators. In the Dirac relation, we shall consider n = 1 and e = electron charge. Searches for "classical" monopoles have been performed at every new high-energy accelerator.[3,4] (ii) GUT monopoles (or GUM) predicted by the Grand Unified Theories (GUT) of weak, electromagnetic and strong interactions. The monopoles are produced at the phase transition when a gauge group breaks spontaneously into subgroups, one of which is U(1). The monopole mass is connected to the vector boson mass m, which defines the unification scale

$$m_M \simeq 2 \sqrt{2} \pi \, m/G^2 \qquad\qquad\qquad (2)$$

where G^2 is the dimensionless coupling constant. If $m \simeq m_X \sim 10^{14}$ GeV one obtains $m_M \sim 10^{16}-10^{17}$ GeV. This is an enormous mass; therefore, monopoles cannot be produced at any existing (and even at any conceivable) accelerator. They could only be made at the extremely high energies available in the first instants of the Universe, immediately after ($\sim 10^{-35}$ seconds) the Big Bang. We shall assume that the stable monopoles have a mass $m_M = 10^{16}$ GeV $\simeq$ 0.02 µg. The poles could possess also an electric charge (dyons) or could be electrically neutral. In the following, we shall consider the lowest mass monopoles as electrically neutral. A large number of searches for GUT monopoles are now in progress.[4-8]

We shall not discuss other possibilities such as the tachyon monopoles. But we shall try to give limits also for (iii) monopoles like (ii) but with a mass of 10^4 GeV, that is, poles which could be connected with the electroweak unification

$$(m_W \sim 10^2 \text{ GeV}, \, m_M \sim 10^4 \text{ GeV}).$$

There are many questions open on the magnetic monopoles; for instance: (a) Is the monopole connected to the electron (in which case its magnetic charge is g = 67.5 e) or to the quark (in which case g = 3 × 67.5 e)? If[0] free quarks exist, it is almost unavoidable that monopoles are connected to the non-integral charge (and thus have a larger magnetic charge). If quarks are absolutely confined, then the basic electric charge is the electron charge.

Conversely, the detection of a pole with the Dirac charge would probably prove that quarks are absolutely confined. (b) How many types of stable monopoles could be consistent with GUT theories? Are there stable poles with electric charge? Is 10^{16} GeV the only important mass to consider? (c) Does the monopole possess (an open) color charge? (d) What is the exact structure of a monopole? Does it catalyze proton decay with a sizable cross section?

For the practical purpose of detection it is very important to establish the precise energy loss for poles with low velocities.

Because of all these uncertainties (and many others), it is important that experimenters keep an open mind.

This paper is arranged as follows: the second section describes the properties of magnetic monopoles relevant to their detection; the third and fourth sections are devoted to the searches for "classical" and GUT monopoles respectively. The fifth section discusses the monopole catalysis of proton decay. The conclusions, as well as an outlook of the future, are given in the last section.

PROPERTIES OF MAGNETIC MONOPOLES RELEVANT TO THEIR DETECTION

In this section will be summarized the main features of magnetic monopoles, which can be obtained from Eq. (1), assuming n = 1 and that the elementary charge is that of the electron:

Magnetic charge

$$g_0 = 1/2 \frac{\hbar c}{e} = \frac{137}{2} e = 3.29 \times 10^{-8} \text{ cgs units} \tag{3}$$

If the elementary electric charge would be that of quarks, with charge 1/3, one would have an elementary magnetic charge 3 times larger and ionization losses 9 times larger (Eq. (6)). A similar situation arises if $|n| > 1$.

Coupling constant

$$\frac{g_0^2}{\hbar c} = \frac{e^2}{\hbar c} \left(\frac{g_0}{e}\right)^2 = \frac{1}{137} \left(\frac{137}{2}\right)^2 = 34.25 \tag{4}$$

Energy acquired in a magnetic field B

$$W(eV) = g H l = 2.06 \text{ MeV/gauss.m} \tag{5}$$

Thus, because of the large g-value, monopoles acquire large energies in even modest magnetic fields.

Energy loss by ionization. The moving monopole creates an electric field which ionizes the medium. The relation between the ionization energy loss of a moving magnetic charge and that of an electric charge with the same β is

$$\left(\frac{dE}{dx}\right)_g = \left(\frac{dE}{dx}\right)_e \left(\frac{g_0}{e}\right)^2 (n \beta)^2 \tag{6}$$

where

$$\left(\frac{g_0}{e}\right)^2 = \left(\frac{137}{2}\right)^2 = 4700 \tag{7}$$

The formula is reliable down to $\beta \sim 10^{-3}$. In Eq. (6) are missing some terms due to the interaction of the traveling magnetic monopole with the magnetic dipoles of the medium (for instance, of the electrons). These terms have a lower β-dependence and therefore become important at lower values of β For $\beta > 5 \times 10^{-2}$ Eq. (6) may be approximated with the formula[9]

$$\left(\frac{dE}{dx}\right)_{g,ioniz.} = 0.72 (8.18 + \ln \beta^2 \gamma^2) (GeV \ g^{-1} \ cm^2) \tag{8}$$

For $\beta < 10^{-3}$ there are considerable differences in the predictions of the ionization loss made by different authors.[9-13] Taking into account the minimum energy gap required for ionization (4 eV in Carbon, 13 eV in Argon and Hydrogen) Ritson[12] obtained the following formula for $\beta < 10^2$

$$\left(\frac{dE}{dx}\right)_{g,ioniz.} \geq \frac{1}{4} \frac{\beta}{5 \times 10^{-3}} \frac{\overline{v}_f^2}{\alpha^2 c^2} \left(\frac{dE}{dx}\right)_{protons} \tag{9}$$
$$at \ \beta = 5 \times 10^{-3}$$

where V_f in the Fermi velocity of electrons with $v = \alpha c$; the presence of the energy gap means that the effective V_f decreases with β.

Figure 1 shows the monopole ionization energy loss in hydrogen[11] and in carbon.[12] Relativistic monopoles ionize 4700 times the value of minimum ionizing particles. The monopole ionization energy loss is equal to the minimum ionization of electric particles for β equal to 10^{-3} in hydrogen and in argon, $\beta \simeq 3 \times 10^{-4}$ in carbon. It is possible that, for low velocities,

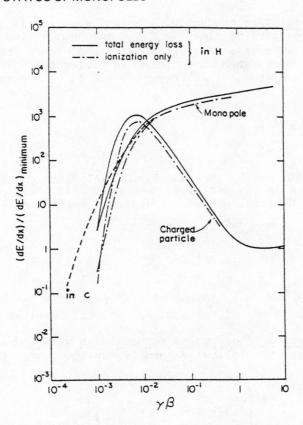

Fig. 1. Total energy loss and ionization energy loss of unit
         magnetic monopoles and of unit electric charges in atomic
         hydrogen[11] plotted versus the $\beta\gamma$. For $\beta\gamma < 10^{-2}$ the
         ionization loss in carbon[12] (which should approximately be
         equal or smaller than the dE/dx in scintillators) is also
         shown.   For $\beta\gamma < 10^{-3}$ the dE/dx is still somewhat
         controversial.

spin interactions could increase these values; on the other hand,
there are uncertainties in the above formulae at low β. The limits
of detection of slow monopoles by ionization losses are still
uncertain, but probably may be placed at β-values around 2×10^{-4}
for scintillators, 5×10^{-4} in proportional counters and 5×10^{-3}
in plastic CR39. (In scintillators, it would be important to
analyze in more detail the light emitted in processes of molecular
and atomic de-excitation (without ionization). These could become
relevant at very small β.)

Energy loss in a conductor. A monopole moving in a conductor produces a time-varying electric field and thus a time varying eddy current. The effective interaction with the conduction electrons is possible only when the pole is moving more slowly than the electrons in the atom. The energy loss is approximately given by[9,13,14]

$$\frac{dE}{dx} \simeq \frac{4\pi^2 \, ng^2 \, e^2}{m_e \, c^2 \, v_f^c} \, \beta \tag{10a}$$

where $v_f^c \simeq 10^8$ cm/s is the Fermi velocity of the conduction electrons and n is the number of conduction electrons per cm^3. For copper, one has approximately (higher values have also been quoted)[14]

$$\frac{dE}{dx} \simeq 2 \times 10^2 \, \beta \, (GeV/cm) \tag{10b}$$

which should saturate at losses of the order of 1 GeV/cm for $\beta >$ few 10^{-3}. This is one of the dominant energy losses at low β.

Energy losses of monopoles in collision with atoms. Ahlen and Kinoshita give the following formula for the energy loss on atoms of Si:[9]

$$\frac{dE}{dx} \, (Si) = \frac{0.79}{4} \left[1.31 + \ln \beta - \ln \left(\frac{\beta}{137 \, e} \right) \right] \tag{11}$$

For $\beta = 10^{-3}$ Eq. (11) gives an energy loss which is about 7% of the stopping power due to ionization. The relative contribution of (11) to the total energy loss increases as β decreases.

Energy losses in ferromagnetic materials. A slow moving monopole will align the magnetic domains of ferromagnetic materials. The energy loss connected with this mechanism becomes relatively large at very low monopole velocities, when they can efficiently align magnetic domains.[5,13]

Goebel[15] has estimated the cross section for nuclear capture (via the emission of a bremstrahlung photon of sufficient energy to allow capture). He obtains a mean free path for nuclear capture

$$\lambda = (N \, \sigma_{int} \, P_\gamma)^{-1} = 1.2 \, Km \tag{12}$$

for monopoles with $\beta = 5 \times 10^{-3}$. If this is correct, all monopoles arriving at the earth surface from below have captured some nuclei.

   Magnetic monopoles may be trapped in bulk paramagnetic and ferromagnetic materials by an image force, which, in ferromagnetic materials, may reach the value of ~10 eV/Å (11 eV/Å in iron, 3.5 in magnetite).[16] (This value has to be compared with the force of gravity at the surface of the earth ~0.1 eV/Å).

   An electromotive force and thus a <u>current</u> (Δi) is <u>induced</u> when a monopole passes through a coil. In particular, for a superconducting coil with N turns and inductance L, one has

$$\Delta i = 4 \pi N \, ng/L = 2 \, \Delta i_0 \qquad (13)$$

where Δi_0 is the current change corresponding to a change of one unit of the flux quantum of superconductivity, $\phi_0 = hc/2e$. The change in current will occur with a characteristic time given by $b/\gamma v$, where b = radius of coil, v = velocity of monopole. The change in the current and thus in the field may be observed with a SQUID magnetometer.

SEARCHES FOR CLASSICAL MONOPOLES

   In the early 1970's the "classical" monopole was considered to be a member of the family of "well known undiscovered objects." Searches were made at every new higher energy accelerator, in cosmic rays and in bulk matter.[3,4,17,18] One thought that monopoles could be produced in high-energy collisions, the simplest production reaction being of the type

$$p+p \rightarrow p+p+g+\overline{g} \qquad (14)$$

where $\overline{g}$ is an antimonopole.

   The main lines of searches will now be briefly mentioned. It may be worthwhile to recall that the methods of detection are based on the large ionizaton energy losses of fast monopoles.

Accelerator searches

   Broadly speaking, the searches may be classified into two groups: (i) Direct detection of monopoles, immediately after their production in high-energy collisions, and (ii) indirect searches, where monopoles are searched for a long time after their production.

   Examples of direct searches are those performed at SLAC,[19] at the CERN-ISR[20,21] and at the $\overline{p}p$ collider[22] with plastic detectors. A set of thin plastic sheets of CR39 or of Nitrocellulose and/or of Makrofol E (Lexan) surrounded an intersection region. Heavily ionizing magnetic poles produced in e^+e^-, pp or $\overline{p}p$ collisions should have crossed some of the plastic sheets and should have left a sort of dislocation along their paths in the plastic sheets. When properly developed, the sheets should show holes along monopole tracks. The experiment at SLAC at the PEP e^+e^- storage ring placed an upper limit cross section of 10^{-37} cm^2, which is about three orders of magnitudes smaller than the QED cross section for point particles.[19] Thus this experiment would exclude poles with masses up to 14 GeV. The new experiment at the KERN $\bar{p}p$ collider, using Kapton foils inside the vacuum chambers and CR39 outside, established an upper limit of $\sigma \leq 3 \times 10^{-32}$ cm^2 for monopole masses up to 150 GeV.[22]

Examples of indirect searches at high-energy accelerators are those performed at the CERN-ISR, Fermilab, CERN-SPS and at other accelerators, using ferromagnetic materials.[4,17,18] For instance, in the experiment at the CERN-SPS, the 400 GeV protons interacted (before reaching a beam dump) in a series of ferromagnetic targets. The poles produced in high-energy pp and pn (and also πN) collisions should lose their energy quickly and be brought to rest inside the target, where they are assumed to be bound. Later on, the pieces of the material were placed in front of a pulsed 200 Kgauss magnetic field; this should have been large enough to extract and accelerate the poles, which should have been detected in nuclear emulsions and in plastic sheets. In this sort of experiment, one can obtain very good cross section upper limits, since one integrates the production over long time intervals. But one must believe in several hypotheses on the behavior of monopoles in matter.

Figure 2 summarizes, as a function of the magnetic charge, the production cross section upper limits (at the 95% C.L.) in pN collisions. Figure 3 summarizes the same limits as a function of monopole mass. Solid lines refer to "direct" measurements; dashed lines to "indirect" measurements at high-energy accelerators; dotted lines refer to "indirect" cosmic-ray experiments.

Cosmic-ray searches

Searches for a flux Φ of fast magnetic monopoles were made using counters, plastic detectors and nuclear emulsions.

Most of the searches with electronic detectors were aimed at detecting lowly ionizing quarks at sea level and at mountain altitude.[4,17,18,23] The information on magnetic poles is only indirect, from a reanalysis of the data. The experimental upper limits are $\Phi < 3 \times 10^{-2}$ cm^{-2}yr^{-1}.

Searches were conducted with nuclear emulsions and plastic detectors. In 1975 was reported a monopole candidate from a high altitude, balloon-born stack of 18 m^2 plastic detectors, nuclear

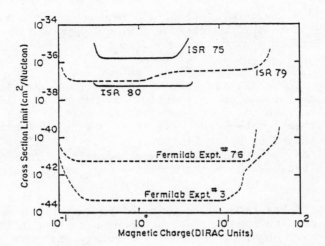

Fig. 2.   Compilation of some upper limits for monopole production
          (at the 95% C.L.) in pp and p-nuclei collisions at
          accelerators plotted versus magnetic charge. Solid and
          dashed lines refer to "direct" and "indirect" measurements
          (see text).

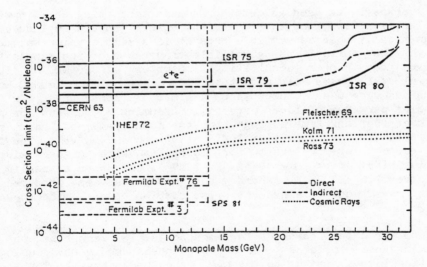

Fig. 3.   Compilation of upper limits for magnetic monopole
          production plotted versus monopole mass. Solid and dashed
          lines refer to "direct" and "indirect" measurements at
          accelerators; dotted lines refer to cosmic-ray
          experiments. The new limit from the SPS $\bar{p}p$ collider (σ
          $\leq 3 \times 10^{-32}$ cm^2) extends up to 150 GeV.

emulsions and a Cherenkov detector.[24] Because of inconsistencies in the various detector readings, the authors later excluded a monopole. From this exposure, and from two subsequent ones, they obtained an upper limit $\Phi < 10^{-4}$ cm^{-2}yr^{-1}.[23,24]

Searches for ancient tracks which traversed samples of mica and obsidian were also conducted.[4] It has to be remembered that mica is full of tracks from natural radioactivity. As far as monopoles, these detectors have high thresholds ($\beta n > 2$). Within this limitation, flux upper limits of $\Phi < 10^{-11}$ cm^{-2}yr^{-1} in mica and $\Phi < 10^{-9}$ cm^{-2}yr^{-1} in obsidian were reported. These good limits are obtained because the rocks had ages of approximately 2×10^{8} years.

Magnetic poles from outer space or produced by cosmic rays at the top of the atmosphere could be brought to rest if their energy is not extremely high. If they have a relatively low mass, they would then drift slowly in the earth magnetic field. These poles could be sucked and accelerated by the magnetic lines of force of solenoids. Their detection would have been performed by counters or emulsions. The estimated upper limit found corresponds roughly to $\Phi < 10^{-6}$ cm^{-2}yr^{-1}.[4]

Searches in bulk matter

An experiment used as detector a superconducting coil in which an electric field, and thus a current change, should be induced by a magnetically charged particle present in a sample which was moved through the coil.[25] Using multiple traversals of the sample they achieved the proper sensitivity. Samples of 20 Kg of lunar material, kilograms of magnetite from earth mines and 2 Kg of meteorites were used. They placed a limit of less than 2×10^{-4} monopoles per gram of lunar material. Assuming a constant monopole flux over the long time during which the moon remained unaltered, they estimated a pole flux $\Phi < 3 \times 10^{-9}$ poles cm^{-2}yr^{-1}. The flux limit becomes less relevant for monopoles with higher kinetic energies (K), $\Phi < 3 \times 10^{-18}$ K^{2} (GeV^{2}), and drops to negligible values for K $> 10^{8}$ GeV. Assuming that monopoles are produced by cosmic rays, one obtained the cross section upper limits shown in Fig. 3.

Another group searched for monopoles in magnetite and in ferromanganese nodules (from deep ocean sediments) using a detecting layout similar to the one described above.[4] The poles should have been extracted, accelerated and sent towards a detector by a large magnetic field. They obtained $\Phi < 10^{-10}$ cm^{-2}yr^{-1}.

Multi-γ events

In nuclear plates exposed to high-altitude cosmic rays, five

events, characterized by a very energetic narrow cone of tens of
γ-rays, without any incident charged particle, were found.[26] One of
the possible explanations of these events could be the following: a
high energy γ-ray, with energy of the order of 10^{12} eV, produces in
the plate a monopole-antimonopole pair by a mechanism similar to
e^+e^- production. The pair then suffers bremsstrahlung and
annihilation, producing the final multi-γ events. Experiments
performed at the ISR and at Fermilab failed to observe them. A
recent ISR experiment at $\sqrt{s}$ = 53 GeV placed an upper limit cross
section of 10^{-37} cm^2.[27] It has been estimated by the authors that
this is about 10 times smaller than the cross section predicted by
a reasonable extrapolation of the higher energy data.

Discussion

The various searches for classical monopoles yielded null
results, whose significance is shown in terms of production upper
limit cross sections in Figs. 2 and 3.

G.U.T. POLES

Properties of G.U.T. poles

As stated in the introduction, the Grand Unified Theories of
strong and electroweak interactions predict the existence of
magnetic monopoles with large masses. In the following, we shall
assume that stable monopoles have masses $m_M \simeq 10^{16}$ GeV.

The GUT pole "has grown in complexity" and is now intuitively
pictured, as sketched in Fig. 4. The pole should have a core of
10^{-29} cm, a region of $r \sim 10^{-16}$ cm typical of the unified
electroweak interaction, a "confinement region" up to 1 Fermi and
finally a condensate of fermion-antifermion pairs which should
extend (in an exponential way) to distances of the order of r
$\sim m_f^{-1}$, that is, up to few Fermis. One may think that going through
the monopole one sees a "small universe", with different regions
full of different virtual particles (from the outside:
fermion-antifermion-pairs, quanta of non-unified forces, particles
typical of the unified electroweak interactions and finally the
core with X-particles).

Another interesting and exotic object is the "monopolonium",
that is, a state monopole-antimonopole analogous to the e^+e^-
positronium; the monopolonium should have an extremely long
lifetime.[28]

As far as the experimental searches which will be described,
the relevant properties of the GUT poles are the mass and the
magnetic charge (discussed in the second section). The extremely

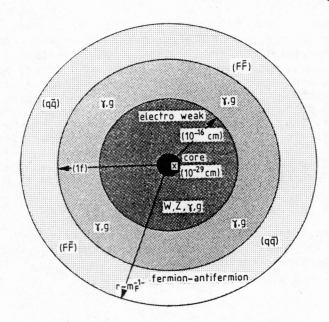

Fig. 4. Illustration of the GUT monopole structure. The sketch
         illustrates various regions, corresponding to: (i) grand
         unification (r ~10^{-29} cm; inside this core, one may find
         virtual X-mesons); (ii) electroweak unification (r ~10^{-16}
         cm; inside one may find virtual W,Z,γ,g); (iii) the
         confinement region (r ~10^{-13} cm; inside one may find
         virtual γ,   g,   f$\bar{\text{f}}$ pairs); (iv) condensate   of
         fermion–antifermion virtual pairs (extending to r ~m_f^{-1}).

large mass implies that poles would have nonrelativistic velocities
and at the same time large kinetic energies.

The history of GUT poles

        Using the "standard model" of the Big Bang, most of the Grand
Unified Theories predict a large production of magnetic monopoles
at the cosmic time t ~10^{-35} seconds, at the phase transition when
the grand unified interaction separated into strong and electroweak
interactions (at which time a subgroup U(1) appeared).   Many
theoreticians tried to find ways to reduce the large number of
produced monopoles.[1]

        As the Universe expanded, the monopoles should have lost
kinetic   energy   and   should   have eventually reached thermal
velocities, with β ~10^{-10} and kinetic energies K ~1 MeV, at the
beginning of the matter-dominated era (at t ~10^{11} sec after the Big
Bang). As matter started to condense into galaxies (at t ~10^{15}

seconds), galactic magnetic fields should have developed. These
fields acted as monopole accelerators.[29,31]

Few of the original monopoles should have been lost from birth
to the time of formation of the galaxies. Possible losses could be
due to monopole-antimonopole annihilation, which could have played
a role only in the early universe, when matter density was large;
its effect is expected to be important only if the number of
produced monopoles was really very large.[31,32] Some pole-antipole
pairs should also have been trapped into monopolonium, which in
most cases should have a lifetime larger than the age of the
universe.

Magnetic poles inside the galaxies should have been
accelerated, preferentially in the plane of the galaxy, by magnetic
fields of the order of $B \simeq 5 \times 10^{-6}$ gauss acting over distances
comparable to the radii of the galaxies ($r \sim 5 \times 10^{22}$ cm).
Monopoles would thus spiral outward in the galaxies and after times
of the order of 10^6-10^7 years would be ejected with velocities

$$\beta_{ejected} \sim 2 \ g \ Br/m \ c^2 \sim 3 \times 10^{-2} \qquad\qquad (15)$$

These relatively fast poles would have had the time to encounter
many galaxies, where they could be accelerated or deaccelerated.
On the average, this would have no net change of energy in the
monopole, nor in the field of the galaxies. The poles ejected from
galaxies would give rise to an intergalactic flux of high-energy
monopoles. An upper limit on the number of these monopoles was
computed in the following way.[30] The regeneration time of the
galactic magnetic field from the dynamo effect is of the order of τ
$\sim 10^8$ years, which is longer than the average time required by a
monopole to escape from the galaxy. Thus if the number of escaping
poles is too large, they would extract energy from the field faster
than it could be replenished and the field would be destroyed. One
may assume that the energy in all the poles should be equal to the
energy stored in the galactic fields. From this equality, one
obtains the number of monopoles per unit volume, from which one has
an upper limit for the flux of $\Phi \lesssim 3 \times 10^{-8}$ cm^{-2} yr^{-1} sr^{-1} (Parker
bound). This number has large uncertainties; it may be raised by
one-two orders of magnitude.[31,32] This limit does not apply to
poles accelerated by possible primordial magnetic fields. Other
alternatives have also been considered; for instance, it has been
suggested that the galactic magnetic fields themselves could arise
from plasma oscillations of a monopole-antimonopole gas. If this
were true, the Parker bound would become irrelevant.

These fast monopoles may be considered to be uniformly
distributed in the universe; therefore, their flux is low. But
like matter, which likes to concentrate in galaxies, stars and

planets, monopoles of lower velocities should probably also
cluster. One may expect that monopoles with velocities $\beta \sim 10^{-3}$ are
bound to the galaxy and may be particularly abundant in the galaxy
halo on the outside borders of the galaxy. Similarly, monopoles
with $\beta \sim 10^{-4}$ may be bound to the solar system and should be
travelling like little asteroids; some monopoles could also orbit
around the earth. For the monopoles with medium and low
velocities, the Parker bound does not apply and higher local fluxes
could be expected. In fact, a fraction of the monopoles impinging
on the sun could be captured and eventually find its way into solar
orbits.[33]

On the basis of what was said before, one could anticipate
that on earth arrives a flux of monopoles with a velocity spectrum
as that sketched in Fig. 5. Notice the values of the escape
velocities from the various astrophysical systems; these values are
also given in Table 1, together with the monopole kinetic energies
with which they correspond.

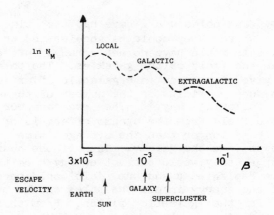

Fig. 5. Sketch of the possibly expected flux of cosmic monopoles
         versus their β. The various peaks correspond to poles
         trapped locally (to the sun and the earth), to poles
         trapped in the galaxy and to extragalactic poles. The
         extragalactic flux should be isotropic, while the others
         should be mainly in the plane of the orbits (in the plane
         of the galaxy and in the plane of the orbit around the
         sun). There could be also a flux of very low energy poles
         surrounding the earth. When poles fall to the earth, they
         acquire kinetic energy; therefore, one does not expect a
         flux of monopoles with $\beta < 3 \times 10^{-5}$ (except for those
         having traversed a certain fraction of the earth).

Table 1. Escape velocities from typical astrophysical systems. The table also gives the values of the monopole kinetic energies (for m = 10^{16} GeV).

System	Escape velocity β	Kinetic energy (GeV)
Cluster of galaxies	3×10^{-3}	5×10^{10}
galaxy	10^{-3}	5×10^{9}
solar system	10^{-4}	5×10^{7}
earth	3.7×10^{-5}	7×10^{6}

Note that poles with $\beta = 10^{-4}$ may be stopped by the earth; poles with $\beta = 10^{-3}$ may be stopped by a star.

Searches for GUT poles--generalities

It has been stated that a flux of cosmic monopoles could reach the earth; its velocity spectrum could be of the type shown in Fig. 5. One has also to remember: pole mass equals $m_M = 10^{16}$ GeV, kinetic energy acquired in earth gravitational field equals 1.2 GeV/m, and gravitational binding to earth is 0.1 eV/Å.

From Fig. 5, and from the considerations made, one may conclude that the velocity range from 3×10^{-5} c to 5×10^{-2} c is the experimentally interesting range for monopole searches. Searches for cosmic poles may be classified as: (i) direct searches for a flux of poles now reaching the earth; (ii) searches for poles which have been trapped in earth ferromagnetic materials; (iii) searches for tracks left in certain materials over the ages by passing poles.

The present searches do not differ in principle from the searches of classical poles and one may ask what is the relevance of those searches for searches for cosmic poles (clearly neglecting the searches at the present accelerators):

i) Experiments performed with counters (more generally with electronics devices) were tuned to fast particles and were therefore insensitive to slow particles (because of long flight times). Experiments of this type may instead play an important role in the future. The experiment performed using lexan plus emulsion detectors had a global threshold of $\beta n > 0.3$ (fixed by lexan). They would therefore be okay only

for high velocities and/or large n-values. The use of CR39
will improve the situation.

ii) The experiments which looked for fossil tracks in mica and
obsidian had too high thresholds, $\beta n > 2$; one can thus repeat
more strongly the same comment made for the search 2b.

iii) It is improbable that heavy poles are stopped at the surface
of the moon, at the surface of the earth or in meteorites.
These last two cases could nevertheless be important for
future searches (using lots of material). For the moon, one
has also to remember that the lunar material was taken to the
earth, experiencing high ($\sim 10^2$ g) decelerations. Monopoles
trapped in all materials except ferromagnetic would have been
lost. There would be partial loss also of monopoles trapped
in iron meteorites (at least in those parts which melt). As
detector one must use a detector sensitive to all monopole
velocities; the superconducting coil is thus okay, but not the
use of a strong magnetic field because the velocities acquired
by the poles would certainly not be sufficient to ionize in
the detector.

Table 2 summarizes the relevance of "classical" searches to
the searches for GUT poles as well as to poles with m $\sim 10^4$ GeV.
None of the searches for "classical" monopoles was really relevant
to the question of the existence of massive poles. The searches
have instead some relevance for possible monopoles with m $\sim 10^4$ GeV.
(In fact, the sensitivity of previous experiments decreases with
increasing pole mass).

Detection of a monopole flux using superconducting coils

The technique of looking for monopoles using small
superconducting coils was first used by the Berkeley group.[25] Since
then, the technique has been improved considerably and one should
now be able to detect the single passage of one magnetic monopole.
This method of detection with a superconducting ring is based
solely on the long-range electromagnetic interactions between the
magnetic charge and the macroscopic quantum state of the
superconducting ring. A passage of a monopole with the smallest
Dirac charge and with any velocity would be observed as a jump of
two flux quanta (fluxons).

Early in 1982 a velocity and mass independent search for
moving monopoles performed by monitoring the current in a 20 cm^2
area superconducting loop was presented.[34] In a run of 151 days the
author found one step current variation, which could be consistent
with that due to the passage of a magnetic monopole according to
Eq. (13) with N = 4 and n = 1. The author stated that "although a
spontaneous and large mechanical impulse seems highly unlikely in

Table 2. Relevance of "classical" monopole searches to searches for GUT poles and poles with $m \approx 10^4$ GeV. In general, the experiments are irrelevant (or have very low sensitivity) for GUT poles, while they have some relevance (that is, detection with low efficiency or marginal capture efficiency) for poles with 10^4 Gev.

	SEARCH TYPE	βn	FLUX $cm^{-2} y^{-1}$	RELEVANCE 10^{16} GeV	FOR 10^4 GeV	REASONS
1.	At accelerators			None	None	Energies too low
2.	Cosmic ray fluxes					
	Counters	>.3	$<10^{-2}$	None	Some	Time of flight – OK for future
	Lexan	>.3	$<10^{-4}$	None	Some	OK for fast poles and large n (CR39 is better)
	Ancient tracks in mica	> 2	$<10^{-11}$	None	Some	OK for fast poles and large n
	Drifting poles		$<10^{-6}$	None	None	Improbable and not detectable
3.	Bulk matter					
	Lunar material ⎫ coil		~1	Doubtful	Some	$m=10^{16}$:capture low;lost when coming to earth $m=10^4$:relevant
	Meteorites ⎭			Possible	Some	Small capture probability (Need mass production) 10^4: better capture probability
	Ferromagnetic			Possible	Some	Small capture probability (Need mass production)
	Ferromagnetic (solenoid)		–	None	Some	Not enough acceleration-Lexan →too high threshold
4.	Multi-γ			None	Some	Energies too low?

an unoccupied laboratory, the evidence presented by this single
event does not preclude that possibility." Thus he considered that
the experiment set an upper limit of 0.53 $m^{-2}d^{-1}sr^{-1}$ for an
isotropic distribution of any moving particle with magnetic charge
greater than 0.06 g_0.

Since then, the experiment continued running and no other
candidate was detected.[6] The author has set up a new larger
detector with 3 large coils mounted one perpendicular to the other.
A monopole should pass through two of the coils, thus reducing
possible background. Also this new detector did not indicate new
candidates during a two month run. Therefore, the present limit
may be estimated to be $\Phi \lesssim 0.06 \ m^{-2}d^{-1}sr^{-1}$.[6,7]

Numerous other superconducting coil detectors have been or are
being set up in different places. Table 3 gives a rough summary of
the present situation. For more details the reader is referred to
Ref. 7. It is clear that there will be results from several
first generation coils (with surface areas of S ~20 cm^2) as well as
from more elaborate second generation coils (with S ~100 cm^2 and
two or more independent coils in order to have effective two-fold
coincidences). Some of these detectors have been surrounded by
scintillation counters in order to have further information.[7]

In the future one may expect further improvements for
rejecting spurious background, larger coils as well as an increase
in their number. It may be worthwhile to mention that one of the
induction coil detectors uses a coil at room temperature.

<u>Counter searches for cosmic monopoles</u>

Several searches for a cosmic flux of magnetic poles have been
performed with scintillation counter arrays. Table 4 lists these
experiments, together with some relevant properties of the
apparatuses (the product $S\Omega$ = area × solid angle, the minimum dE/dx
detectable, the β-range). No monopole was detected; the experiments
can thus place only upper limits (usually at the 90% confidence
level) for the β-range covered. A few comments on some of the
layouts will now be made; more details are given by Loh.[8]

Ullmann[35] (BNL) performed a search with a proportional counter
array having $S\Omega$ = 1.9 m^2sr. With this system, he established upper
limits at the level of 0.03-0.06 $m^{-2}d^{-1}sr^{-1}$ for monopoles with
velocities between 100 and 350 Km/sec (see Fig. 6).

The Bologna search used the existing apparatus of a cosmic-ray
station located on the roof of the physics building.[36] This simple
apparatus, enlarged to reach a value $S\Omega$ = 33 m^2sr, yielded an upper
limit $\Phi < 2\times10^{-4} \ m^{-2}d^{-1}sr^{-1}$ for poles with 0.005 < β < 0.5; the
limit becomes 5 × 10^{-4} for 0.002 < β < 0.005 (see Fig. 6).

Table 3. List of experiments searching for cosmic monopoles using superconducting devices (for more details, see Ref. 7).

LABORATORY	AUTHORS	AREA (cm^2)
1. Stanford	Cabrera[3,4]	20
Stanford	Cabrera[6]	~100
2. Fermilab-Chicago	R. Carrigan, H.J. Frisch, et al.	~140
3. Texas A & M	H. Armbuster, W.P. Kirk	
4. IBM	P. Chaudan, J. Chi et al.	~100
5. Imperial College	A.D. Caplin, J.G. Park, et al.	
6. South Caroline	T. Datta	~20
7. NBS	Clark	
8. Virginia	B.S. Deaver	
9. LBL	P.H. Eberhard, D.E. Morris	(non superconducting)

Table 4. List of experiments which searched or are searching for a flux of cosmic monopoles with scintillation counters, proportional tubes and plastic detectors.

LABORATORY	LOCATION	DETECTOR	$S\Omega$ $(m^2 sr)$	dE/dx (minimum)	β-RANGE	FLUX UPPER LIMIT $(m^{-2} d^{-1} sr^{-1})$	REFERENCE
1. BNL	Building	Proportional	1.9	2.0	$3 \times 10^{-4} - 1.2 \times 10^{-3}$	0.03	35
2. Bologna	Building,roof	Scintillator	33.0	20.0	$2 \times 10^{-3} - 0.6$	1.5×10^{-4}	36
3. Tokyo	Building	Scintillator	2.0	1.2	$10^{-2} - 10^{-1}$	1.3×10^{-2}	37
	Building	Scintillator	6.0	0.025	$2 \times 10^{-4} - 5 \times 10^{-3}$	2.6×10^{-3}	
	Kamioka mine	Scintillator	22.0	0.062	$3 \times 10^{-4} - 1$	3×10^{-3}	
4. Michigan	Building	Scintillator	11.0	0.01	$3 \times 10^{-4} - 10^{-2}$	0.07	38
5. Utah-Stanford	Mayflower mine	Scintillator	3.2	0.03	$10^{-4} - 10^{-2}$	8×10^{-3}	39
6. Minnesota-Argonne	Soudan mine	Proportional	60.0	0.5	$3 \times 10^{-4} - 3 \times 10^{-2}$	2×10^{-4}	40
7. URSS	Baksan mine	Scintillator	large	0.25	$4 \times 10^{-3} - 5 \times 10^{-2}$	1.3×10^{-5}	42
8. India-Japan	Kolar mine	Proportional	218	2.5	$2 \times 10^{-3} - 0.9$	3×10^{-5}	*
9. Berkeley		CR39	150	$z/\beta \geqslant 30$	$0.02 - 1$	1.3×10^{-4}	43

* Communication at this workshop

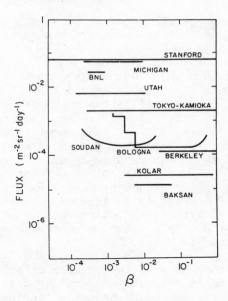

Fig. 6.   Compilation of upper limits for a flux of cosmic GUT
          monopoles plotted versus the β of the monopoles. The
          Stanford experiment corresponds to one candidate event;
          the   other   experiments   are   upper   limits at the 90%
          confidence level.   Most limits   were   obtained   with
          scintillation or gas tube detectors (see Tables 3 and 4):
          the Berkeley experiment was performed with CR39 plastics.

        The first Tokyo layout was a prototype for a larger layout,
which is being installed in the Kamioka mine.[37] The limits given in
Table 4 and Fig. 6 were obtained from the prototype, the equipment
in the Kamioka mine and from a third layout. All these layouts are
characterized by a large number of counter hodoscopes.

        The Michigan[38] and Utah-Stanford[39] layouts were specifically
designed to detect very low ionization losses (down to 0.01 times
minimum ionization).

        The   Minnesota[40]   and   India-Japan[41]   layouts   employed
proportional gas counter arrays.   The India-Japan apparatus is
situated in the Kolar mine in India and was designed for proton
decay studies.

        The largest apparatus in operation uses liquid scintillators
and is located in the Baksan mine in the Soviet Union.[42] The
apparatus was designed as a high-energy neutrino detector.   It
yielded an upper limit $\Phi < 1.3 \times 10^{-5}$ $m^{-2}d^{-1}sr^{-1}$ for poles with
$4 \times 10^{-3} < \beta < 5 \times 10^{-2}$.

A search has been performed by a Berkeley group using CR39 plastic detectors.[23,43] They quote an upper limit $\Phi < 1.3 \times 10^{-4}$ $m^{-2}d^{-1}sr^{-1}$ for $\beta > 0.02$.

The limits obtained by the above mentioned detectors are quoted in Table 4 and illustrated graphically in Fig. 6. There is a general consensus that scintillation counters detect with 100% efficiency monopoles with $\beta > 10^{-3}$; they very probably have some efficiency for $2 \times 10^{-4} < \beta < 10^{-3}$. The proportional gas counter detectors are probably reliable for $\beta >$ few 10^{-3}. One may thus conclude that for $\beta > 10^{-3}$ there already exist limits at the level of few 10^{-5} $m^{-2}d^{-1}sr^{-1}$. For $3 \times 10^{-5} < \beta < 2 \times 10^{-4}$ there is only the value from the superconducting coil experiment, while for $2 \times 10^{-4} < \beta < 10^{-3}$ there is an intermediate limit. The counter limits apply to poles with the g charge and n = 1; if n > 1 or e = 1/3, and thus g = 3g , some of the counter limits will certainly be okay down to $\beta^0 \simeq 10^{-4}$.

In the near future some of the counter arrays for high-energy neutrino experiments at CERN and at Fermilab will probably also be devoted to searches for magnetic monopoles (during beam off times).

MONOPOLE CATALYSIS OF PROTON DECAY

It was known for some time that GUT monopoles would give rise to interactions with the nucleons of ordinary matter which would violate baryon number conservation:

$$pM \rightarrow Me^+ \pi^0$$

$$\rightarrow M \mu^+ K^0 \qquad\qquad\qquad (16)$$

$$\rightarrow Me^+ \mu^+ \mu^-, \text{ etc.}$$

One thought that the cross section would be very small, comparable to the geometric cross section of the core ($\sim 10^{-58}$ cm^2), where may be found the X-mesons which mediate the $\Delta B \neq 0$ interaction. Recently Rubakov et al.[44] suggested that the cross section could be comparable to the cross section of ordinary strong interactions σ $\sim 10^{-26}$ cm^2) because, as sketched in Fig. 4, the monopole should be surrounded by a condensate of fermion-antifermion pairs. A proton (neutron) hitting the fermion condensate is probably "captured" and eventually should reach the core, where would happen the proton decay (monopole catalysis of proton decay). This possibility has stirred up considerable theoretical interest and many controversies.[45-47] At this time, one should in fact consider the cross section to be uncertain by several orders of magnitude.

If the $\Delta B \neq 0$ cross section for monopole catalysis of the proton decays (16) would be large, then a monopole would trigger a chain of baryon "decays" along its passage through a large detector, such as those designed to study baryon decay or in a large bubble chamber. The mean free path λ between two successive monopole-induced proton decays would be[45]

$$\lambda(cm) = \frac{43}{\rho(g \ cm^{-3})} \ \frac{\beta}{\sigma} \qquad\qquad (17)$$

where σ is the cross section in units of the typical strong interaction cross section ($\sigma_0 = 4 \times 10^{-26}$ cm^2), βc is the velocity of the pole and ρ the density (in g cm^{-3}) of the material traversed.

Equation (17) predicts that in a large deuterium bubble chamber one would have a "proton decay" along the monopole track every 30 cm (3 cm) for monopoles with β equal to 10^{-3} (10^{-4}). The Big European Bubble Chamber (BEBC) at CERN and the 15 ft bubble chamber at Fermilab, filled with hydrogen, deuterium and Ne/H, have taken more than 20 million pictures. These events have not been scanned for; assuming nevertheless an efficiency of the order of 10% for having seen them coming from off-beam tracks with any direction, and assuming a sensitive time of 15 msec, one probably already has an upper limit, which could be interpreted either as a limit on the monopole-induced proton decay cross section ($\sigma < (10^{-1}-10^{-2})$ σ strong) or as an upper limit on the flux $\Phi < 10^{-2}$ m^{-2}d^{-1}sr^{-1} of cosmic poles with $3 \times 10^{-5} \lesssim \beta \lesssim 10^{-3}$. (For $\beta < 3 \times 10^{-4}$ one probably does not see anymore the monopole track, but one can see the string of close by decays.)

Ellis et al.[45,46] quote another upper limit based on the present limit for the proton lifetime: if $\tau_p > 10^{31}$ years and if $\sigma = \sigma_0$, then $\Phi < 3 \times 10^{-3} \beta$ m^{-2}d^{-1}sr^{-1}.

From the above reasoning, one concludes either that the flux of monopoles is low or that the Rubakov cross section is much smaller than σ_{strong}.

Other informations for monopole induced proton decays come from the study of neutron stars.[47-49]

Magnetic monopoles should be stopped and trapped in neutron stars. In fact, because of their strong magnetic fields the effective surface of the star is much larger ($\sim 10^6$) than its geometrical dimensions (10 Km). Moreover, the material of the neutron star should be an exceptionally good conductor (with 10^9 times better conductivity than copper). Therefore eddy current losses should be high, and neutron stars should stop very efficiently low velocity monopoles. Monopoles will keep

accumulating in the neutron star, as time goes on. They will heat
the star to high temperatures, increasing considerably the emitted
x-ray flux, which could easily overwhelm the general x-ray
background. Kolb et al.[49] obtained a limit $\Phi \sigma < 5\times10^{-49}$ $s^{-1}sr^{-1}$.

What would happen if a monopole also has a negative electric
charge? Its long-range electric field could capture protons at
atomic distances.[50] They could quickly go to lower orbits and
eventually hit the fermion condensate giving rise to $\Delta B \neq 0$
reactions. Thus the effective Rubakov cross section would be much
larger (of the order of $(r_{at}/r_{nu})^2 \sim 10^{10}$). Thus a small number of
negative dyons captured by a star could produce significantly large
energies. The limit of the flux Φ for the negative dyons can thus
be considerably improved. (As a result of catalyzed proton decays,
one should produce a significant number of muon neutrinos with
energies of the order of 100 MeV.)

OUTLOOK

As it was stated in the introduction, it is evident that the
search for magnetic monopoles has evolved into a fascinating
interdisciplinary field of physics, with implications in
fundamental theories, particle physics, astrophysics and cosmology.
In the future, one may expect many new mathematical and theoretical
analyses and many elaborate experiments.

Searches for "classical monopoles" will continue to be
performed at each new accelerator, even if this is not the hottest
topic on the subject of monopoles. One experiment is going on at
the CERN $\bar{p}p$ collider ($E_{c.m.}$ = 540 GeV).[22] A proposal has been
approved for the tevatron $\bar{p}p$ collider at Fermilab (it should reach
a c.m. energy of 2000 GeV).[51] These experiments will be pertinent
to monopoles with masses up to 800 GeV. The accessible cross
sections will be limited to $\sigma \sim 10^{-34}$ cm^2 by the available
accelerator luminosities. Better limits could be obtained by
higher luminosity colliders.[52]

The new searches for a flux of cosmic GUT monopoles will be
performed by more elaborate and more sensitive "second generation
experiments." In the case of searches with superconducting
devices, the "second generation" will imply coils of larger areas
($\gtrsim 100$ cm^2) and of greater sophistication, for instance, using two
or more coils in "coincidence" to reduce the background.[6,7]

Scintillation counters are probably sensitive to monopoles
with β as low as 10^{-4}. It would clearly be important to establish
with more certainty the energy loss (and the photon yield) at low
values. There are plans to use the large neutrino detectors at

CERN (Experiment WA 18) and at Fermilab (additional electronics have to be installed in order to detect monopoles with low velocities). "Second generation" counter layouts will probably involve large arrays, hundred of square meters in size, with several layers. In order to reduce the background, the detectors will probably be used underground.[8,41,42]

Proton decay experiments are adding electronics in order to be able to detect strings of monopole induced-proton decays.[41,53] Thus some proton decay experiments will be connected with monopole searches.

Different types of new detectors are and will be studied.[54,55] They may essentially complement the present searches.

There is at least one proposal to perform a large scale experiment of the bulk matter type, using ferromagnetic material and a superconducting device.[56] The experiment will be located at a steel plant in Wisconsin, where are processed ferromagnetic materials which were formed several hundred million years ago. Monopoles trapped in the material will leave it when it is heated above the melting point (or above the Curie point, when the material stops being ferromagnetic). The poles would freely fall towards the earth and will be detected when they pass through the superconducting device.

A cosmic monopole has a small probability of stopping and being captured at the surface of the earth. It is to offset this small probability that very large quantities of ferromagnetic materials have to be processed if one wants to reach high sensitivities.

The present astrophysical analyses and limits on the monopole abundance could be improved; it should be possible to find completely new ones. An example is the study of x-rays from neutron stars to see if one finds indications of proton decays catalyzed by monopoles.

An analysis of the magnetic field of the sun has revealed a monopole component in the field.[57,58] It is difficult to see how this could come about from the trapping of magnetic poles, since it would indicate preference for one type of poles. It is still interesting to improve the measurements of the magnetic field of the sun and its analysis.

A search was performed for absorption of 42 cm electromagnetic waves in neutral galactic hydrogen.[59] If this absorption existed it could indicate the presence of atoms for which the magnetic moment of the proton could arise from a distribution of magnetic charges rather than from a distribution of circulating currents.

Several analyses of where the monopoles could have been concentrated have been performed; monopoles could have concentrated inside the earth, the stars, in the halo of the galaxies, in black holes, and so on.[55,60] Also, these analyses may be improved.

CONCLUSION

The problem of the existence of magnetic monopoles is still an open one. It would be desirable to improve the theoretical predictions and make them more quantitative; from the experimental point of view, it is clear that future searches will involve major experimental efforts. The need of more quantitative predictions and less extreme speculation was well emphasized by a solid state physicist who quoted Mark Twain: "I adore science; nowhere else is it possible to reap such a magnificent harvest of speculation from such a trifling investment in fact."

ACKNOWLEDGEMENTS

I would like to acknowledge many colleagues for discussions and for sending material before publication. In particular, I would like to thank Drs. E. Amaldi, R. Bonarelli, P. Buford Price, B. Cabrera, P. Capiluppi, R. A. Carrigan, Jr., D. Cline, M. Koshiba, E. Loh, M. J. Longo, G. Mandrioli, P. Musset, P. Trower, A. M Rossi, C. C. Tsuei and C. N. Yang. The skillful help of Miss R. Genasi is gratefully acknowleged.

REFERENCES

1. See the proceedings of the conference on "Monopoles in Quantum Field Theory," Trieste World Scientific (1981).
2. P.A.M. Dirac, Proc. Roy. Soc. 133, 60 (1931); Phys. Rev. 74, 817 (1948); Int. J. Theor. Phys. 17, 235 (1978).
3. G. Giacomelli, Searches for missing particles, invited paper at the 1978 Singapore meeting on "Frontiers of Physics," Proceedings of the Conference (1978).
4. G. Giacomelli, Review of the experiment status (past and future) of monopole searches, invited paper at the Conference on "Monopoles in Quantum Field Theory," Trieste (1981). Revised for the 1982 Zuoz Spring School of Physics.
5. D. B. Cline, The search for magnetic monopoles, invited talk at the meeting on Experimental Test of Unified Theories, Venezia (1982).
6. B. Cabrera (Stanford superconducting monopole detectors), invited paper at this workshop.
7. C. C. Tsuei, Review of other induction experiments, invited paper presented at this workshop.

8. E. C. Loh, Electronic cosmic-ray searches, invited paper at
    this workshop.
9. S. P. Ahlen and K. Kinoshita, Calculation of the stopping
    power of very low velocity magnetic monopoles, Berkeley,
    preprint (1982).
10. S. P. Ahlen, Monopole energy loss and detector excitation
    mechanism, invited paper presented at this workshop.
11. S. Geer and W. G. Scott, Calculation of the energy loss for
    slow monopoles in atomic hydrogen, CERN $\bar{p}p$ Note (1981).
12. D. Ritson, Fermi-Teller theory of low velocity ionization
    losses applied to monopoles, SLAC-PUB-2950 (1982).
13. V. P. Martem'yanov, et al., Slowing down of a Dirac monopole
    in metals and ferromagnetic substances, Soviet Phys. JETP
    35, 20 (1972).
14. M. J. Longo, Phys. Rev. D 25, 2399 (1982).
15. C. A. Goebel, private communication to P. M. McIntyre.
16. E. Goto, Progr. Theor. Phys. 30, 700 (1963).
17. R. A. Carrigan, Jr., Magnetic monopole bibliography, FERMILAB
    77/42 (1977); R. E. Craven et al., FERMILAB 81/37 (1981).
18. Review of Particle Properties, Magnetic Monopole Searches,
    Phys. Lett. B 111, 1 (1982).
19. K. Kinoshita, Search for highly ionizing particles in e^+e^-
    collisions at PEP, Ph.D. Thesis, Berkeley (1981).
20. G. Giacomelli et al., Nuovo Cimento 28A, 21 (1975).
21. H. Hoffman et al., Nuovo Cimento Lett. 23, 357 (1978).
22. B. Aubert et al., Search for magnetic monopoles at the CERN $\bar{p}p$
    collider, Experiment UA3, Annecy-CERN Collaboration; P.
    Musset, Search at the CERN-SPS collider, invited paper
    presented at this workshop.
23. P. B. Price, Searches for exotic particles, LBL preprint
    (1982).
24. P. B. Price et al., Phys. Rev. Lett. 35, 487 (1975); Phys.
    Rev. D 18, 1382 (1978).
25. R. R. Ross, P. H. Eberhard, L. W. Alvarexz and R. D. Watt,
    Phys. Rev. D 8, 689 (1973); Phys. Rev. D 4, 3260 (1971).
26. M. Schein, D. M. Haskin and R. G. Glasser, Phys. Rev. 95, 855
    (1954); Phys. Rev. 99, 643 (1955).
27. G. F. Dell et al, Multigamma-ray events at the CERN-ISR,
    presented at this workshop, to be published (1982).
28. G. T Hill, Monopolonium, invited paper presented at this
    workshop.
29. M. S. Turner, Monopoles and astrophysics, invited paper
    presented at this workshop.
30. E. Purcell, Monopoles and galactic magnetic fields, invited
    paper presented at this workshop.
31. T. Kibble, Monopoles in the early universe, 1981 Trieste
    Meeting.
32. N. Straumann, Cosmological production of magnetic monopoles,
    lectures given at the SIN Spring School, Zuoz, Switzerland,
    April (1982).

33. S. Dimopoulos et al., Nature 298, 824 (1982).
34. B. Cabrera, Phys. Rev. Lett. 48, 1378 (1982).
35. J. D. Ullmann, Phys. Rev. Lett. 47, 289 (1981).
36. R. Bonarelli et al., Phys. Lett. 112B, 100 (1982); and paper
     presented at this workshop.
37. T. Moshimo, K. Kawagoe and M. Koshiba, Journal of the Phys.
     Soc. of Japan 51, 3065 (1982); M. Koshiba and S. Orito,
     Particle physics at LICEPP, present and future, LICEPP
     report, Tokyo and communication to this workshop.
38. J. K. Sokolowski and L. R. Sulak, A search for
     lightly-ionizing slow magnetic monopoles, Univ. of
     Michigan, preprint. Submitted to the 21st Int. Conf. on
     High Energy Physics, Paris (1982).
39. D. E. Groom, E. C. Loh and D. M. Ritson, A search for a
     slow-moving flux of massive magnetic monopoles, contributed
     to the XXI Conf. on High Energy Physics, Paris and UHEP
     82/2 (1982).
40. J. Bartelt et al., A new limit on magnetic monopole flux,
     preprint, University of Minnesota (1982).
41. India-Japan proton decay experiment in the Kolar mine,
     presented at this workshop.
42. E. N. Alexeyev, Search for superheavy magnetic monopoles at
     the Baksan underground telescope, Moscow Institute for
     Nuclear Research, preprint, submitted to Nuovo Cimento
     Lettere (1982).
43. P. Buford Price, Searches at accelerators and elsewhere,
     invited paper presented at this workshop.
44. V. A. Rubakov and M. S. Serebryakov, Anomalous baryon number
     non-conservation in the presence of SU(5) monopoles, Moscow
     preprint (1982).
45. J. Ellis, D. V. Nanopoulos and K. A. Olive, Baryon number
     violation catalyzed by Grand Unified Monopoles, TH 3323
     CERN (1982).
46. F. A. Bais et al., More about baryon number violation
     catalyzed by Grand Unified Monopoles, Preprint TH 3383 CERN
     (1982).
47. G. C. Callan, Jr., Monopole catalysis of baryon decay, invited
     paper presented at this workshop.
48. S. Dimopoulos, Catalyzed nucleon decay in neutron stars,
     NSF-TP-82-91 (1982).
49. E. W. Kolb, et al., Phys. Rev. Lett. 49, 1373 (1982).
50. L. Bracci and G. Fiorentini, Monopolic atoms and monopole
     catalysis of proton decay, IFUP TH 82-18 (1982).
51. K. Kinoshita et al., Proposal to search for highly ionizing
     particles for the DO Area at Fermilab, Fermilab experiment
     E-713.
52. G. Giacomelli and G. Kantardjian, Magnetic monopole searches
     at Isabelle, Proceedings of the 1981 Isabelle Summer
     Workshop, BNL 51443.

53. E. Fiorini et al., Proton decay experiment under Mt. Blanc, private communication.
54. B. L. Barish, Acoustical detection of monopoles, invited paper presented at this workshop.
55. C. W. Akerlof, Limits on the thermoacoustic detectability of electric and magnetic charges, preprint, Univ. of Michigan (1982).
56. D. Cline, Trapped monopoles in matter, invited paper presented at this workshop.
57. J. M. Wilcox, Why does the sun sometimes look like a magnetic monopole, Comments on Astrophysics and Space Physics $\underline{6}$, 141 (1974).
58. L. W. Alvarez, Some thoughts on monopoles, LBL-Physics Notes - Memo 932 (1982).
59. J. J. Broderick et al., Phys. Rev. D $\underline{19}$, 1046 (1979).
60. R. A. Carrigan, Jr., Nature $\underline{288}$, 348 (1980).

MONOPOLE COMPATIBILITY WITH COSMOLOGY

G. Lazarides

Department of Physics
Rockefeller University
New York, New York 10021

GRAND UNIFIED MONOPOLES

Grand Unified Theories (GUTs)[1,2] of strong, weak and electromagnetic interactions predict the existence of superheavy magnetic monopoles.[3,4] The exact value of the mass of these monopoles, their production rate in the early universe, their subsequent history and their abundance at present depend on the specific GUT model we adopt as well as on some astrophysical details. <u>But the existence of magnetic monopoles is the most general consequence of the very idea of Grand Unification</u>. As soon as we embed $U(1)_{em}$ into a simple grand unifying gauge group G, we are bound to have magnetic monopoles.

Let us now see what a GUT monopole looks like. For definiteness, we will consider the simplest GUT based on the gauge group $SU(5)$.[1] The symmetry breaking goes as follows:

$$SU(5) \xrightarrow[24]{M_x \sim 10^{15} \text{GeV}} [SU(3) \times SU(2)] \otimes U(1)_y \xrightarrow[5]{M_w \sim 10^2 \text{GeV}} SU(3) \otimes U(1)_{em}.$$

$$(1)$$

The <u>fundamental</u> monopole of this theory is a complicated configuration where all the mass scales appear. It contains a superheavy core of radius M_x^{-1}, which is made out of superheavy gauge and Higgs particles. At the center of this core the VEV of the Higgs 24-plet is zero, whereas outside the core, this field

71

takes values in its vacuum manifold. The magnetic field around
this monopole, for distances $\lesssim M_w^{-1}$, consists of one unit of
chromomagnetic flux, one unit of $SU(2)$-magnetic flux and a minimal
amount of $U(1)_y$-magnetic flux.[5,6] For distances between M_w^{-1} and
Λ^{-1} ($\Lambda \simeq 100$ MeV), we have one unit of chromomagnetic flux and a
minimal amount of ordinary magnetic flux. The chromomagnetic flux
is screened by the QCD vacuum at distances $\sim \Lambda^{-1}$ ($\Lambda \simeq 100$ MeV) and
the only magnetic field that appears at even larger distances is
the ordinary magnetic field of the monopole.

At non-zero temperature T, the structure of the monopole is
different. For $M_w \lesssim T \lesssim M_x$, the chromomagnetic and the
$SU(2)$-magnetic flux are screened by thermal fluctuations at a
distance $\sim (g^2 T)^{-1}$.[7] The ordinary magnetic flux is long ranged. For
$\Lambda \lesssim T \lesssim M_w$, the electroweak core of radius $\sim M_w^{-1}$ is formed and the
chromomagnetic flux is still screened by thermal fluctuations at
distances $\sim (g^2 T)^{-1}$. For $T \lesssim \Lambda$, confinement sets in and the
chromomagnetic flux is screened by the QCD vacuum at distances
$\sim \Lambda^{-1}$. The structure of the monopole is then essentially its
zero-temperature structure.

MONOPOLE PRODUCTION IN THE EARLY UNIVERSE

Monopoles are produced at the phase transition where the
unifying gauge symmetry $SU(5)$ breaks down to $[SU(3) \times SU(2)] \otimes$
$U(1)_y$. This transition takes place at a critical temperature T_c
$\sim (M_x/g) \sim 10^{15}$ GeV and, for the time being, we will assume that it
is a second-order transition. The finite temperature effective
potential $V(\phi)$ of the Higgs 24-plet ϕ has its absolute minimum at
$\phi = 0$ at any $T \geq T_c$ and $SU(5)$ is unbroken. For $T < T_c$, $\phi = 0$
becomes a local maximum. The absolute minimum of $V(\phi)$ then appears
at $\langle\phi\rangle \neq 0$, and $SU(5)$ breaks down to $SU(3) \times SU(2) \times U(1)$. The
magnitude of $\langle\phi\rangle$ grows continuously from zero to $\langle\phi\rangle$ (T=0) $\sim M_x/g$ as
the universe cools

$$\langle\phi\rangle \ (T) \simeq \langle\phi\rangle \ (T=0) \left(1 - \frac{T^2}{T_c^2} \right)^{1/2} , \ T \leq T_c . \tag{2}$$

The Higgs mass $m_H \sim h\langle\phi\rangle$ (h^2 is some combination of the quartic
Higgs self-couplings) also grows continuously from zero to its
zero-temperature value.

At temperatures of the universe below T_c, the difference in
free energy density between the $SU(5)$ symmetric maximum at $\phi = 0$
and the asymmetric minimum at $\phi = \langle\phi\rangle$ (T) is given by

$$\Delta V \sim h^2 <\phi>^4 \; . \tag{3}$$

For temperatures just below T_c, ΔV is very small and thermal fluctuations of the Higgs field ϕ back and forth across the local maximum at $\phi = 0$ are very common. The condition for this to happen is the following. <u>The free energy needed for ϕ to fluctuate from the minimum of $V(\phi)$ back to $\phi = 0$ in a sphere of radius equal to the Higgs correlation length $\xi = m_H^{-1}$ is not greater than the temperature T</u>, i.e.,

$$\left(\frac{4\pi}{3}\right)\xi^3 \; \Delta V \leq T \; . \tag{4}$$

The T at which equality holds is called the <u>Ginzburg</u> <u>temperature</u> T_G.[8]

For $T<T_G$, ϕ may fluctuate near the minimum but it is very unlikely to fluctuate back to $\phi = 0$. So, below T_G, $<\phi>$ lies essentially on the vacuum manifold M almost everywhere in space. The choice of $<\phi>$ is otherwise random and $<\phi>$ is expected to take different values in M in different regions of space. Magnetic monopoles are effectively frozen in for $T<T_G$. They are topologically stable extended structures with $<\phi> = 0$ at their center. On any sphere around the monopole with radius bigger than the size of the monopole, $<\phi> \in M$, and S^2 is mapped <u>onto</u> a closed 2-dimensional surface in M which is homotopically nontrivial. This means that this image of S^2 in M cannot be deformed <u>continuously</u> down to a point because of some topological "obstacles" in M.

The Higgs field is correlated up to distances $\sim\xi = m_H^{-1}$. Thus, at T_G, we may imagine that space splits into regions of linear dimension of the order of the Higgs correlation length at T_G,

$$\xi_G \sim \frac{1}{h^2 T_c} \; . \tag{5}$$

Within each such region, the Higgs field is essentially aligned but its values in different regions are uncorrelated. (At the boundaries of these regions, ϕ varies smoothly from one value to the other.) At the corners where several regions meet we will occasionally have monopoles.[9] Thus, the number density of monopoles produced at T_G is

$$n_M \sim p \; \xi_G^{-3} \sim p h^6 \; T_c^3 \; . \tag{6}$$

Here p is a geometric factor of order 1/10. Instead of n_M, we customarily use the "relative monopole density"

$$r = \frac{n_M}{T^3} .$$
(7)

The above estimate, then, gives for the <u>initial</u> relative monopole density,

$$r_{in} \sim 10^{-6} .$$
(8)

This estimate depends on various assumptions about the monopole production mechanism and it is also considerably uncertain because of uncertainties in the values of h^2 and T_G. It is thus important to find a lower bound on r_{in} which is more general and does not depend on so many details. Whatever the details of the transition are, the Higgs field cannot be correlated over distances greater than the particle horizon 2t (t is the cosmic time) at T_G. So, the linear dimension of the regions in which ϕ is aligned cannot be bigger than 2t at T_G. We thus derive the causality bound[10] on the initial monopole density

$$n_M \gtrsim \frac{p}{(4\pi/3)(2t)^3}$$
(9)

or, equivalently,

$$r_{in} \gtrsim 10^{-10} .$$
(10)

SUBSEQUENT MONOPOLE ANNIHILATION

After monopole production at T_G, the monopole number density follows the equation[11]

$$\frac{dn_M}{dt} = -Dn_M^2 - 3 \frac{\dot{R}}{R} n_M .$$
(11)

The second term in the RHS of this equation describes the dilution of monopoles due to the cosmological expansion. The first term describes monopole-antimonopole annihilation. The annihilation process is characterized by the coefficient D and has been studied by Preskill.[11] The monopoles diffuse towards antimonopoles through the plasma of light charged particles, capture each other in Bohr orbits and finally annihilate. This process goes on as long as the mean free path of the monopole is shorter than the capture distance, i.e.,

$$T \gtrsim T_f \sim 10^{12} \text{ GeV}. \tag{12}$$

For $T < T_f$, monopoles and antimonopoles can capture each other only by emitting radiation. This mechanism turns out to be very ineffective and does not lead to any appreciable annihilation, for $T < T_f$. The final result is summarized as follows. If

$$r_{in} \gtrsim 10^{-9}$$

then

$$r_{fin} \sim 10^{-9},$$

If

$$r_{in} \lesssim 10^{-9}$$

then

$$r_{fin} \sim r_{in}. \tag{13}$$

This result together with the causality bound on the initial monopole density implies

$$r_{fin} \gtrsim 10^{-10}. \tag{14}$$

OBSERVATIONAL BOUNDS ON r

We can obtain a bound on the present value of r from the requirement that the mass density due to monopoles does not exceed the limit on the mass density of the universe imposed by the observed values of the Hubble constant and the deceleration parameter. This bound reads[11]

$$r_{now} \lesssim 10^{-24}. \tag{15}$$

So, there is a big discrepancy between this constraint and our previous theoretical estimates.

However, one may argue that, since we do not know the details of monopole history in the late universe, the above observational bound on r cannot be compared with our theoretical estimates. (Monopoles may have been accumulated in special places such as cores of galaxies or massive stars where they have been annihilated.)

There is another limit on r which does not depend on the details of the late universe. The standard scenario of nucleosynthesis requires that the monopoles do not dominate the energy density of the universe at $T \simeq 1$ MeV. This yields the cosmological bound[11]

$$r(T \simeq 1 \text{ MeV}) \lesssim 10^{-19} , \tag{16}$$

which is still in great disagreement with our theoretical predictions.

GUTs are in trouble. They predict too many magnetic monopoles in the universe, at least within the standard hot big bang cosmology.

CAN THE PREDICTED MONOPOLE DENSITY BE REDUCED?

Many people have suggested possible solutions to this problem. I will try to review some of the suggestions. My list is by no means complete.

Bais and Rudaz[12] have argued that, at temperatures between T_c and T_G, single monopoles can be created or destroyed by the fluctuations of the Higgs field back and forth across the maximum at $\phi = 0$. They also argued that thermal equilibrium is maintained in this process and thus the number of monopoles that freeze in at T_G is given by the Boltzman law

$$r_{in} = \left(\frac{M(T_G)}{2\pi\, T_G} \right)^{3/2} \exp\left[-\frac{M(T_G)}{T_G} \right] . \tag{17}$$

Here M is the monopole mass. This formula gives an acceptably small r_{in}, provided $m_H \gtrsim 2.5 m_V$. The main difficulty of this approach is the following. It is not easy to see how, by using thermodynamic arguments, one can avoid the causality limit on r_{in}.

Langacker and Pi[13] have proposed a model where $U(1)_{em}$ remains spontaneously broken at temperatures $> T_c \gtrsim 1$ TeV. $U(1)_{em}$ is restored at $T_c \gtrsim 1$ MeV and monopoles are produced at this temperature. At such low T's the particle horizon is big and the causality limit on r_{in} not so stringent. The Kibble[9] mechanism for monopole production is also not effective, since we have restoration rather than breaking of symmetry at T_c. Moreover, any thermal production of monopoles will be strongly suppressed since $M(T_c) \gg T_c$. This proposal appears to be viable but it is rather inelegant since it requires a complicated ad hoc set of Higgs fields.

Lazarides and Shafi[14] have suggested that the cosmological monopole problem may be solved in models where $SU(2)_L$ and $U(1)_y$ do not intersect. (In such models all combinations of $SU(2)_L$ and $U(1)_y$ quantum numbers are allowed.) It turns out that the monopoles of these theories do not carry $SU(2)_L$ magnetic fields. At the Weinberg-Salam transition, these monopoles may acquire a tube carrying Z^0-magnetic flux. They are then connected to antimonopoles and annihilate. Unfortunately, it is not easy to construct realistic models where this mechanism is realized.

Linde[15] has argued that at high temperatures the plasma of quarks, leptons, etc. causes the non-abelian part of the magnetic flux of monopoles to squeeze into tubes of thickness $\sim(g^2T)^{-1}$. Monopoles and antimonopoles are then joined by flux tubes and annihilate. I do not consider this mechanism plausible for the following reasons: i) Monte Carlo calculations by Billoire, Lazarides and Shafi[7] have shown that non-abelian magnetic flux at high temperatures is screened rather than squeezed into tubes, and ii) squeezing of non-abelian magnetic flux would require infinite conductivity of non-abelian electric currents (compare with Meissner effect) which is very unlikely in a high temperature plasma.

Another very interesting suggestions was made by Guth and Tye.[16] They introduced a cubic term into the Higgs potential for the SU(5) 24-plet ϕ. This leads to a __first-order__ transition. It was also noticed that the Coleman-Weinberg (CW) potential leads to a first-order transition too. Such a transition proceeds as follows. At $T > T_{c1}$, $V(\phi)$ has a single minimum at $\phi = 0$. At $T = T_{c1}$, a secondary minimum appears. At the critical temperature T_c the two minima become degenerate. For $T < T_c$, $\phi = 0$ becomes metastable and tends to decay to the true vacuum by quantum tunnelling. Since this process is very slow, we can have a prolonged supercooling of the universe in the false vacuum $\phi = 0$. The transition proceeds by nucleation of bubbles of the true vacuum $\phi = <\phi>$. The Higgs field ϕ is aligned within each bubble and monopoles may be produced by the Kibble mechanism as the bubbles coalesce to fill the space. If the tunnelling rate is low enough, the monopole production is __suppressed__. At the completion of the transition, latent heat is released and reheats the universe back to a temperature $T_R \lesssim T_c$. The thermal production of $M\bar{M}$ pairs during reheating is also suppressed. Unfortunately, this scenario has many problems. As the universe supercools in the false vacuum, the energy density of this vacuum dominates over the radiation energy density and leads to exponential expansion (inflation) of the parts of the universe that are in the false vacuum. Then the bubbles of the true vacuum never coalesce and the transition is not completed.[17,18] Even if we achieve completion of the transition, we end up with a very inhomogeneous universe.

Linde[19] as well as Albrecht and Steinhardt[20] have suggested a very clever variant of the above scenario (new inflation). They consider SU(5) with the CW potential and they observe that ϕ does not take its vacuum value $<\phi>$ immediately after tunnelling away from $\phi = 0$. It actually takes a value ϕ_1 very close to $\phi = 0$. The roll over of the field ϕ from ϕ_1 to $<\phi>$ can be very slow. During this roll over, there is appreciable vacuum energy density which may lead to exponential expansion. A single bubble can expand exponentially (inflate), so that it becomes much bigger than our Universe. We are then within a single bubble, and there is no monopole creation by the Kibble mechanism. One can also estimate the monopole density due to thermal fluctuations after reheating. This turns out to be utterly negligible.[21] So, we conclude that (new) inflation leads to a complete absence of monopoles.

This improved inflationary scenario also has its problem. It leads to unacceptably large density perturbations in the universe.[22,23]

Finally, Lazarides and Shafi[24] have recently shown that the monopole problem can in principle be solved in the context of supersymmetric GUTs of the O'Raifeartaigh[25]-Witten[26] type. In these theories there is only one fundamental mass scale, namely the supersymmetry (SUSY) breaking scale $\sqrt{F} \gtrsim 10^9$ GeV. The GUT scale is given by

$$M \sim \sqrt{F} \exp (d/g^2) >> \sqrt{F}, \; d>0 \; . \tag{18}$$

The simplest SU(5) model of this variety contains two Higgs 24-plets A and Y and a Higgs singlet X. $<A>$ and $<Y>$ are parallel and break SU(5) to SU(3) $\times$ SU(2) $\times$ U(1), $<A>$ $\sim\sqrt{F}$ and $<Y>$, $<X>$ $\sim M$.

At a $T_c \sim\sqrt{F}$, SUSY breaks and A acquires a VEV $\sim\sqrt{F}$. So SU(5) breaks to SU(3) $\times$ SU(2) $\times$ U(1) but only by a VEV $\sim\sqrt{F}$. Superheavy monopoles of mass $\sim(\sqrt{F}/\alpha)$ are then produced. After Preskill's annihilation, the relative monopole density is

$$r \simeq 10^{-6} \frac{\sqrt{F}}{M_p} \; . \tag{19}$$

We assume that the potential for the X-field is of the Coleman-Weinberg type. The universe supercools in the X = 0 phase and exponential expansion takes place. The X = 0 phase is destabilized at its Hawking temperature $\sim F/M_p$ and the transition to $<X>$ $\sim M$ takes place. The mass of the monopoles increases to (M/α). Assuming that after this transition, the universe reheats back to $T_R \sim\sqrt{F}$, we get a further dilution of monopoles, and

$$r_{fin} \sim 10^{-6} \left(\frac{F}{M_p^{\,2}} \right)^2 . \tag{20}$$

This is acceptably small if $\sqrt{F} \lesssim 10^{16}$ GeV. The difficulty of this approach is that it is not easy to get reheating since the curvature of the potential at X = M tends to be very low.

CONCLUSIONS

So far, we do not know of any simple and elegant GUT model which solves the cosmological monopole problem and at the same time is consistent with all phenomenological and cosmological requirements.

But we should not be pessimistic. After all, we do not know the ultimate model yet. We have many possible mechanisms for reducing the monopole density and it is reasonable to expect that some combination of these mechanisms (as well as other mechanisms yet to be discovered) will solve the problem in the context of the ultimate theory.

ACKNOWLEDGEMENT

This work was supported in part by the U.S. Department of Energy under Contract No. DE-AC02-82ER40033.B000.

REFERENCES

1. H. Georgi and S. Glashow, Phys. Rev. Lett. 32, 438 (1974).
2. J. G. Pati and A. Salam, Phys. Rev. Lett. 31, 661 (1973).
3. G. 't Hooft, Nucl. Phys. B 79, 276 (1974).
4. A. M. Polyakov, JETP Lett. 20, 194 (1974).
5. M. Daniel, G. Lazarides and Q. Shafi, Nucl. Phys. B 170, 156 (1979).
6. C. P. Dokos and T. N. Tomaras, Phys. Rev. D 21, 2940 (1980).
7. A. Billoire, G. Lazarides and Q. Shafi, Phys. Lett. 103B, 450 (1981).
8. V. L. Ginzburg, Sov. Phys.-Solid State 2, 1824 (1960).
9. T.W.B. Kibble, J. Phys. A 9, 1387 (1976).
10. M. B. Einhorn, "Proceedings of Europhysics Conference on Unification of the Fundamental Interactions," Erice (1980).
11. J. P. Preskill, Phys. Rev. Lett. 43, 1365 (1979).
12. F. Bais and S. Rudaz, Nucl. Phys. B 170, 507 (1980).

13. P. Langacker and S.-Y. Pi, Phys. Rev. Lett. $\underline{45}$, 1 (1980).
14. G. Lazarides and Q. Shafi, Phys. Lett. $\underline{94B}$, $\overline{1}49$ (1980).
15. A. D. Linde, Phys. Lett. $\underline{96B}$, 293 (1980).
16. A. H. Guth and S. H. Tye, Phys. Rev. Lett. $\underline{44}$, 631 (1980).
17. M. B. Einhorn, D. L. Stein and D. Toussaint, Phys. Rev. D $\underline{21}$,
     3295 (1980).
18. A. H. Guth and E. Weinberg, Phys. Rev. D $\underline{23}$, 876 (1981).
19. A. D. Linde, Phys. Lett. $\underline{108B}$, 389 (1982).
20. A. Albrecht and P. Steinhardt, Phys. Rev. Lett. $\underline{48}$, 1220
     (1982).
21. G. Lazarides, Q. Shafi and P. Trower, Phys. Rev. Lett. $\underline{49}$,
     1756, (1982).
22. A. H. Guth and S.-Y. Pi, unpublished, (1982).
23. J. Bardeen, P. Steinhardt and M. Turner, unpublished, (1982).
24. G. Lazarides and Q. Shafi, unpublished, (1982).
25. L. O'Raifeartaigh, Nucl. Phys. B $\underline{96}$, 331 (1975).
26. E. Witten, Phys. Lett. $\underline{105B}$, 267 (1981).

REDUCING THE MONOPOLE ABUNDANCE

Alan H. Guth

Laboratory for Nuclear Science and Department of Physics
Massachusetts Institute of Technology
Cambridge, Massachusetts 02139

When a typical grand unified theory (GUT) is considered in the context of standard cosmology, one is led immediately to the conclusion that far too many magnetic monopoles would have been produced in the very early history of the universe.[1,2] In this talk I will discuss two possible mechanisms which might suppress monopole production. The first is the idea of a high temperature superconductor, as proposed by Langacker and Pi.[3] The second is the inflationary universe.[4,5,6] Befitting my biases, I will devote most of my time to the latter.

MONOPOLE PRODUCTION IN THE STANDARD SCENARIO

I will begin by briefly explaining why standard assumptions lead to an overproduction of monopoles. First, one must recall that the 't Hooft-Polyakov magnetic monopoles are in fact topologically stable knots in the Higgs field expectation value.[7] If the Higgs field has a correlation length ξ, then one would expect a density of monopoles given roughly by the Kibble relation[8]

$$ n_M \approx 1/\xi^3 . \tag{1} $$

When the universe cools below the critical temperature[9] $T_c \sim 10^{14}$ GeV of the GUT phase transition, it becomes thermodynamically favorable for the Higgs field to align uniformly over large distances. However, it takes time for these correlations to be established. Note that the horizon length, defined as the distance which a light pulse could have traveled since the initial singularity, is given in the radiation-dominated era by

$$\ell_H = 2t \; , \tag{2}$$

where t is the age of the universe. Thus, causality alone implies[10] that

$$\xi \lesssim 2t \; . \tag{3}$$

To relate the time t to the temperature T, one writes the mass density of the radiation of effectively massless particles (i.e., $M \ll T$) as

$$\rho = cT^4 \; , \tag{4}$$

where

$$c = \frac{\pi^2}{30} \; (N_b + \frac{7}{8} \; N_f) \; . \tag{5}$$

Here N_b denotes the number of effectively massless bosonic spin degrees of freedom (e.g., the photon contributes two units), and N_f denotes the corresponding number for fermions (e.g., electrons and positrons together contribute four units). For the minimal SU(5) theory,[11] $c \approx 52.9$ at the highest temperatures. The temperature is then given by[12]

$$T^2 = \left(\frac{3}{32\pi c}\right)^{1/2} \frac{M_P}{t} \; , \tag{6}$$

where $M_P \equiv G^{-1/2} \approx 1.2 \times 10^{19}$ GeV is the Planck mass. It is convenient to compare the monopole density to the entropy density s, given by

$$s = \frac{4}{3} \; cT^3 \; . \tag{7}$$

If we assume that the phase transition occurs instantaneously when the temperature falls to T_c, then these relationships imply that immediately after the phase transition,

$$n_M/s \gtrsim 18.2 \; c^{1/2} \; (T_c/M_P)^3 \; . \tag{8}$$

According to Preskill,[1] the rate of monopole-antimonopole annihilation would be insignificant, and thus n_M/s would be essentially unchanged today. Assuming a photon temperature today $T_\gamma \approx 2.7°$K, with three species of massless neutrinos at

$T_\nu = (4/11)^{1/3}T_\lambda$, one finds $s \approx 2.8 \times 10^3$ cm^{-3}. For a monopole mass of 10^{16} GeV, this would give a monopole mass density $\rho_M \gtrsim 5 \times 10^{-18}$ gm cm^{-3}. Taking an estimate of the current Hubble "constant" $H \approx (10^{10}$ yr$)^{-1}$, the critical mass density (which gives a precisely flat (k=0) universe) is given by $\rho_c = 3H^2/8\pi G \approx 1.8 \times 10^{-29}$ gm cm^{-3}. Thus, this density of monopoles would yield

$$\Omega \equiv \rho/\rho_c \gtrsim 3 \times 10^{11}. \tag{10}$$

The current age of the universe would then be given by

$$t_0 = \frac{\pi H^{-1}}{2\sqrt{\Omega}} \left[1 + O(\Omega^{-1/2})\right] \lesssim 30,000 \text{ yr}. \tag{11}$$

I am told that an age of this order is considered acceptable in some circles; but I have checked with my friends, and they all believe that such an age can be ruled out. Thus, some mechanism must be found to suppress the production of magnetic monopoles.

THE HIGH-TEMPERATURE SUPERCONDUCTOR

    First, I will summarize the solution proposed by Langacker and Pi.[3] By introducing extra Higgs fields (they use three Weinberg-Salam doublets) and by carefully choosing the couplings, they have shown that the phase structure of an SU(5) GUT might behave as shown in Fig. 1. At the lowest temperatures the unbroken symmetry is $SU(3)_c \times U(1)_{EM}$, the standard symmetry of quantum chromodynamics and electromagnetism. When the temperature is raised to ~10^3 GeV, one has a Weinberg-Salam phase transition of an unusual sort--while ordinarily the SU(2) × U(1) symmetry would be

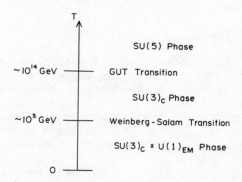

Fig. 1: The phase diagram for the high-temperature superconductor mechanism of monopole suppression.

restored in this case U(1)$_{EM}$ symmetry is broken. (Of course the SU(2) × U(1) symmetry does become valid at high energies and zero temperature, in complete agreement with the standard model. The Langacker-Pi model takes advantage of the fact that high temperature physics, which is governed by thermal expectation values of the Higgs field, can sometimes be quite different from high energy physics.) Above the GUT phase transition at ~10^{14} GeV, the full SU(5) symmetry is restored (as in the standard model). Monopole production is suppressed because the intermediate phase, with SU(3)$_c$ symmetry, does not contain any stable monopoles. Monopoles do exist in the ground state phase as always, but the breaking of U(1)$_{EM}$ at the Weinberg-Salam phase transition causes space to act like a high temperature superconductor. The magnetic flux becomes confined to flux tubes which join monopoles to antimonopoles. Thus, any monopoles which form at the GUT phase transition are confined by flux tubes and they rapidly annihilate. Monopoles might in principle be produced thermally at temperatures below the Weinberg-Salam phase transition, but the Boltzmann factor is so small that one would expect no monopoles in the observed universe.

Technically I believe that the Langacker-Pi mechanism is quite sound. The model does require large scalar self-couplings (in order to overcome the effects of the gauge couplings, which tend to restore the SU(2) × U(1) symmetry), so the perturbation theory used to calculate the phase structure is not totally reliable. Nonetheless, I think one would expect that the proposed phase structure will hold at least for some range of parameters. I still find it somewhat implausible that this is the actual mechanism by which monopole production was suppressed. My general feeling is that whenever a complicated mechanism is constructed solely to solve a single problem, there is only a small chance that the mechanism is correct. Of course, my feelings on this could easily be wrong.

THE INFLATIONARY UNIVERSE

I now turn to what I regard as the most plausible mechanism of monopole suppression: the inflationary universe. The basic idea is very simple--the prediction of the monopole excess was based on the smallness of the horizon length at the time of the phase transition, and thus this prediction can be avoided if the phase transition occurs much later, after extreme supercooling. Under these circumstances, the universe expands exponentially.

There are two versions of the inflationary scenario, so I would like to begin by explaining the distinction. The difference hinges on the mechanism by which the belated phase transition occurs. In my original work[4] I was aware of only the

Coleman-Callan "fate of the false vacuum" mechanism,[13] in which bubbles of the new phase nucleate randomly in the background of the old phase and then start to grow. However, it was soon found[4,14,15] that the randomness of this process would lead to gross inhomogeneities. The bubbles would have a huge range of sizes, extending over at least 29 orders of magnitude. They would asymptotically fill the volume, but they would be forever contained within finite clusters which would never coalesce to fill the space uniformly.[4,14] A typical cluster would be dominated by a single bubble, and most of the energy in the cluster would remain unthermalized in the wall of the dominant bubble. Thus, although the inflationary universe scenario solved several important problems (which I will come back to), it was known to contain a crucial flaw.

This crucial flaw was overcome by a proposal made by Linde[5] and independently by Albrecht and Steinhardt,[6] known as the "new inflationary universe." These authors proposed that by fine-tuning the Higgs field effective potential to be of the Coleman-E. Weinberg form,[16] it is possible to arrange for a single bubble (or fluctuation) to expand enough to encompass the entire observed universe. The coalescence of bubbles is then irrelevant. The crucial flaw of the original model is avoided, and all of its successes are preserved.

Unfortunately, a totally successful version of the new inflationary universe has not yet emerged. The version[5,6] which has been studied in detail, and which will be discussed below, is based on an underlying SU(5) particle theory. It has the serious flaw that it predicts density fluctuations with an amplitude about 10^5 times larger than the value desired for galaxy formation. Nonetheless, the general idea that the early universe supercooled and underwent a period of exponential expansion still seems very promising; I think one can realistically hope to find a modification which suppresses these residual fluctuations.

THE HORIZON AND FLATNESS PROBLEMS

Thus, let me continue to discuss the virtues of the new inflationary universe. While the inflationary scenario grew out of my work with Henry Tye on monopole suppression,[17] it soon became apparent that the scenario resolves two other perplexing problems of classical cosmology: the horizon and flatness problems.

The horizon problem was first pointed out by Rindler[18] in 1956. The observational basis for this problem is the uniformity of the cosmic background radiation, which is known to be isotropic to about one part in 10^3. This fact is particularly difficult to understand when one considers the existence of the horizon length,

the maximum distance that light could have traveled since the
beginning of time. Consider two microwave antennas pointed in
opposite directions. Each is receiving radiation which is believed
to have been emitted (or "decoupled") at the time of hydrogen
recombination, at $t \simeq 10^5$ yr. At the time of emission, these two
sources were separated from each other by over 90 horizon
lengths.[19] The problem is to understand how two regions over 90
horizon lengths apart came to be at the same temperature at the
same time. Within the standard model this large scale homogeneity
is simply assumed as an initial condition.

The flatness problem was first pointed out by Dicke and
Peebles[20] in 1979. The basis of this problem is the fact that
today the ratio between the actual mass density and the critical
mass density (i.e., $\Omega \equiv \rho/\rho_c$) is conservatively known to lie in the
range

$$0.01 < \Omega < 10 \ . \tag{12}$$

No one is surprised by how narrowly this range brackets $\Omega = 1$.
However, within the evolution of the standard model, the value
$\Omega = 1$ is an unstable equilibrium point. Thus, to be near $\Omega = 1$
today, the universe must have been very near to $\Omega = 1$ in the past.
When T was 1 MeV, Ω had to equal one to an accuracy of one part in
10^{15}. At the time of the GUT phase transition, when T was about
10^{14} GeV, Ω had to be equal to one to within one part in 10^{49}! In
the standard model this precise fine-tuning is simply put in as an
initial condition. However, I feel that the model is believable
only if a mechanism to create this fine-tuning can be found.

Since the horizon and flatness problems are peculiarities of
the initial conditions required in the standard model, it must be
pointed out that both problems could conceivably be solved during
the era of quantum gravity, when $T \gtrsim 10^{19}$ GeV. Since quantum
gravity is only poorly understood, such a solution would be highly
speculative at the present time. Inflation, on the other hand,
relies only on rather well-accepted physical mechanisms. The fact
that inflation can solve these problems and the monopole problem as
well, must (in my opinion) be considered strong evidence in its
favor.

THE NEW INFLATIONARY UNIVERSE

I will now describe the new inflationary universe scenario[5,6]
in some detail. The scenario is cast in the context of the SU(5)
GUT.[11] In this theory the SU(5) gauge symmetry is broken to SU(3)$_c$
× SU(2) × U(1) by means of a Higgs field Φ in the adjoint
representation. Φ can be represented as a traceless hermitian
5 × 5 matrix. The symmetry breaking is achieved by a vacuum

expectation value of the form

$$\Phi = \sqrt{\frac{2}{15}} \; \phi \; \text{diag}[1,1,1,-3/2,-3/2] \; . \qquad (13)$$

(The factor of $\sqrt{2/15}$ normalizes the trace of the square of the matrix to unity.) The new inflationary scenario requires that the effective potential $V(\phi)$ be adjusted to obey the Coleman-E. Weinberg condition,[16] $\partial^2 V/\partial\phi^2 = \partial^3 V/\partial\phi^3 = 0$ at $\phi = 0$. Including the one-loop gauge field quantum corrections (calculated in the flat-space approximation), the potential takes the form

$$V(\phi) = \frac{25}{16} \; \alpha^2 [\phi^4 \ln(\phi^2/\sigma^2) + \frac{1}{2}(\sigma^4 - \phi^4)] \; , \qquad (14)$$

where $\alpha \equiv g^2/4\pi \approx 1/45$ is the gauge coupling and $\sigma \approx 1.2 \times 10^{15}$ GeV. The form of this potential is shown in Fig. 2. The minimum lies at $\phi = \sigma$, corresponding to the true vacuum. The point $\phi = 0$ is an equilibrium point which is just barely unstable. I will refer to the field configuration $\phi = 0$ as the false vacuum, even though this term is traditionally reserved for configurations which are classically stable. For any temperature $T > 0$, the false vacuum is stabilized by a bump in the finite temperature effective potential. For small T this bump has a height of order T^4, and a width of order T. There is a phase transition with critical temperature $T_c \sim 10^{14}$ GeV, above which the $\phi = 0$ configuration is the minimum of the potential (and is hence the thermal equilibrium state).

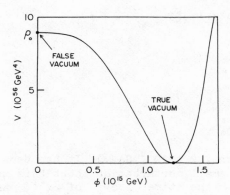

Fig. 2:   The Coleman-Weinberg potential for the SU(5) adjoint Higgs field.

    The starting point of a cosmological scenario is somewhat a
matter of taste and philosophical prejudice. Some physicists find
it plausible that the universe began in some highly symmetrical
state, such as a de Sitter space. I prefer to believe that the
universe began in a highly chaotic state; one advantage of the
inflationary scenario, from my point of view, is that it appears to
allow a wide variety of starting configurations. I require only
that the initial universe is hot ($T > 10^{14}$ GeV) in at least some
places, and that at least some of these regions are expanding
rapidly enough so that they will cool to T_c before gravitational
effects reverse the expansion.

    In these hot regions, thermal equilibrium would imply $<\Phi> = 0$.
(Actually, though, the universe has not had time at this point to
thermalize.[21] Thus, I need to assume that there are some regions of
high energy density with $<\Phi> \approx 0$, and that some of these regions
lost energy with Φ being trapped in the false vacuum.) These
regions will cool to T_c, and nucleation rate calculations[22]
indicate that they will continue to supercool well below T_c. The
energy density ρ will approach $\rho_0 \equiv V(\phi = 0)$. Since this false
vacuum state is Lorentz-invariant, the energy-momentum tensor must
then have the form

$$T_{\mu\nu} = \rho_0 g_{\mu\nu} . \tag{15}$$

It follows that the false vacuum has a large and negative pressure
$p = -\rho_0$.

    As long as the energy-momentum tensor has the form of
Eq. (15), any region which is approximately described by a
Robertson-Walker metric will evolve into a de Sitter space, which
can be described by the metric

$$ds^2 = -dt^2 + R^2(t) \, d\vec{x}^2 , \tag{16}$$

where

$$R(t) = e^{\chi t} \tag{17}$$

and

$$\chi^2 = (8\pi/3) \, G\rho_0 . \tag{18}$$

(For our parameters, $\chi \approx 10^{10}$ GeV.) It has been shown[23] that any
locally measurable small perturbation about this metric is damped
exponentially on the time scale of χ^{-1}, and it has been
conjectured[24] that a "cosmic no-hair theorem" implies a similar
behavior for large perturbations. Thus, a smooth de Sitter metric
arises naturally, without any need to fine-tune the initial

conditions.   (Advocates of an initial de Sitter space are welcome
to join the scenario at this point.)

   As the space continues to supercool and exponentially expand,
the energy density is fixed at ρ_0. Thus, the total matter energy
(i.e., all energy other than gravitational) is increasing!   This
seems to violate our naive notions of energy conservation, but we
must remember that the gravitational field can exchange energy with
the matter fields. The energy-momentum tensor for matter obeys a
covariant conservation equation, which in the Robertson-Walker
metric reduces to

$$\frac{d}{dt} (R^3 \rho) = -p \frac{d}{dt} (R^3) \ . \tag{19}$$

For $\rho = -p = \rho_0$, the above equation is satisfied identically, with
the energy of the expanding gas increasing due to the negative
pressure. If the space were asymptotically Minkowskian it would be
possible to define a conserved total energy (matter plus
gravitational).   However, in the Robertson-Walker metric, a global
conservation law of this type does not exist, except perhaps in a
trivial version in which the total energy vanishes identically.

   Let me digress a moment to discuss the contents of the
observed universe. I have just explained that matter energy is not
conserved, but can be increased dramatically by exponential
expansion.   If baryon number is also nonconserved (as in GUTs),
then the universe is, so far as we know, devoid of any conserved
quantities.[25] In that case, it is very tempting to believe that the
universe began from nothing, or from almost nothing.   The
inflationary model illustrates the latter possibility. It is often
said that there is no such thing as a free lunch. However, one of
the most exciting implications of modern cosmology is the
possibility that the universe is a free lunch.

   Now let me return to my region of space which is supercooling
into a de Sitter phase. It is a striking fact about de Sitter
space that quantum fluctuations mimic thermal effects at a Hawking
temperature[26]

$$T_H = \chi/2\pi \ . \tag{20}$$

In our case, $T_H \approx 10^9$ GeV. Calculations on the decay of the
supercooling phase have been carried out mainly in Minkowski space,
omitting the effects of gravitation (including the Hawking
temperature).   These calculations[22,5,6] indicate that the
supercooling continues until $T \approx 10^{7-8}$ GeV, at which point the
calculations break down. (They break down when the energy of a
critical size bubble becomes comparable to T, resulting in a

failure of the steepest descent approximation. The situation becomes even more difficult to analyze when T falls to ~10^6 GeV, at which point the SU_5 gauge coupling becomes strong.) Since gravitational effects apparently prevent the temperature from ever falling below T_H[24], these Minkowski space calculations are illustrative but inconclusive. The gravitational effects have been studied by several authors,[27] but important questions appear to be still unresolved.

It is clear that within the supercooling region, the Higgs field will be undergoing fluctuations due to thermal and/or quantum effects. One would expect that some fluctuations would begin to grow, and at some point these fluctuations will become large enough so that their subsequent evolution can be described by the classical equations of motion. I will used the term "fluctuation region" to denote any region in which this has occurred. The size and shape of the fluctuation regions are uncertain--they are presumably irregular,[6] and are presumably homogeneous on a length scale of order χ^{-1} (which appears to be the only relevant length scale in the problem).

The Higgs field ϕ then "rolls" down the potential of Fig. 2, obeying the classical equations of motion

$$\ddot{\phi} + 3 \frac{\dot{R}}{R} \dot{\phi} = - \frac{\partial V}{\partial \phi} . \tag{21}$$

If the initial fluctuation is small, then the flatness of the potential for $\phi \approx 0$ will imply that the rolling begins very slowly. Note that the second term on the left-hand-side of Eq. (21) is a damping term, helping to further slow down the speed of rolling. As long as $\phi \approx 0$, the energy density ρ remains about equal to ρ_0, and the exponential expansion continues. The exponential expansion occurs on a time scale χ^{-1} which is short compared to the time scale of the rolling.

For the scenario to work, it is necessary for the length scale of homogeneity to be stretched from χ^{-1} to at least about 10 cm before the Higgs field ϕ rolls off the plateau in Fig. 2. This corresponds to an expansion factor of about 10^{25}, which requires about 58 times constants (χ^{-1}) of expansion. It can be shown[28] that such an expansion would result if the initial fluctuation in ϕ were less than about $1.4T_H$ (with no initial fluctuation $\dot{\phi}$). One would expect fluctuations of roughly this magnitude, but a more detailed understanding is required to determine if the initial fluctuations are small enough or not.

When the ϕ field reaches the steep part of the potential, it falls quickly to the bottom and oscillates about the minimum. The time scale of this motion is a typical GUT time of $(10^{14}$ GeV$)^{-1}$, which is very fast compared to the expansion rate. The Higgs field oscillations are then quickly damped by the couplings to the other fields, and the energy is rapidly thermalized.[29] (The Higgs field oscillations correspond to a coherent state of Higgs particles; the damping is simply the decay into other species.) The release of this energy (which is just the latent heat of the phase transition) raises the temperature to the order of 10^{14} GeV. (More precisely, the reheating temperature is about one sixth of the vector boson mass.)

From here on the standard scenario ensues, including the production of a net baryon number. The length scale of homogeneity increases to $\geq 10^{10}$ light-yr by the time T falls to 2.7°K.

SOLUTION OF THE HORIZON AND FLATNESS PROBLEMS

The horizon problem is clearly avoided in this scenario, since the entire observed universe evolves from a single fluctuation region. This region had a size of order χ^{-1} at the time when the fluctuation began to grow classically--it was thus causally connected. As I explained earlier, the Higgs field is expected to be homogeneous on this length scale. The exponential expansion causes this very small region of homogeneity to grow to be large enough to encompass the observed universe.

The flatness problem is avoided by the dynamics of the exponential expansion of the fluctuation region. As ϕ begins to roll very slowly down the potential, the evolution of the metric is governed by the energy density ρ_0. Assuming that the fluctuation region (or a small piece of it) can be approximated locally by a Robertson-Walker metric, then the scale factor evolves according to the standard equation

$$\left(\frac{\dot{R}}{R}\right)^2 = \frac{8\pi}{3} G\rho_0 - \frac{k}{R^2} , \qquad (22)$$

where $k = +1$, -1, or 0 depending on whether the region approximates a closed, open, or flat universe, respectively. (There could also be perturbations, but these would die out quickly, as discussed following Eq. (18).) In this language, the flatness problem is the problem of understanding why the second term on the right-hand-side is so extraordinarily small. But as the fluctuation region expands exponentially, the energy density ρ remains very nearly constant at ρ_0, while the k/R^2 term falls off as the square of the exponential factor. Thus the k/R^2 term is suppressed by at least a factor of 10^{50}, which provides a "natural" explanation of why its value immediately after the phase transition is less than 10^{-49} times the value of the other terms.

Except for a very narrow range of parameters, this suppression of the curvature term will vastly exceed that required by present observations. For example, the known fact $\Omega \sim 1$ today can be explained if the initial fluctuations in ϕ are about $1.4T_H$. Suppose, however, that the initial fluctuations are just 10% smaller. It then turns out[28] that today Ω would be equal to one to within one part in 10^{10}. This leads to the prediction that the value of Ω today is expected to be equal to one to within an extraordinary degree of accuracy.

MONOPOLE SUPPRESSION IN THE NEW INFLATIONARY UNIVERSE

I now turn to the important question of magnetic monopole suppression. First, it is easy to see that the Kibble production mechanism, which produced a large excess of monopoles in the standard scenario, is totally ineffective here. The Higgs field is correlated throughout the initial fluctuation region, and the exponential expansion increases the Higgs correlation length ξ to greater than 10^{10} light-yr. Thus, this mechanism would produce $\lesssim 1$ monopole in the observable universe.

One must also consider the possibility of thermal production of monopoles immediately after the GUT phase transition. I will here follow the analysis of Lazarides, Shafi, and Trower,[30] with some minor changes in their numerics. The thermal production will be suppressed by a Boltzmann factor

$$f \equiv e^{-M_M/T_r} \, , \qquad\qquad\qquad\qquad (23)$$

where M_M is the monopole mass and T_r is the temperature to which the universe reheats imediately after the phase transition. These numbers depend on the details of the underlying particle theory, but they may be estimated in the minimal SU(5) model as follows. The monopole mass is bounded by the value it has in the Bogomol'nyi- Prasad-Sommerfield limit,[31] in which the Higgs potential approaches zero with $\langle\Phi\rangle$ fixed. This implies[17] that

$$M_M \geq M_X/\alpha \, , \qquad\qquad\qquad\qquad (24)$$

where M_X is the mass of the superheavy vector bosons. α is the SU(5) gauge coupling constant, and is almost certainly smaller than $1/40$. The reheating temperature T_r is estimated by using conservation of energy--the energy density of the false vacuum is equated with cT^4 (see Eq. (4)), with a value of c reflecting the appropriate number of effectively massless degrees of freedom. One finds that $T_r \approx M_X/6$. Thus,

$$f \lesssim e^{-240} \approx 10^{-104} \ . \tag{25}$$

If monopoles had time to come into thermal equilibrium at the temperature T_r, then the ratio n_M/s would be proportional to f. But as Preskill[1] has shown, monopoles do not have time to come into thermal equilibrium--if they did, we would not need a suppression mechanism. Thus, we must estimate the rate at which monopoles are produced, when the other particles are in thermal equilibrium. Since the incoming particles must have an energy exceeding $2M_M$ to produce a monopole-antimonopole pair, these rates are proportional to f^2. Thus, the thermally produced contribution to n_M/s is proportional to f^2, resulting in no monopoles in the observable universe. (The thermal contribution to n_M/s is further suppressed by the cross section σ for monopole-antimonopole pair production, which is believed[32] to be suppressed by a factor exp(-constant/α).)

Thus, if the new inflationary scenario is correct, and the underlying particle theory is similar to the minimal SU(5) model, then one would conclude that the number of magnetic monopoles in the observed universe is vastly exceeded by the number of physicists at this conference.[33] However, the calculations of density fluctuations tell us that the new inflationary universe based on the SU(5) model is not completely correct; thus, the previous sentence is actually irrelevant. Turner[34] emphasizes that the calculation of f depends exponentially on M_M and T_r, each of which depends on the details of the underlying theory. Thus it is conceivable (though I would imagine unlikely) that one could find an inflationary model which would predict an observable magnetic monopole flux.[39]

FLUCTUATIONS IN THE NEW INFLATIONARY UNIVERSE

Finally, now, let me summarize the results which have been obtained regarding the density fluctuations in the new inflationary universe. The calculation of these fluctuations was a popular sport[35] at the Nuffield Workshop on the Very Early Universe, held last summer in Cambridge, England. The goal was to calculate $\delta\rho/\rho$ on a given comoving length scale, at the time when the physical length is comparable to the Hubble length H^{-1}. At the outset, all parties agreed that the value of $\delta\rho/\rho$ was roughly independent of length scale, in accord with the spectrum proposed by Harrison[36] and Zeldovich[37] to account for galaxy production. However, the initial estimates of the amplitude ranged from 10^{-16} to 10^2, while the desired value is $10^{-4} - 10^{-2}$. By the end of the workshop, all of the groups converged to a value near 10^2. The fluctuations arise mainly from the quantum fluctuations in the Higgs field ϕ, which cause it to fall off the plateau of Fig. 2 and release the false vacuum energy density ρ_0 at a time which varies from one position to another.

It was of course disappointing to discover that the new inflationary universe could not correctly account for the density fluctuations in the universe. However, I would like to argue that this result should be viewed not as a discouraging failure, but rather as an encouraging near miss. First, one should remember that this is the first time we have ever had a scenario with enough detail to even allow a calculation of $\delta\rho/\rho$ from first principles. Traditionally, an ad hoc initial spectrum of fluctuations has simply been assumed. Second, I claim that the new inflationary universe does quite a bit better than a toy model which I would regard as a reasonable elaboration of the standard model. In this toy model, I will assume that the mysterious effects of quantum gravity leave the universe in thermal equilibrium at a perfectly uniform temperature $T \sim M_p$. However, I will insist that the gas which fills the universe exhibit the usual fluctuations of Poisson statistics. Due to the incredible instabilities of the early universe, it turns out that the fluctuations on galactic scales then grow to be 10^{12} times larger than those predicted by the new inflationary universe.[28] And third, recall that the fluctuation results are for the particular case of an underlying SU(5) model. The result is model-dependent, and in a certain class of supersymmetric models the amplitude of the fluctuations is about right.[38] (Unfortunately, these supersymmetric models do not produce enough reheating.) Thus, it is possible that a model will be found which will solve all of these problems.

In conclusion, I want to say that I began working on cosmology only because I was pressured into it by Henry Tye. He had to do a lot of arm-twisting, because at that time I very strongly believed that cosmology was the kind of field in which a person could say anything he wanted, and no one could ever prove him wrong. I am sure that truth does not change with time, but after three years of working in cosmology my prejudices about the subject have completely reversed. It now appears that it is very easy to show that a cosmological scenario is wrong, and far more difficult than I had ever imagined to develop a totally consistent picture.

ACKNOWLEDGEMENTS

I would like to acknowledge the support of the U. S. Department of Energy (DOE) under contract number DE-AC02-76ER03069, and also an Alfred P. Sloan Fellowship.

REFERENCES

1. J. P. Preskill, Phys. Rev. Lett. 43, 1365 (1979).
2. Ya. B. Zeldovich and M. Y. Khlopov, Phys. Lett. 79B, 239 (1978).

3. P. Langacker and S.-Y. Pi, Phys. Rev. Lett. $\underline{45}$, 1 (1980). See also F. A. Bais and P. Langacker, Nucl. Phys. $\underline{B197}$, 520 (1982).

4. A. H. Guth, Phys. Rev. D $\underline{23}$, 347 (1981).

5. A. D. Linde, Phys. Lett. $\underline{108B}$, 389 (1982).

6. A. Albrecht and P. J. Steinhardt, Phys. Rev. Lett. $\underline{48}$, 1220 (1982).

7. G. 't Hooft, Nucl. Phys. B $\underline{79}$, 276 (1974); A. M. Polyakov, Pis'ma Zh. Eksp. Teor. Fiz. $\underline{20}$, 430 (1974) [JETP Lett. $\underline{20}$, 194 (1974)].

8. T. W. B. Kibble, J. Phys. A $\underline{9}$, 1387 (1976).

9. I set $\hbar = c = k = 1$, and I take the GeV as my fundamental unit. Then 1 GeV = 1.16×10^{13} °K = 1.78×10^{-24} gm, and 1 GeV^{-1} = 1.97×10^{-14} cm = 6.58×10^{-25} sec.

10. J. P. Preskill (private communication); A. H. Guth and S.-H. Tye, Phys. Rev. Lett. $\underline{44}$, 631, 963 (1980); M. B. Einhorn, D. L. Stein, and D. Toussaint, Phys. Rev. D $\underline{21}$, 3295 (1980).

11. H. Georgi and S. L. Glashow, Phys. Rev. Lett. $\underline{32}$, 438 (1974).

12. For a general background in cosmology, see S. Weinberg, "Gravitation and Cosmology," Wiley, New York (1972). At a less technical level, see J. Silk, "The Big Bang," W. H. Freeman & Co., San Francisco (1980), or S. Weinberg, "The First Three Minutes," Bantam Books, New York (1977).

13. S. Coleman, Phys. Rev. D $\underline{15}$, 2929 (1977); C. G. Callan and S. Coleman, Phys. Rev. D $\underline{16}$, 1762 (1977); see also S. Coleman in "The Whys of Subnuclear Physics," Ettore Majorana, Erice, 1977, A. Zichichi, ed., Plenum, New York (1979).

14. A. H. Guth and E. J. Weinberg, Nucl. Phys. B $\underline{212}$, 321 (1983).

15. S. W. Hawking, I. G. Moss, and J. M. Stewart, Phys. Rev. D $\underline{26}$, 2681 (1982).

16. S. Coleman and E. J. Weinberg, Phys. Rev. D $\underline{7}$, 1888 (1973).

17. A. H. Guth and S.-H. Tye, Ref. 10.

18. W. Rindler, Mon. Not. R. Astron. Soc. $\underline{116}$, 663 (1956). See also S. Weinberg, "Gravitation and Cosmology," Ref. 12, pp. 489-490 and 525-526.

19. A. H. Guth, to be published in "Asymptotic Realms of Physics: Essays in Honor of Francis E. Low," A. H. Guth, K. Huang, and R. L. Jaffe, eds., MIT Press, Cambridge, Massachusetts (1983).

20. R. H. Dicke and P. J. E. Peebles, in "General Relativity: An Einstein Centenary Survey," S. W. Hawking and W. Israel, eds., Cambridge University Press, Cambridge, U.K. (1979).

21. G. Steigman, Proceedings of the Europhysics Study Conference: Unification of the Fundamental Interactions II, Erice, October 6-14, 1981.

22. M. Sher, Phys. Rev. D $\underline{24}$, 1699 (1981).

23. J. A. Frieman and C. M. Will, Ap. J. $\underline{259}$, 437 (1982); J. D. Barrow, to appear in the "Proceedings of the Nuffield

Workshop on the Very Early Universe," G. W. Gibbons,
S. W. Hawking, and S. T. C. Siklos, eds., Cambridge
University Press, Cambridge, U.K. (1983).

24. S. W. Hawking and I. G. Moss, Phys. Lett. 110B, 35 (1982).
25. D. Atkatz and H. Pagels, Phys. Rev. D 25, 2065 (1982) and
    references therein.
26. G. W. Gibbons and S. W. Hawking, Phys. Rev. D 15, 2738 (1977).
27. S. W. Hawking and I. G. Moss, Ref. 24; A. Vilenkin and L. H.
    Ford, Phys. Rev. D 26, 1231 (1982); A. Vilenkin, Phase
    Transitions in de Sitter Space, Tufts Univ. preprint
    TUTP-82-10 (1982); P. Hut and F. R. Klinkhamer, Phys. Lett.
    104B, 439 (1981).
28. A. H. Guth, to appear in the "Proceedings of the Nuffield
    Workshop on the Very Early Universe," Ref. 23.
29 A. Albrecht, P. J. Steinhardt, M. S. Turner, and F. Wilczek,
    Phys. Rev. Lett. 48, 1437 (1982); L. F. Abbott, E. Farhi,
    and M. B. Wise, Phys. Lett. 117B, 29 (1982); A. D. Dolgov
    and A. D. Linde, Phys. Lett. 116B, 329 (1982).
30. G. Lazarides, Q. Shafi, and W. P. Trower, Phys. Rev. Lett. 49,
    1756 (1982).
31. E. B. Bogomol'nyi, Sov. J. of Nucl. Phys. 24, 449 (1976);
    M. K. Prasad and C. M. Sommerfield, Phys. Rev. Lett. 35,
    760 (1975).
32. E. Witten, Nucl. Phys. B 160, 57 (1979); A. K. Drukier and S.
    Nussinov, Phys. Rev. Lett. 49, 102 (1982).
33. This quip was shamelessly stolen from Curt Callan.
34. M. S. Turner, Phys. Lett. 115B, 95 (1982).
35 A. A. Starobinsky, manuscript in preparation; A. H. Guth and
    S.-Y. Pi, Phys. Rev. Lett. 49, 1110 (1982); S. W. Hawking,
    Phys. Lett. 115B, 295 (1982); J. M. Bardeen, P. J.
    Steinhardt, and M. S. Turner, Spontaneous Creation of
    Almost Scale-Free Density Perturbations in an Inflationary
    Universe, Univ. of Penna. preprint UPR-0202T (1982). See
    also the "Proceedings of the Nuffield Workshop on the Very
    Early Universe," Ref. 23.
36. E. R. Harrison, Phys. Rev. D 1, 2726 (1970).
37. Ya. B. Zeldovich, Mon. Not. R. Astr. Soc. 160, 1P (1972).
38. See the contributions by P. J.Steinhardt and by M. S. Turner
    in the "Proceedings of the Nuffield Workshop on the Very
    Early Universe, Ref. 23.
39. Since the Magnetic Monopole Workshop, A. S. Goldhaber, S.-Y.
    Pi and I have begun to think about the production of
    magnetic monopoles during the early stages of inflation,
    while $\phi \approx x$. At this time magnetic monopoles are relatively
    light, and can be produced abundantly. These monopoles are
    then diluted by the subsequent expansion, but we think that
    a conceivably observable flux of monopoles could remain.
    (The possibility that monopoles could be produced in this
    way was suggested by M. S. Turner.)

CATALYSIS OF BARYON DECAY

Curtis G. Callan, Jr.

Joseph Henry Laboratories
Princeton University
Princeton, New Jersey 08544

INTRODUCTION

Recent studies of the interaction between fermions and grand unification monopoles have shown that the cross-sections for monopole-catalysed proton decay reactions such as $M + p \rightarrow M + e^+ + \pi^0$ are independent of the unification scale, M_x, and in fact of typical strong interaction size.[1,2] Other speakers at this conference have discussed the phenomenological implications of this result, so I will concentrate on trying to explain <u>why</u> it happens. Since fairly subtle field theory phenomena are involved, I will not be able to give an exhaustive discussion in the short time available to me. Instead, I will present a line of argument which, while incomplete, brings out what I believe to be the essential physical mechanisms underlying this rather paradoxical phenomenon. I hope to convince the baffled experimentalist as well as the skeptical theorist that sensible physics, not magic, is responsible for monopole catalysis of baryon decay.

SU5 MONOPOLE SUMMARY

In order for baryon decay catalysis to occur, it turns out to be essential that the underlying theory violate baryon number at the Lagrangian level. For definiteness, then, I choose to work with the standard SU_5 model with one generation of quarks and leptons. The least massive monopole of this theory has the following structure: The magnetic charge is confined to a core of radius M_x^{-1} in which specifically SU_5 (and in fact baryon-number-violating) gauge fields are excited. Outside this

97

core, only color magnetic and ordinary magnetic fields survive. A
gauge can be chosen so that the color magnetic field is
proportional to λ_8, or color hypercharge. The long-range fields
are then the superposition of two-single-Dirac-unit Abelian
monopoles, one for ordinary charge, Q, and one for color
hypercharge, Y_c. Because of confinement, the color magnetic field
is eventually screened (at distances of order one fermi) and the
Abelian treatment of color breaks down. Since confinement seems to
have nothing essential to do with baryon decay catalysis, I will
ignore its effects for most of this talk. As a result, I will be
able to make only the crudest estimates of actual reaction rates.

The quarks and leptons carry ordinary charge and color
hypercharge and respond to both the long-range fields of the SU_5
monopole. Outside the monopole core, they behave in the same way
as particles of total charge $\bar{Q} = Q + Y_c$ in the field of a simple
Abelian Dirac monopole. The right-handed Weyl fermions of the
first $5 + \overline{10}$ generation divide as follows according to their
effective charge $\bar{Q}$:

$$\bar{Q} = +1: \quad e_R^+ \ \bar{d}_{3R} \ u_{1R} \ u_{2R}$$

$$\bar{Q} = -1: \quad e_R^- \ d_{3R} \ \bar{u}_{1R} \ \bar{u}_{2R}$$

$$\bar{Q} = 0: \quad d_{1R} \ d_{2R} \ u_{3R} \ \bar{d}_{1R} \ \bar{d}_{2R} \ \bar{d}_{3R} \ \bar{\nu}_R$$

The subscript on the quark field identifies the color quantum
number--the 3 direction is singled out by the gauge choice which
makes the monopole's color magnetic field proportional to λ_8. The
$\bar{Q} = 0$ fermions in effect do not see the monopole at all. The
others behave in exactly the same way as electrons or positrons in
the field of an Abelian Dirac monopole so that, as far as motion
outside the core is concerned, the complications of SU_5 simply
reduce to having four equivalent flavors of Dirac fermion (one each
for e^+, $\bar{d}_3$, u_1, u_2).

Since baryon-number-violating gauge fields are to be found
only inside the monopole core, it would appear that
baryon-number-violating scattering can only occur for those
fermions which have a high probability of being found inside the
core. Standard quantum-mechanical wisdom has it that if the
fermion energy E is small compared to the inverse core radius, M_x.
then the relative probability of finding the fermion inside the
core scales as $(E/M_x)^3$. Consequently, baryon-number-violating
scattering should be utterly negligible at any energy accessible in
the laboratory. This is the argument underlying the feeling that
strong monopole catalysis of baryon decay is counterintuitive. The
flaw in the argument is the hidden assumption that there is no

enhancement of the fermion wave function at the monopole core location. In fact, as I will discuss in the next section, the magnetic field of a monopole has precisely the effect of concentrating the wave function of a spin-1/2 particle at the monopole core in a way which ruins the naive quantum mechanics argument.

S-WAVE PECULIARITIES

To discuss spin-1/2 quantum mechanics, we will have to solve the Dirac equation in the background field of a magnetic monopole. To begin with, let us consider a massless electron moving in the field of a point Abelian Dirac monopole. There is a conserved total angular momentum (taking on integer values because of a contribution peculiar to the electron-monopole system) which we may use to partial-wave-analyze the Dirac equation. The S-waves (J = 0) have certain features, first studied by Kazama and Yang[3] and Goldhaber,[4] which are very relevant to our problem.

The J = 0 solutions of the Dirac equation are explicitly

$$\psi_+ = \frac{e^{iEr}}{r} \begin{pmatrix} \eta \\ \eta \end{pmatrix}, \quad \psi_- = \frac{e^{-iEr}}{r} \begin{pmatrix} \eta \\ -\eta \end{pmatrix}$$

where $E > 0$ is the energy, $\eta(\theta,\phi)$ is a special two-component angular function and the standard Pauli-Dirac representation of the gamma matrices is being used.[5] Notice that both solutions are singular at $r = 0$ and that no linear combination of them can be found which eliminates the singularity in all components of ψ. Since the wave function blows up as r^{-1}, the S-wave contribution to bilinear current densities (such as the electric charge current) can blow up as r^{-2}, corresponding to a __finite__ flux of the corresponding charges into the __point__ $r = 0$. Notice also that both solutions are helicity eigenstates ($\gamma_5 \psi_+ = \pm\psi_+$), as might be expected, since the massless Dirac equation conserves helicity. However, since ψ_+ (ψ_-) contains only outgoing (ingoing) waves, scattering events will be very peculiar: if I make a negative helicity in-state, the entire S-wave part (a finite fraction of the initial wave packet) will eventually disappear into the origin! This is yet another way of seeing that the Dirac equation allows physical quantities to disappear into the origin at a finite rate.

None of the above peculiarities occur for $J \neq 0$ partial waves: the wave functions are finite at the origin, helicity eigenstates have both in- and out-going waves and there is no flux of physical quantities into the origin. The peculiar doings in the S-wave really show that the S-wave Dirac equation is not, by itself, a

well-defined system and of course raise the question of how it must
be extended to make it well defined. At the same time, the fact
that S-wave states, no matter what their energy, have finite fluxes
into and out of the point center of the monopole is a concrete
example of naive quantum mechanical intuition about reaction rates
being falsified by wave function enhancements. In the end, this
enhancement or "sucking in" of the wave function is due to the very
singular attractive interaction between the magnetic moment of the
S-wave electron and the monopole magnetic field.

BOUNDARY CONDITIONS

The problem with the S-wave Dirac equation is not that the
wave functions are singular at the origin--all charge and energy
integrals are nonetheless finite. The real problem is that the
Dirac equation is not Hermitean (does not conserve probability)
precisely because flux leaks into and out of the point monopole
core. This problem can be eliminated by imposing a bag-like
boundary condition

$$\gamma_0 \, \psi(0) = e^{i\delta\gamma_5} \, \psi(0)$$

(where δ is an arbitrary constant) on ψ at $r = 0$. The J = 0 wave
function is then uniquely determined to be[3,4,6]

$$\psi_\delta = \frac{e^{i(Er+\delta/2)}}{r} \binom{\eta}{\eta} + \frac{e^{-i(Er+\delta/2)}}{r} \binom{\eta}{-\eta}.$$

The boundary condition or the explicit form of ψ_δ guarantees that
the flux of electric charge into the point r = 0 vanishes. As is
apparent from the form of ψ_δ, this happens because incoming waves
of negative helicity are reflected off r = 0 as outgoing waves
(with relative phase $e^{i\delta}$) of positive helicity. The boundary
condition conserves charge and probability at the price of
violating chirality. (The phase $e^{i\delta}$ can introduce CP-violating
effects which I will not have time to discuss.[6]) Even though the
Dirac equation conserves chirality, this chirality-violating
boundary condition is the most general boundary condition which
makes the Dirac equation Hermitean by forbidding electric charge
flow into the monopole core.

There remains the question why this, or any, boundary
condition is an adequate approximation to the full field theory
physics of fermions interacting with the monopole. Before facing
up to that issue I want to point out an interesting feature of the

symmetry violation that has been introduced into the system by the boundary condition. Consider the scattering of fermions by the monopole. There will now be a helicity flip cross-section coming entirely from the S-wave which can easily be shown to have the value[4,5]

$$\sigma_{\Delta h} = \pi/E^2 \ .$$

This cross-section is of the size of what one might loosely call the unitarity limit precisely because all the incoming S-wave flux of one helicity is converted by the boundary condition to outgoing flux of the opposite helicity. As a result, the symmetry breaking effect of the point boundary condition has a scale set by the energy of the scattering particle, not by the size of the region where the boundary condition is applied. When we come to the SU_5 problem we will find baryon-number-violating boundary conditions producing large baryon-number-violating cross-sections via precisely the same mechanism.

I now want to return to the SU_5 monopole in order to discuss a case in which boundary conditions on S-wave functions can be shown to be a simple representation of the correct physics. Consider the Dirac equation for massless fermions of the first generation moving in the field of the SU_5 monopole described earlier. Outside the (almost pointlike) core the fields are such that the e^-, d_3, $\bar{u}_1$ and $\bar{u}_2$ behave like Dirac electrons moving in the field of an ordinary point Dirac monopole (while the d_1, d_2, u_3 and ν_L decouple) and the exterior solution of the Dirac equation for these particles must have the properties described in the last section: for each positive (negative) helicity S-wave particle there exists an outgoing (ingoing) wave solution.

Since the SU_5 monopole has a finite core with finite (albeit large) fields inside it, it will be possible to continue the exterior solutions of the Dirac equation all the way in to the origin and successfully impose the condition that the wave function be finite everywhere, without ever inducing any violation of chirality. The result of this exercise will be to construct exterior solutions of the full Dirac equation which are linear combinations of the solutions discussed in the last paragraph. The solutions will contain both ingoing and outgoing waves so that no conservation of probability difficulty need arise. They can be most conveniently described if we remember that the four relevant Dirac fermions can be thought of as having eight different particle states of a given helicity (L for definiteness): e_L^+, e_L^-, $\bar{d}_{3L}$, d_{3L}, u_{1L}, $\bar{u}_{1L}$, u_{2L}, $\bar{u}_{2L}$. There are half as many S-wave particle solutions of the full Dirac equation. Their structure outside the core may be symbolically indicated as

$$e_L^+ \qquad \bar{d}_{3L} \qquad u_{1L} \qquad u_{2L}$$

$$d_{3L} \qquad e_L^- \qquad \bar{u}_{2L} \qquad \bar{u}_{1L}$$

where the upper symbol indicates the particle identity of the outgoing wave and the lower symbol that of the incoming wave pieces of the wave function. To demonstrate this one has to do the detailed mathematics of solving the Dirac equation in the full monopole field.[7] I am just quoting the results in their simplest form.

Note that although our solutions conserve helicity, they do not conserve charge--ingoing and outgoing waves differ in all cases by the charges of the X_3 gauge boson of SU_5! This can only make sense if, in a scattering event, the monopole absorbs the difference--i.e., changes its <u>electric</u> charges by those of the X_3 boson. Precisely such electrically charged states of the monopole, called dyons, are in fact expected to exist and the Dirac equation is simply telling us that the dyon degree of freedom is necessarily excited when S-wave fermions scatter from the monopole.

The same result could have been obtained by ignoring the monopole core structure and imposing a boundary condition at $r = 0$ on the exterior solutions. The boundary condition has to express the result that an ingoing e_L^+ is completely reflected into an outgoing d_{3L}, that an ingoing u_{1L} is completely reflected into an outgoing $\bar{u}_{2L}$, and so on. In the Abelian case, the only possible boundary condition violated helicity conservation and led to large helicity flip scattering cross-sections; by the same token, in the SU_5 case, the boundary condition must violate charge conservation and we will find large charge-exchange cross-sections such as

$$\sigma \ (e_L^+ \ M \rightarrow d_{3L} \ M') = \pi/E^2.$$

Once again, the symmetry-breaking effect of a boundary condition imposed at a point monopole core has a scale set by the energy of the scattering particle, not by the size of the monopole core.

Before we can convert these observations into a theory of monopole catalysis of baryon decay, there is one important physical fact, not contained in the Dirac equation, which must be faced. According to the Dirac equation, S-wave scattering at any energy involves excitation of the dyon degree of freedom of the monopole. However, the dyon is expected to be more massive than the neutral monopole by approximately $\delta M = e^2 M_x$ (the Coulomb energy of charge e confined to a region of radius M_x^{-1}).[8] Since e^2 is not all that

small, projectiles of ordinary energy will not be energetic enough
to actually excite the dyon degree of freedom and something else
must happen. The simplest way out of this impasse is to imagine
that some sort of fermion pair creation event allows the charge
which would have been stuck on the dyon to radiate away to infinity
before an energetic harm is done. This sort of mechanism is a
non-trivial field theory process (it takes us out of the
one-particle Hilbert space of the Dirac equation and into the full
state space of field theory) and we will need a very efficient
formalism to deal with it in any sort of quantitative fashion. I
will present such a method in the next section.

BOSONIZATION

From all that I have said, it is apparent that only the S-wave
fermions can be directly influenced by the symmetry-breaking
boundary conditions which replace the details of the monopole core.
This suggests that a reasonable model field theory to study is one
which contains only the fermionic S-waves and their interactions
with each other and the monopole core.[1,2] With only one partial
wave, the problem is essentially one-dimensional (the only spatial
coordinate is the radial coordinate) and we can use some of the
tricks developed to simplify and solve one dimensional field
theories.

The trick which proves most useful is the bosonization
strategy of Mandelstam[9] and Coleman.[10] It turns out that a one
space dimension Fermi field is equivalent in a certain sense to a
one space dimension Bose field. The equivalence means that certain
fermion bilinears can be replaced as operators by simple local
functions of the Bose field: For example,

$$\bar{\psi}\gamma\cdot\partial\psi \simeq \frac{1}{2}\sum(\partial_\mu\phi)^2$$

$$\bar{\psi}\gamma_\mu\psi \simeq \frac{1}{\sqrt{\pi}}\partial_\mu\phi$$

$$\bar{\psi}\psi \simeq \mu\cos\sqrt{\pi}\,\phi .$$

This type of relation allows one to rewrite the Hamiltonian for a
fermion as a completely different Hamiltonian for a boson. The
boson Hamiltonian usually makes clear what the qualitative features
of the ground state are and sometimes reveals that the system is
soluble.

We have four flavors of S-wave fermions and we must introduce one scalar field for each flavor:

$$(e^+_L, e^-_L) \rightarrow \Phi_{e^+} \qquad\qquad (d_{3L}, \bar{d}_{3L}) \rightarrow \Phi_{\bar{d}^3}$$

$$(e_{1L}, \bar{u}_{1L}) \rightarrow \Phi_{u_1} \qquad\qquad (u_{2L}, \bar{u}_{2L}) \rightarrow \Phi_{u_2}$$

The interpretation of the scalar fields is roughly that $1/\sqrt{\pi}\ \Phi_a(r)$ is equal to the number of particles of type a interior to r. In the case of massless fermions the Lagrangian for the scalar fields which is equivalent to the original fermi system is just a sum of free massless scalar kinetic energies:

$$L_k = \int_0^\infty dr \left\{ \frac{1}{2} \sum_{a\mu} (\partial_\mu \Phi_a)^2 \right\}$$

$$\mu = t,r \ ; \ a = e^+, \bar{d}_3, u_1\ u_2 \ .$$

If the fermions have masses, it turns out that we have to add a sum of Sine-Gordon-like interaction terms:

$$L_M = \int_0^\infty dr \left\{ - \sum_a \mu_a^2 \cos 2\sqrt{\pi}\ \Phi_a \right\}$$

The total Lagrangian, $L = L_k + L_M$, allows for the free propagation of solitons for each type of field (in these solitons ϕ_a interpolates between $N\sqrt{\pi}$ and $(N\pm1)\ \sqrt{\pi}$). The solitons are essentially the same thing as the single fermion and antifermion states with which we started.

Because our world terminates at $r = 0$ we need boundary conditions to specify what happens to the scalar fields there. If we want to realize the charge-exchange boundary condition described in the previous section, it turns out that we must impose the following set of four boundary conditions:

$$\Phi_{e^+}(0) = \Phi_{\bar{d}_3}(0) \qquad\qquad \Phi_{u_1}(0) = \Phi_{u_2}(0)$$

$$\Phi'_{e^+}(0) = -\Phi_{\bar{d}_3}(0) \qquad\qquad \Phi'_{u_1}(0) = -\Phi'_{u_2}(0)$$

It is easy to obtain the solutions to the free scalar field equations, augmented by these boundary conditions. One finds for example (see Fig. 1) that an incoming Φ_{e^+} soliton (ingoing e^+) is

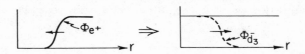

Fig. 1. An e^+ soliton scattering by the boundary conditions into a $\bar{d}_3$ antisoliton.

reflected into an outgoing $\Phi_{\bar{d}_3}$ antisoliton (outgoing d3). The entire content of the charge-changing boundary condition on the original Fermi fields turns out to be reproduced by this simple set of conditions on the equivalent Bose fields.

A slight modification of these boundary conditions will allow me to deal with the problem discussed at the end of the last section: the large dyon excitation energy associated with charge exchange scattering from the monopole. To see how that works, it is necessary to be more explicit about the dynamical meaning of Bose field boundary conditions.

In general, boundary conditions should be derivable directly from the Lagrangian and should tell us about the nature of the dynamical variables (if any) on the boundary surface. Consider the simple action

$$ L = \int_0^\infty dr \left[\frac{1}{2} \dot{\Phi}^2 - \frac{1}{2} \Phi'^2 \right] + \frac{1}{2} A \Phi(0)^2 . $$

The variation of L is, after some integration by parts,

$$ \delta \int dt\, L = \int dt\, \delta\Phi(0,t)\, [\Phi'(0,t)+A\Phi(0,t)] + \int dt \int_0^\infty dr\, \delta\Phi(r,t)\, \Box\, \Phi(r,t) $$

The action principle then yields both the equation of motion $\Box \Phi = 0$ and the boundary condition

$$ 0 = \Phi'(0,t) + A\Phi(0,t) . $$

The two extreme boundary conditions, $\Phi'(0) = 0$ and $\Phi(0) = 0$, can be achieved by setting $A = 0$ and $A = \infty$, respectively. These two choices correspond in an obvious way to having no extra dynamical variable on the boundary ($\Phi'(0) = 0$ or "free") or to having an infinitely massive variable on the boundary ($\Phi(0) = 0$ or "fixed").

This means that we can include the boundary conditions for the bosonized S-wave fermi fields by adding a set of quadratic terms in the surface variables, $\Phi_a(0)$, to L_k. One easily sees that the correct choice is

$$L_k' = \frac{1}{2} \int_0^\infty dr \sum_a \left(\partial_\mu \Phi_\alpha\right)^2$$

$$+ a_1 \left(\Phi_{e^+}(0) - \Phi_{\bar{d}_3}(0)\right)^2 + a_2 \left(\Phi_{u_1}(0) - \Phi_{u_2}(0)\right)^2$$

$$+ b_1 \left(\Phi_{e^+}(0) + \Phi_{\bar{d}_3}(0) + \Phi_{u_1}(0) + \Phi_{u_2}(0)\right)^2$$

$$+ b_2 \left(\Phi_{e^+}(0) + \Phi_{\bar{d}_3}(0) - \Phi_{u_1}(0) - \Phi_{u_2}(0)\right)^2$$

with

$$a_1 = a_2 = \infty$$
$$b_1 = b_2 = 0.$$

I have emphasized that the surface terms in L_k' tell about the dynamics of extra surface variables located at $r = 0$. Since the dyon degree of freedom really is located at $r = 0$, we might hope to account for its dynamics by a suitable modification of the surface terms in L_k.

The charge which counts the degree of excitation of the dyon is the charge $\bar{Q} = Q + Y_c$ which I introduced in discussing the interaction of quarks and leptons with the SU5 monopole; the charge-changing scattering events always involve the conversion of a $\bar{Q} = +1$ fermion into a $\bar{Q} = -1$ fermion, or vice-versa. The excitation energy of a dyon carrying net charge $\bar{Q}_D$ will, by the Coulomb energy argument sketched earlier, be

$$E_{DYON} = C e^2 M_x \bar{Q}_D^2 .$$

Since I interpret $1/\sqrt{\pi}\, \Phi_a(r)$ as the number of a-type particles contained within radius r, I must interpret $1/\sqrt{\pi}\, \Phi_a(0)$ as the number of such particles residing at $r = 0$ or on the monopole core. Since e^+, $\bar{d}_3$, u1 and u2 all carry $\bar{Q} = 1$, I can identify the net $\bar{Q}$ carried by the monopole/dyon as

$$\bar{Q}_D = \frac{1}{\sqrt{\pi}} \left(\Phi_{e^+}(0) + \Phi_{\bar{d}_3}(0) + \Phi_{u_1}(0) + \Phi_{u_2}(0)\right) .$$

But, to include dyon energetics in L'_k, all I have to do is replace $b_1 = 0$ by $b_1 = C e^2 M_x$. Since M_x is superlarge, this amounts to replacing $b_1 = 0$ by $b_1 = \infty$, thus changing one of the original boundary conditions (the one controlling the flow of $\bar{Q}$ into the monopole) from free to fixed!

The net result is that all the relevant dynamics of the interaction of massless S-wave fermions with the monopole is summarized by free propagation of the bosonized fields Φ_{e^+}, $\Phi_{\bar{d}_3}$, Φ_{u_1}, and Φ_{u_2} subject to the boundary conditions

$$\Phi_{e^+}(0) - \Phi_{\bar{d}_3}(0) = \Phi_{u_1}(0) - \Phi_{u_2}(0) = 0$$

$$\Phi_{e^+}(0) + \Phi_{\bar{d}_3}(0) + \Phi_{u_1}(0) + \Phi_{u_2}(0) = 0$$

$$\Phi'_{e^+}(0) + \Phi'_{\bar{d}_3}(0) - \Phi'_{\bar{u}_1}(0) - \Phi'_{u_2}(0) = 0 .$$

The real virtue of the bosonization trick is that it allows us to account for the genuinely complicated field theory effects of the dyon energy by a simple change of boundary condition. I shall show in the next section that this simple change leads to baryon number violating particle production processes.

BARYON NUMBER NON-CONSERVATION

The boundary conditions I extracted from the Dirac equation had baryon number violation built into them because they arose from the SU_5 fields in the monopole core. These boundary conditions had to be modified in order to account for the energetics of exciting the dyon degree of freedom. I now want to show that the new boundary conditions still violate baryon number.

Let met first examine the properties of the monopole ground state. Since the scalar Lagrangian is quadratic, we can find the quantum ground state just by finding the classical ground state. In the classical ground state the fields Φ must all be independent of r (to make the kinetic energy term vanish). For purposes of interpretation, it is convenient to include a fermion mass term

$$L_M = \int_0^\infty dr \sum_a \mu_a^2 \cos 2\sqrt{\pi}\, \Phi_a .$$

To minimize the associated energy the Φ_a must be integer multiples of $\sqrt{\pi}$: $\Phi_a = N_a \sqrt{\pi}$. To satisfy the boundary conditions there must be

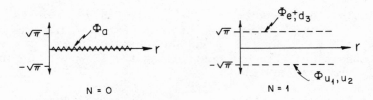

Fig. 2. Degenerate monopole ground state configurations.

relations among the N_a which are satisfied if we choose

$$\Phi_{e^+} = \Phi_{\bar{d}_3} = N\sqrt{\pi} \qquad \Phi_{u_1} = \Phi_{u_2} = -N\sqrt{\pi}$$

$$N = \ldots, -1, 0, 1, \ldots \ .$$

Thus, the ground state is degenerate! The ground state configurations for $N = 0$ and $N = 1$ are displayed in Fig. 2.

The interpretation of this degeneracy is very simple: $\Phi_a(0) = N\sqrt{\pi}$ means that N a-type particles have been swallowed by the monopole core. Thus in the Nth ground state the monopole appears to have swallowed N times the combination $(e^+\bar{d}_3\bar{u}_1\bar{u}_2)$. This combination is completely neutral (it has neither ordinary nor color electric charges) but does have baryon number -1. Therefore the different degenerate monopole ground states are distinguished only by baryon number. The monopole has the remarkable ability to swallow hydrogen atoms at no cost in energy!

When external fermions scatter from the monopole, transitions between different N-states will occur, leading to a change in the net baryon number of the ingoing versus outgoing fermions. Since fermions appear as solitons in the scalar field language, we have only to scatter a soliton from the $r = 0$ boundary conditions to examine the details of this process. A typical scattering event, obtained by solving the classical equations of motion subject to the boundary conditions, is shown in Fig. 3. The particle interpretation of this event is

$$u_1 + (N = 0) \rightarrow (N = -1) + \bar{u}_2 + \bar{d}_3 + e^+$$

As advertised, the baryon number (though not B-L) of the fermions changes by -1 while the baryon number of the monopole core changes by +1. Since there is no observable difference between the $N = 0$ and the $N = 1$ ground state we have to interpret this process as monopole catalysis of baryon number violation:

$$u_1 + M \rightarrow M + \bar{u}_2 + \bar{d}_3 + e^+ \ .$$

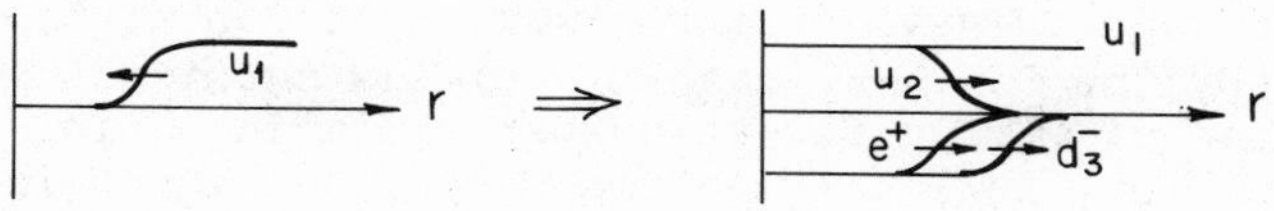

Fig. 3. A scattering event in which the $N = 0$ ground state is transformed into the $N = -1$ ground state.

For precisely the same reasons as given in the discussion of spin-flip scattering from the Abelian monopole, the cross-section for this S-wave process must be of order the S-wave "unitarity limit":

$$\sigma_{\Delta B} = C \cdot \frac{\pi}{E^2} .$$

The constant C could be calculated and is a number of order one.

The above discussion was entirely in terms of free quarks. To obtain a cross-section for monopole catalysis of <u>baryon</u> decay, the baryon number-violating quark cross-section has to be integrated into a picture of hadrons as confined quarks. I cannot go into detail here, but a simple bag model calculation allows us to estimate the cross-section for

p + M → M + (anything with B ≠ 1)

as

$$\sigma_{\Delta B} \simeq (10^{-27} \text{ cm}^2)\ \beta_p^{-1} \cdot F \cdot \overline{F}$$

where β_p is the relative velocity of proton and monopole. The fudge factor F contains some nuclear physics effects and could be as small as 10^{-4} for small β_p but is of order 1 for relativistic β_p. The fudge factor $\overline{F}$ contains the effect of the long-range interaction between the monopole and the proton's magnetic moment and gives a further enhancement of order β_p^{-1} for very small β_p [11]. Even under the most extreme assumptions, the baryon decay catalysis cross-section is truly a strong cross-section.

CONCLUSION

My intention in this talk was to explain the fundamental physical mechanisms responsible for strong monopole catalysis of baryon decay. The essential ingredients are: (a) a large enhancement of the S-wave electron wave function at the monopole core, produced by the $SU_3 \times U_1$ magnetic fields outside the core and (b) the existence of strong baryon-number violating fields inside the core. As long as a unification theory has monopoles at all, requirement (a) will be met one way or another. What will vary from one model to another are the fermion flavors which actually have S-wave states relative to the monopole (in the standard model studied here the neutrino and half the quark states do not). On the other hand, requirement (b) need not be met. For example, in the Pati-Salam $SU_4 \times SU_2 \times SU_2$ unification models, baryon number is conserved in all the gauge interactions. Therefore no baryon-number violation follows from the strong interaction of external fermions with the superheavy gauge mesons in the monopole core.[12] It is quite conceivable that one can construct theories intermediate between the two extremes of SU_5 (strong baryon decay catalysis) and $SU_4 \times SU_2 \times SU_2$ (no baryon decay catalysis) in the sense of having a baryon decay catalysis cross-section which is some small fraction of strong interaction size. This question has not been explored in detail.

What I have described obviously just scratches the surface of this subject. Evidently monopoles present a new and challenging arena in which to do strong interaction physics. The greatest lesson of this development is that the low energy interactions of the monopole give us a direct and detailed glimpse of details of the physics going on at the distance scale of the monopole core itself. Since we shall not be building accelerators to explore these scales for a long time to come, we have yet another reason to persevere in the great search for monopoles.

REFERENCES

1. V. Rubakov, Nucl. Phys. B 203, 311 (1982); JETP Lett. 33, 644 (1981).
2. C. G. Callan, Jr., Phys. Rev. D 25, 2141 (1982); Phys. Rev. D 26, 2058 (1982); Nucl. Phys. B 203, 311 (1982).
3. Y. Kazama, C. N. Yang and A. S. Goldhaber, Phys. Rev. D 15, 2287 (1977).
4. A. S. Goldhaber, Phys. Rev. D 16, 1815 (1977).
5. Y. Kazama, Int. Jour. Theo. Phys. 17, 262 (1978).
6. C. Besson, Ph.D. Thesis, Princeton University, June 1982.
7. R. Jackiw and C. Rebbi, Phys. Rev. D 13, 3398 (1976).
8. C. Dokos and T. Tomaras, Phys. Rev. D 21, 2940 (1980).
9. S. Mandelstam, Phys. Rev. D 11, 3026 (1975).
10. S. Coleman, Phys. Rev. D 11, 2088 (1975).
11. I am indebted to C. Goebel for discussions on this point.
12. I am indebted to T. Tomaras for discussions on this subject.

FRACTIONAL CHARGE AND MAGNETIC MONOPOLES

John Preskill

Lyman Laboratory of Physics
Harvard University
Cambridge, Massachusetts 02138

INTRODUCTION

Last spring, when word began to circulate that Blas Cabrera had detected something interesting, many of us recognized the opportunity to make an ironic remark (see, for example, Refs. 1-3): Two experiments, both performed at Stanford, had obtained results which were flatly contradictory. Cabrera[4] had observed a candidate magnetic monopole carrying the Dirac magnetic charge $g_D = 1/2e$. Fairbank and collaborators[5] had observed fractional electric charge $q = e/3$ on matter. Together, these two observations violate a very general consistency condition derived many years ago by Dirac.[6]

I first met Bill Fairbank at the Vanderbilt Conference on Particle Physics last May, and could not resist the temptation to make this ironic remark. Although, of course, he had heard it before, Bill responded kindly and patiently. He suggested that, instead of trying to be ironic, I should give some serious thought to the possibility that both results are correct. They are good experiments. What if both are right? This struck me as a provocative question.

So I thought about it a little.[7] Others[8-11] have done the same, presumably without any prompting from Bill Fairbank. For the remainder of this paper, I will assume without further apology that both experiments are correct; that is, I will assume that Cabrera has really discovered a monopole with the Dirac magnetic charge, and that Fairbank and collaborators have really discovered objects carrying fractional electric charge. I do not mean to endorse either result. I merely ask, if both experiments are right, how

111

can we explain the apparent violation of the Dirac consistency condition? This question leads us into some desperate speculations.

THE DIRAC QUANTIZATION CONDITION

First, let us recall Dirac's[6] reasoning, which indicates that the results of Fairbank and Cabrera are incompatible. (For a more detailed review, see Ref. 12.)

Dirac envisaged a magnetic monopole as a semi-infinitely long, infinitesimally thin solenoid. One end of the solenoid, viewed in isolation, appears to be a magnetic charge. But it makes sense to identiy this object as a magnetic monopole only if no conceivable experiment can detect the solenoid, in the limit in which it is infinitesimally thin.

We might imagine trying to detect the solenoid by doing an electron interference experiment; such an experiment gives a null result only if the phase picked up by the electron wave function, when the electron is transported along a closed path enclosing the solenoid, is trivial. Suppose a monopole with magnetic charge g sits at the origin, so that the magnetic field is

$$\vec{B} = g \frac{\hat{r}}{r^2} \,, \tag{1}$$

and that the solenoid lies on the negative z-axis. Then the vector potential can be written in polar coordinates as

$$\vec{A} \cdot d\vec{r} = g(1-\cos\theta) \, d\phi \,. \tag{2}$$

The electron interference experiment fails to detect the solenoid if

$$\text{phase} = \exp[-ie\oint\vec{A} \cdot d\vec{r}] = \exp[-4\pi ieg] = 1 \,, \tag{3}$$

where $-e$ is the electron charge. Hence, we require the magnetic charge g to satisfy the (Dirac) quantization condition

$$g = n/2e \,, \tag{4}$$

where n is an integer. The minimal allowed charge $g_D = 1/2e$ is called the Dirac magnetic charge.

It may disturb you that, to derive the Dirac quantization condition, Eq. (4), I have used the electron charge $-e$. We would like to believe that quarks exist, and the electric charge of a down quark, for example, is $-e/3$. Will not the same argument as before, applied to a down quark instead of an electron, lead to the conclusion that the minimal allowed magnetic charge is $3g_D$ instead of g_D?

No, not if quarks are confined.[13] For if quarks are permanently confined in hadrons, it makes sense to speak of performing a quark interference experiment only over distances less than 10^{-13} cm, the size of a hadron. It is true that, when the down quark is transported around Dirac's solenoid, its wave function acquires the nontrivial phase

$$\exp[-i(e/3)\oint \vec{A}_{em} \cdot d\vec{r}] = e^{-i2\pi/3} \neq 1 \tag{5}$$

due to the coupling of the down quark to the electromagnetic vector potential, if the monopole carries the Dirac magnetic charge g_D. But we must recall that the down quark carries another degree of freedom, color. The solenoid is not detectable if the monopole also has a <u>color-magnetic</u> <u>field</u>, such that the phase acquired by the down quark wave function due to the color vector potential compensates for the phase due to the electromagnetic vector potential, or

$$\exp[ie_c\oint \vec{A}_{color} \cdot d\vec{r}] = e^{i2\pi/3} , \tag{6}$$

where e_c is the color gauge coupling.

The correct conclusion, then, if quarks are confined, is not that the minimal magnetic charge is $3g_D$, but rather that the monopole carrying magnetic charge g_D must also carry a color-magnetic charge. The color-magnetic field of the monopole becomes screened by nonperturbative strong-interaction effects at distances greater than 10^{-13} cm.

How do we know that the color-magnetic field of the monopole is "screened"? Because there are no physical massless particles in QCD, the color-magnetic field can have no long-range effects. (In this respect, QCD differs from QED; long-range effects are mediated by the massless photon.) Either the color-magnetic field decays exponentially with distance r like $\exp(-Mr)$, where M is a hadron mass, or else the color-magnetic field is confined to flux tubes with a width of order $M^{-1} \sim 10^{-13}$ cm. But 't Hooft[14] has persuasively argued that color-magnetic flux tubes cannot form if quarks are exactly confined. The remaining alternative is that the color-magnetic field is screened.

Now we can state precisely the manner in which the observation of fractional electric charge is inconsistent with the observation of the Dirac magnetic charge. Suppose that there exists an isolated particle with electric charge q = e/3. By "isolated", I mean that it is possible to separate this particle from other charged particles by distances large compared to the color-magnetic screening distance M^{-1}. Is is legitimate to apply Dirac's argument to this particle, and we conclude that the minimal allowed magnetic charge of a magnetic monopole is $3g_D$ rather than g_D. It is for this reason that the results of the Fairbank experiment and the Cabrera experiment appear to be incompatible.

I now wish to reformulate the above discussion of the color-magnetic field of the monopole using a different notation which will prove useful in the ensuing discussion. The vector potential of a magnetic monopole which, like that considered above, carries more than one type of magnetic charge, can in general be written[15]

$$\sum_a e_a T^a \vec{A}^a \cdot d\vec{r} = \frac{1}{2} M (1-\cos\theta) d\phi , \tag{7}$$

where M is a constant matrix. The sum on a on the left-hand side of Eq. (7) runs over all unbroken gauge generators--for example, the electric charge and the eight color generators. The gauge couplings e_a have been absorbed into M. By an argument similar to that invoked above, we can derive the generalized Dirac quantization condition

$$\exp (2\pi i M) = 1 . \tag{8}$$

That is, M must have integer eigenvalues.

For example, in the SU(5) grand unified model, the electric charge generator may be written as a 5 × 5 matrix

$$Q_{em} = \text{diag} \left(\frac{1}{3}, \frac{1}{3}, \frac{1}{3}, 0, -1\right) , \tag{9}$$

where the diag(, ... ,) notation denotes a diagonal matrix with the indicated eigenvalues. The eigenvalues of Q_{em} are the electric charges, in units of e, of the elements of the defining 5 representation of SU(5)--antidown quarks in three colors, the neutrino, and the electron. The $SU(3)_c$ color generators are traceless 3 × 3 matrices acting only on the quarks; one of these is

$$Q_{color} = \text{diag} \left(- \frac{1}{3}, - \frac{1}{3}, \frac{2}{3}, 0, 0\right) . \tag{10}$$

With the electric charge generator Q_{em} given by Eq. (9), it is clear that objects which carry trivial $SU(3)_c$ triality have integer electric charge (in units of e), even though objects with nontrivial triality have fractional charge. A more mathematical, but useful, restatement of this observation is that $\exp(2\pi i Q_{em})$ is a nontrivial element of Z_3, the center of $SU(3)_c$; that is, it acts trivially on color singlets, and multiplies quarks of all three colors by the same nontrivial phase. For a monopole which satisfies the quantization condition (8), the matrix M may be written as

$$M = nQ_{em} + mQ_{color} ,\qquad\qquad\qquad (11)$$

where

$$\exp[2\pi i n Q_{em}] = \exp[-2\pi i m Q_{color}] = Z ,\qquad\qquad (12)$$

and Z is an element of Z_3; both n and m must be integers. The magnetic charge of this monopole is the coefficient of eQ_{em} in 1/2M, which is $n/2e = ng_D$.

The color-magnetic charge of the monopole must be defined with some care. The integer n is a conserved magnetic charge, but m is not, because the $U(1)_{em}$ and $SU(3)_c$ gauge groups have different topological properties. Color-magnetic monopoles with different values of m which correspond to the same element of Z_3, that is, values of m which are congruent modulo 3, are topologically equivalent to one another. Therefore, while the $U(1)_{em}$ charge can be any integer multiple of g_D, the color-magnetic charge may assume only three distinct values, which correspond to the three distinct elements of Z_3. We say that the monopole carries a conserved Z_3 color-magnetic charge. An assembly of three identical monopoles, each with a nontrivial Z_3 charge, has a trivial Z_3 charge. (How the conserved Z_3 flux can become screened is elucidated in Ref. 12.)

Two examples of monopoles in the SU(5) model are those defined by

$$M = Q_{em} + Q_{color} = \text{diag } (0,0,1,0,-1),\qquad\qquad (13)$$

and

$$M = 3Q_{em} = \text{diag } (1,1,1,0,-3);\qquad\qquad\qquad (14)$$

both are consistent with the quantization condition, Eq. (8). The monopole defined by Eq. (13) has $U(1)_{em}$ magnetic charge g_D and a nontrivial Z_3 color-magnetic charge; the monopole defined by Eq. (14) has $U(1)_{em}$ magnetic charge $3g_D$ and trivial Z_3 color-magnetic charge. In the SU(5) model, the lightest monopole, and therefore the stable one, is the one defined by Eq. (13).

INTERPRETATIONS OF FRACTIONAL ELECTRIC CHARGE

The above argument seems to exclude, on the basis of very general considerations, the possibility that both monopoles with magnetic charge $g_D = 1/2e$ and isolated fractional charges with $q = e/3$ exist. We will now hold this argument up to further scrutiny, discussing three different theoretical interpretations of the observation of fractional electric charge.

Unconfined Quarks

The interpretation of the observation of electric charge $e/3$ which seems to require minimal speculation is that quarks are not exactly confined, and Fairbank has detected free quarks.[16] But we have concluded that if any isolated charge-$e/3$ particles exist, then the minimal allowed magnetic charge of a monopole is $3g_D$, contrary to Cabrera's observation.

Further insight into this conclusion is gained if we recall another result obtained by 't Hooft.[14] I noted earlier that 't Hooft argued that if quarks are exactly confined, then the conserved Z_3 color-magnetic charge is screened rather than confined. The converse is also true! If quarks are not exactly confined, that is, if physical states exist which carry nontrivial color triality, then the Z_3 color-magnetic charge is confined, assuming that there are no long-range color forces (see below).

Consider, for example, how the monopoles of the SU(5) grand unified model would be affected if the $SU(3)_c$ gluons acquired masses of order μ due to the Higgs mechanism.[1] The stable SU(5) monopoles carry $U(1)_{em}$ magnetic charge g_D and nontrivial Z_3 color-magnetic charge. The Z_3 color-magnetic flux emerging from these monopoles would collapse to a tube with a width of order μ^{-1}, and the monopoles would be confined to magnetically neutral "mesons" or to "baryons" with trivial Z_3 color-magnetic charge and $U(1)_{em}$ magnetic charge $3g_D$. The magnetic baryons would thus carry the minimal magnetic charge allowed by the Dirac quantization condition, in the presence of free quarks.

The above remarks require one qualification. The existence of a free quark is consistent with the existence of a free monopole with $U(1)_{em}$ magnetic charge g_D if the monopole carries a long-range Z_3 color-magnetic field; that is, if the color-magnetic field is

not screened. This theoretical possibility must be excluded by
appealing to experiment. There can be long-range color forces only
if there are massless hadrons. If there were massless hadrons,
then they would be copiously produced in hadronic collisions, and
we would surely know about it.

To summarize: If there are no long-range color interactions,
then either quark confinement is exact, in which case we cannot
interpret Fairbank's fractional charge as a free quark, or else
monopoles with magnetic charge g_D are confined, in which case
Cabrera could not have detected and isolated monopole with that
charge. (We can safely rule out the possibility that the monopole
confinement scale μ^{-1} is macroscopic, for in that case the
probability of finding any free quarks would be negligible.) On
the other hand, there can be long-range color interactions only if
there are massless hadrons, which is experimentally excluded. To
reconcile the observations of Fairbank and Cabrera, we must find a
different interpretation of the observed fractional charge.

Dyons

One amusing possibility is that the objects which carry
fractional electric charge are actually "dyons" which also carry
magnetic charge. For two dyons with electric and magnetic charges
(q_1,g_1) and (q_2,g_2), the Dirac consistency condition becomes
generalized to[17]

$$q_1 g_2 - q_2 g_1 = n/2 \tag{15}$$

where n is an integer. This condition is satisfied if the electric
charge q of an object with magnetic charge g is given by

$$q/e = m + eg\theta/\pi \tag{16}$$

where m is an integer, and θ, which may take any value from $-\pi$ to
π, is a CP-violating parameter of electrodynamics which has
observable effects only in the presence of a monopole.[18] The
minimal electric charge of a dyon with the Dirac magnetic charge is
$q = e\theta/2\pi$. Conceivably, Fairbank and Cabrera have discovered the
same object--a dyon; Fairbank has measured its electric charge and
Cabrera has measured its magnetic charge.

Is there any reason to expect $|\theta|/2\pi \sim 1/3$? I know of no such
reason. On the other hand, $|\theta|/2\pi \sim 1/3$ cannot be excluded. In
grand unified theories, the θ appearing in Eq. (16) is related to
another angle,[19] θ_{QCD}, which is known to be very small (less than
10^{-9}).[20] But a natural explanation for the small value of θ_{QCD}, the
Peccei-Quinn mechanism,[21] has been proposed, and this mechanism
permits θ to assume a value which need not be small in general.

In any event, this theoretical possibility appears to be experimentally excluded. Cabrera[22] has measured the magnetic charges of two niobium spheres on which fractional electric charge had been observed; the measured magnetic charges were consistent with zero.

Fractionally Charged Leptons

The remaining possibility is that Fairbank has observed a color-singlet particle with vanishing magnetic charge and fractional electric charge. It could be either a fractionally charged lepton or a hadron containing a colored particle with an exotic charge.

How can we reconcile the existence of such a particle with Cabrera's observation of a monopole with magnetic charge g_D? Our discussion of the generalized quantization condition, Eq. (8), suggests an answer.[7-11] There must be a new long-range gauge interaction, not yet observed, in addition to electromagnetism. This new interaction might be a nonabelian gauge interaction with a macroscopic confinement length,[24] but I will consider the simplest possibility, that it is a U(1) gauge interaction. I will refer to this interaction as extraordinary electromagnetism, and to its gauge group as $U(1)_{ex}$; matter carrying $U(1)_{ex}$ charge will be called extraordinary matter.

Extraordinary electromagnetism may have escaped detection so far because ordinary matter is neutral under $U(1)_{ex}$. But Fairbank's fractionally charged particle and Cabrera's monopole must carry both types of charge. The Dirac quantization condition for an electrically charged particle with $U(1)_{em}$ and $U(1)_{ex}$ charges q and q' and a monopole with magnetic charges g and g' is

$$qg + q'g' = n/2. \tag{17}$$

This condition is satisfied by, for example, a monopole with magnetic charges $g = g_D = 1/2e$ and $g' = g_D' = 1/2e'$, and an extraordinary lepton with electric charges $q = e/3$ and $q' = -e'/3$. The apparent inconsistency with the Dirac quantization condition may have arisen, then, because Fairbank and Cabrera each measured only one of the two (or more) types of charge carried by the lepton and monopole.

We should next consider whether we can construct realistic grand unified models which take advantage of this possibility.

EXTRAORDINARY MODEL BUILDING

Howard Georgi and I have made some preliminary attempts to

construct such models,[25] as have others.[9-11] We seek a model in which the low-energy gauge group

$$G_{LE} = SU(3)_c \times [SU(2) \times U(1)]_{ew} \times U(1)_{ex} \qquad (18)$$

is embedded in some unifying group G. The breakdown of G to G_{LE} might occur all at once at the grand unification mass scale, or there might be a nontrivial sequence of intermediate symmetry-breaking scales. We require that there are at least three generations of ordinary matter, which are neutral under $U(1)_{ex}$ and carry the standard $SU(3)_c \times [SU(2) \times U(1)]_{ew}$ quantum numbers of

$$\begin{pmatrix} u \\ d \end{pmatrix}_L^{1/6}, \quad \bar{u}_L^{-2/3}, \quad \bar{d}_L^{1/3}, \quad \begin{pmatrix} \nu \\ e \end{pmatrix}_L^{-1/2}, \quad \bar{e}_L^{1}. \qquad (19)$$

There must also be a magnetic monopole with $U(1)_{em}$ magnetic charge $g_D = 1/2e$, and a color-singlet charged particle with $U(1)_{em}$ electric charge e/3.

The smallest unifying group for which a model satisfying these criteria is likely to be found is G = SU(7), and the most promising fermion representation to consider in SU(7) is the representation

$$7 + \overline{21} + 35 , \qquad (20)$$

which is obtained by decomposing the spinor representation of SO(14) with respect to SU(7). To begin our search, let us ask whether there is any embedding of G_{LE} in SU(7) such that the representation (20) contains at least one generation of ordinary matter, neutral under $U(1)_{ex}$. This is a variation on a game pioneered by Kim,[26] who investigated the consequences of different embeddings of $SU(3)_c \times [SU(2) \times U(1)]_{ew}$ in SU(7). Kim's motivation was to understand the repeated generation pattern of the quarks and leptons, and we might hope that models which satisfy our criteria will also offer some insight into the generation puzzle.

It turns out that there are only two solutions to the problem of embedding G_{LE} in SU(7) such that the representation (20) contains at least one ordinary generation with conventional quantum numbers. In both solutions, $SU(3)_c \times SU(2)_w$ is embedded in the standard way: $7 \to (3,1) + (1,2) + 2(1,1)$. The generators of $U(1)_{em} \times U(1)_{ex}$, in the first solution, are

$$Q_{em} = \text{diag} (1/3, 1/3, 1/3, 0, -1, 0, 0)$$

$$Q_{ex} = \text{diag} (0, 0, 0, 0, 0, 1/2, -1/2) \qquad (21)$$

where

$$Q_{color} = \text{diag} (-1/3, -1/3, 2/3, 0, 0, 0, 0)$$

This is the trivial solution. The Georgi-Glashow SU(5) has been embedded in SU(7), and $U(1)_{ex}$ commutes with this SU(5). This model does not satisfy our criteria, because all particles with trivial color triality carry integer electric charge. Nevertheless, it is not totally uninteresting from the perspective of the generation puzzle. The representation (20) contains two ordinary left-handed SU(5) generations, neutral under $U(1)_{ex}$, plus two right-handed generations which carry $U(1)_{ex}$ charges. The unbroken $U(1)_{ex}$ gauge symmetry prevents the left-handed and right-handed generations from pairing up and acquiring superheavy masses. Including an unbroken $U(1)_{ex}$ thus allows us to unify two ordinary SU(5) gnerations in the spinor representation of SO(14). This unification of two generations is similar to that achieved by Kim,[26] who considered unconventional embeddings of $U(1)_{em}$ in SU(7).

The other, nontrivial, solution is

$$Q_{em} = \text{diag} (1/6, 1/6, 1/6, 0, -1, 1/2, 0) ,$$

$$Q_{ex} = \text{diag} (1/2, 1/2, 1/2, 0, 0, -1/2, -1) , \qquad (22)$$

$$Q_{color} = \text{diag} (1/3, 1/3, -2/3, 0, 0, 0, 0) .$$

Because $Q_{em} + Q_{ex} + Q_{color}$ has integer eigenvalues, this model has magnetic monopoles with $U(1)_{em}$ magnetic charge $g = g_D = 1/2e$ and $U(1)_{ex}$ magnetic charge $g' = g_D' = \pm 1/2e'$. However, it is also clear that with this embedding of Q_{em}, objects with trivial $SU(3)_c$ triality carry <u>half</u>-integer electric charge, instead of one-third-integer charge.

A surprising feature of this model is that the Weinberg angle comes out right; the unrenormalized value of $\sin^2 \theta_w$ is 3/8 as in the SU(5) model. But, because it contains half-integer charges rather than one-third-integer charges, this model cannot account for the Fairbank experiment. No SU(7) model can, if the fermions are in the representation (20), and the model also contains magnetic monopoles with magnetic charge g_D.

What should we try next? We could consider other representations of SU(7), semisimple unifying groups like SU(5) × SU(5), or simple groups larger than SU(7). It is not hard to construct models based on SU(5) × SU(5) which satisfy our criteria,[9] but I will not discuss this construction here. Instead, I will describe two models based on SU(9).

Both of these models satisfy all our criteria. In the first
model, $SU(3)_c$ and $SU(2)_w$ are minimally embedded in $SU(9)$, and the
generators of $U(1)_{em} \times U(1)_{ex}$ are

Q_{em} = diag (1/3, 1/3, 1/3, 0, -1, -1/3, -1/3, 2/3, 0),

Q_{ex} = diag (0, 0, 0, 0, 0, 1/3, 1/3, 1/3, -1),   (23)

where

Q_{color} = diag (-1/3, -1/3, 2/3, 0, 0, 0, 0, 0, 0).

(This model was also proposed in Ref. 10.) In the second model,
the embedding of $SU(3)_c$ in $SU(9)$ is nonstandard; we have $9 \rightarrow 3 + 3$
$+ 1 + 1 + 1$, and the generators of $U(1)_{em} \times U(1)_{ex}$ are

Q_{em} = diag (1/3, 1/3, 1/3, 0, -1, 0, 0, 0, 0),

Q_{ex} = diag (0, 0, 0, 0, 0, 1/3, 1/3, 1/3 -1),   (24)

where

Q_{color} = diag (-1/3, -1/3, 2/3, 0, 0, -1/3, -1/3, 2/3, 0).

In both models, $Q_{em} + Q_{ex} + Q_{color}$ has integer eigenvalues, and
both models therefore contain magnetic monopoles with magnetic
charges g and g' given by

$$(eg, e'g') = \left(\frac{1}{2}, \frac{1}{2} \right).$$ (25)

Also, in both models, objects with trivial color triality need not
have integer electric charge; instead, it is required that

q/e + q'/e' = n   (26)

where n is an integer and q/e may be one third of an integer.

The fermions may be chosen to lie in the $SU(9)$ representation

$9 + \overline{36} + 84 + \overline{126}$ (27)

which is obtained by decomposing the spinor representation of
$SU(18)$ with respect to $SU(9)$. This representation contains, in
both models, two ordinary generations which are neutral under
$U(1)_{ex}$, plus many extraordinary fermions.   For example, in the

second model, with embedding (24), there are extraordinary leptons with charges

$$(q/e,\ q'/e') = (-1/3,\ 1/3),\ \ (-1/3,\ -2/3),$$
$$(1/3,\ -1/3),\ \ (1/3,\ 2/3),$$
$$(2/3,\ 1/3),\ \ (2/3,\ -2/3), \qquad (28)$$
$$(-2/3,\ -1/3),\ (-2/3,\ 2/3).$$

The model with the embedding (24) appears to have the advantage that the unrenormalized Weinberg angle at the unification scale is $\sin^2\theta_W = 3/8$, as in the SU(5) model. However, because of the nonstandard embedding of SU(3)$_c$, $\sin^2\theta_W$ is renormalized differently, and it is actually necessary in both models to introduce intermediate symmetry-breaking scales in order to get the observed value of $\sin^2\theta_W$ at low energy. This is a heavy price to pay, because our power to compute $\sin^2\theta_W$ is lost. The renormalized value at low energy of the extraordinary gauge coupling e^1 also depends on the intermediate-scale physics, but it is typically of order e. The Higgs structure needed to achieve the desired symmetry-breaking pattern is quite intricate; I will not describe it.

EXTRAORDINARY PHENOMENOLOGY

In all models of the type described in the previous section, there must be at least one stable particle which carries fractional U(1)$_{em}$ and U(1)$_{ex}$ charge. Because this particle has not been produced at Petra, we know that its mass m_{ex} satisfies

$$m_{ex} \gtrsim 20 \text{ GeV}. \qquad (29)$$

Because it mass is forbidden by weak-interaction symmetries, we also know that it cannot be arbitrarily heavy; we expect

$$m_{ex} < G_F^{-1/2} \sim 300 \text{ GeV}. \qquad (30)$$

Apparently, extraordinary quarks and leptons are much heavier than their ordinary counterparts. I do not know why this should be so, but perhaps the question should be phrased differently. The natural value for a mass forbidden by weak-interaction symmetries is of order 300 GeV; what is truly mysterious is not that the extraordinary matter is so heavy, but that the ordinary matter is so light.

If ordinary quarks and leptons are neutral under U(1)$_{ex}$, as we have assumed, then the U(1)$_{ex}$ extraordinary photon couples to ordinary matter only through processes involving virtual extraordinary matter. For example, the extraordinary photon can couple to three gluons through an extraordinary quark loop, but, because of gauge invariance, this coupling is suppressed at low energy by m_{ex}^{-4}. There is no direct experimental evidence, or any astrophysical evidence, against the existence of a massless particle which couples so weakly to ordinary matter.

However, models of this type face a serious difficulty when we consider their cosmological implications. The difficulty, which is encountered by any model which contains fractionally charged color-singlet particles, is that too many such particles are produced in the early universe. Standard estimates indicate that, when the electroweak and extraordinary interactions freeze out and annihilation of extraordinary matter ceases, the relative abundance of extraordinary matter to baryons is given by

$$n_{ex}/n_{baryon} \gtrsim 10^{-5}(m_{ex}/100 \text{ GeV}). \tag{31}$$

Observational bounds on the abundance of fractionally charged particles in the earth's crust are highly uncertain, but it is generally agreed that the abundance must be much smaller than allowed by (31), for $m_{ex} \gtrsim 20$ GeV.

A means of reducing an initially large abundance of fractionally charged leptons to an acceptable level has been proposed by Goldberg.[27] He suggests that most of the leptons annihilated in very massive first-generation stars. The problem with this suggestion is that, to explain the low abundance of fractionally charged particles in the earth's crust, it is required that all but a tiny fraction of the matter in the solar system has been processed in these first-generation stars, which seems implausible.

A more reasonable way to avoid a large cosmological abundance of fractionally charged particles is to construct a model in which all fractionally charged particles are superheavy. In the new inflationary universe scenario,[28] it is possible for the abundance of stable superheavy particles to be both nonnegligible and small. Indeed, this scenario provides the best known way to obtain an acceptable cosmological abundance of magnetic monopoles (for a review, see Ref. 7); we might as well use the same trick to control the abundance of fractionally charged particles.

It is possible to construct models in which all extraordinary matter is superheavy, but not without paying an aesthetic price. Barr, Reiss, and Zee[9] recently constructed such a model based on $SU(5)_{ord} \times SU(5)_{ex}$, in which ordinary matter carries $SU(5)_{ord}$ quantum numbers and extraordinary matter carries $SU(5)_{ex}$ quantum numbers. The two SU(5)'s are treated quite asymmetrically, so that in a sense this model is less unified than the SU(9) model described earlier. But it is nonetheless a quite simple model which satisfies all the criteria specified at the beginning of the previous section, and avoids the cosmological problem as well.

CONCLUSIONS

The observation of a magnetic monopole carrying the Dirac magnetic charge g_D = 1/2e and the observation of isolated electric charge q = e/3 on matter are not necessarily incompatible, but the two results together impose severe constraints on unified model building. Two types of models have been described which can accommodate both observations. One type of model predicts the existence of many new particles with mass less than a few hundred GeV which will turn up in forthcoming accelerator experiments, but runs into cosmological problems. The other type of model avoids the cosmological problems by making all fractionally charged particles superheavy; the new physics predicted by this type of model will not be directly accessible in accelerator experiments.

The model-building exercises outlined here were clearly very ad hoc. Our goal was to construct models which are consistent with the results of both the Cabrera experiment and the Fairbank experiment, and in this we succeeded. But one might have hoped that these models would have other, unexpected, good features; for example, one might have hoped to gain insight into the generation puzzle. Unfortunately, this hope was not realized.

Nevertheless, it is probably wise to bear in mind the implications of the Cabrera experiment and the Fairbank experiment taken together, which are much more far-reaching than the implications of either experiment taken alone. These implications provide us with further motivation to follow with great interest the developing experimental status of fractional charge and magnetic monopoles.

Note added: When I wrote this paper, I had not yet seen the paper by Pantaleone, Ref. 11. His analysis of models incorporating extraordinary electromagnetism is very similar to my presentation here.

REFERENCES

1. G. Lazarides, Q. Shafi, and W. P. Trower, Phys. Rev. Lett. 49, 1756 (1982).
2. M. T. Vaughn, in "Third Workshop on Grand Unification," P. H. Frampton, S. L. Glashow, and H. van Dam, eds., Birkhauser, Boston (1982).

3. G. 't Hooft, Talk at XXI International Conference on High Energy Physics, Paris (1982).
4. B. Cabrera, Phys. Rev. Lett. $\underline{48}$, 1220 (1982).
5. G. S. LaRue, W. M. Fairbank, and A. F. Hebard, Phys. Rev. Lett. $\underline{38}$, 1011 (1977); G. S. LaRue, W. M. Fairbank, and J. D. Phillips, Phys. Rev. Lett. $\underline{42}$, 142, 1019 (E) (1979); G. S. LaRue, J. D. Phillips, and W. M. Fairbank, Phys. Rev. Lett. $\underline{46}$, 967 (1981).
6. P. A. M. Dirac, Proc. Roy. Soc. (London) Ser. A, $\underline{133}$, 60 (1931).
7. J. Preskill, in "The Very Early Universe," S. W. Hawking, G. W. Gibbons, and S. Siklos, eds., Cambridge University Press, Cambridge, to be published (1983).
8. A. E. Strominger, IAS Preprint (1982).
9. S. M. Barr, D. B. Reiss, and A. Zee, Seattle preprint 40048-31 P2 (1982).
10. S. Aoyama, Y. Fujimoto, and Zhao Zhiyong, Trieste preprint (1982).
11. J. Pantaleone, Cornell preprint (1982).
12. S. Coleman, in "Proceedings of the International School of Subnuclear Physics," Ettore Majorana, Erice, 1981, A. Zichichi, ed., Plenum, New York, to be published (1983).
13. G. 't Hooft, Nucl. Phys. B $\underline{105}$, 538 (1976); E. Corrigan, D. Olive, D. Fairlie, and J. Nuyts, Nucl. Phys. B $\underline{106}$, 475 (1976).
14. G. 't Hooft, Nucl. Phys. B $\underline{138}$, 1 (1978); B $\underline{153}$, 141 (1979).
15. P. Goddard, J. Nuyts, and D. Olive, Nucl. Phys. B $\underline{125}$, 1 (1977).
16. A. DeRujula, R. C. Giles, and R. L. Jaffe, Phys. Rev. D $\underline{17}$, 285 (1978); D $\underline{22}$, 227 (1980); R. Slansky, T. Goldman, and G. L. Shaw, Phys. Lett. $\underline{47}$, 887 (1981).
17. J. Schwinger, Phys. Rev. $\underline{144}$, 1087 (1966); D. Zwanziger, Phys. Rev. 176, 1480, 1489 (1968).
18. E. Witten, Phys. Lett. $\underline{86B}$, 283 (1979).
19. G. 't Hooft, Phys. Rev. Lett. $\underline{37}$, 8 (1976); C. G. Callan, R. F. Dashen, and D. J. Gross, Phys. Lett. $\underline{63B}$, 334 (1976); R. Jackiw and C. Rebbi, Phys. Rev. Lett. $\underline{37}$, 177 (1976).
20. V. Baluni, Phys. Rev. D $\underline{19}$, 2227 (1979); R. J. Crewther et al., Phys. Lett. $\underline{88B}$, 123 (1979).
21. R. Peccei and H. Quinn, Phys. Rev. Lett. $\underline{38}$, 1440 (1977); J. E. Kim, Phys. Rev. Lett. $\underline{43}$, 103 (1979); M. Dine, W. Fischler, and M. Srednicki, Phys. Lett. $\underline{104B}$, 199 (1981).
22. B. Cabrera, in "Third Workshop on Grand Unification," P. H. Frampton, S. L. Glashow, and H. van Dam, eds., Birkhauser, Boston (1982).
23. L.-F. Li and F. Wilczek, Phys. Lett. $\underline{107B}$, 64 (1981); H. Goldberg, T. W. Kephart, and M. T. Vaughn, Phys. Rev. Lett. $\underline{47}$, 1429 (1981).
24. L. B. Okun, Nucl. Phys. B $\underline{173}$, 1 (1980).

25. H. Georgi and J. Preskill, unpublished.

26. J. E. Kim, Phys. Rev. Lett. $\underline{45}$, 1916 (1980); Phys. Rev. D $\underline{23}$, 2706 (1981).

27. H. Goldberg, Phys. Rev. Lett. $\underline{48}$, 1518 (1982).

28. A. Guth, Phys. Rev. D $\underline{23}$, 347 (1981); A. D. Linde, Phys. Lett. $\underline{108}$B, 389 (1982); $\underline{114}$B, 431 (1982); A. Albrecht and P. J. Steinhardt, Phys. Rev. Lett. $\underline{48}$, 1220 (1982).

MONOPOLES AND ASTROPHYSICS

Michael S. Turner

Astronomy & Astrophysics Center
The University of Chicago
Chicago, Illinois 60637

INTRODUCTION

The purpose of this workshop is to discuss various aspects of superheavy magnetic monopoles. Coleman and Goldhaber have told us what a grand unified monopole is. It is a topologically-stable, classical configuration of gauge and Higgs fields which is present in the low-energy theory when a semisimple group G breaks down to a group G' which contains a U(1) factor.[1] The long-range fields of this configuration are those of a magnetic monopole with magnetic charge $h = n(2\pi/e) \simeq n(69e)$, $n = \pm1, \pm2, \ldots$, and the energy associated with this configuration is $O(M/\alpha)$ where M is the scale of spontaneous symmetry breaking (SSB) and α is the gauge coupling constant. For SU(5) $M \simeq 10^{14}$ GeV, and the monopole mass $m \simeq 10^{16}$ GeV ($\simeq 10^{-8}$ grams!). In my review I will discuss where the magnetic monopoles with us today came from and how many we should expect. I will trace their path from production until today, discussing where we should expect to find monopoles and what kind of things they are doing today (contributing mass density, devouring nucleons, destroying magnetic fields, etc.) Next I will review the astrophysical constraints on the monopole flux which follow from considering the mass density of the Universe, the survival of large-scale astrophysical magnetic fields, and monopole catalysis of nucleon decay. Finally, I will briefly discuss the possibility that the local flux of monopoles is enhanced (and argue that on kinematic grounds alone it is very unlikely), and comment on the possibility that the origin of the galactic magnetic field is due to magnetic monopoles, rather than electric currents. [This topic will be covered in greater detail by Wasserman.] I will make no contribution to the most important challenging question which this workshop will address: just how does one go about trying to detect these extraordinarily interesting objects? We will hear much about monopole hunting (past, present, and future) from other speakers at this workshop.

PRODUCTION OF MONOPOLES: GLUT OR FAMINE

Since there exist no contemporary sites for producing particles of mass even approaching 10^{16} GeV, the only plausible production site is the early Universe, about 10^{-34} sec after "the bang" when the temperature was $O(10^{14}$ GeV). There are two ways in which monopoles can be produced: (1) as topological defects during the SSB of the unified group G (e.g., SU(5)); (2) in monopole-antimonopole pairs by energetic particle collisions. The first process has been studied by Preskill,[2] and Zel'dovich and Khlopov,[3] and I will briefly review their conclusions here. The magnitude of the Higgs field responsible for the SSB of the unified group G is determined by the minimization of the free energy. However, this does not uniquely specify the direction of the Higgs field in group space. A monopole corresponds to a configuration in which the direction of the Higgs field in group space at different points in physical space is topologically distinct from the configuration in which the Higgs field points in the same direction (in group space) everywhere in physical space (which corresponds to no monopole):


```
    ↑ ↑ ↑                    ↑              → = direction of
    ↑ ↑ ↑                  ← ↓ →                Higgs field
  no monopole                ↓                in group space.
                          monopole
```


In the standard, hot big bang cosmology there are particle horizons, i.e., the distance over which a light signal could have propagated since "the bang" (t = 0) is finite and $\simeq O(ct)$. At the time of SSB (in SU(5) T $\simeq 10^{14}$ GeV and t $\simeq 10^{-34}$ sec) the Higgs field can only smoothly orient itself on scales $\lesssim$ horizon $\simeq$ ct $\simeq 10^{-23}$ cm. This results in $O(1)$ monopole (topological defect) per horizon volume. The horizon volume contains a net baryon number of about $(10^{15}$ GeV/T$)^3$ (corresponding today to that number of baryons). For SU(5) the number of monopoles produced as topological defects is $O(10^{-3})$ per baryon--a seemingly small number when one considers today there are $O(10^{10})$ microwave photons per baryon. Because of their relative scarcity, monopoles and antimonopoles do not annihilate in appreciable numbers, and for the standard cosmology and SU(5) the relic monopole abundance predicted is $O(10^{-3})$ per baryon. This corresponds to a present mass density of about 10^{12} × the critical mass density! This is clearly impossible. [More precisely, when such a Universe cooled to a temperature of O(3K) it would only be O(30,000 yrs) old!]. This catastrophe is known as the monopole problem. A number of possible remedies have been suggested;[4] the most attractive is the inflationary scenario which Guth will discuss. In this scenario our observable Universe today was, at the epoch of SSB, contained well within one horizon volume, so that we would expect less than one monopole in our observable Universe due to this process.

In any case, monopole-antimonopole pairs can be produced in energetic particle collisions, e.g., particle + antiparticle → monopole + antimonopole. The numbers produced are intrinsically small because monopole configurations do not exist in the theory until SSB ocurs ($T_c \simeq M$ = scale of SSB), and have a mass $O(M/\alpha) \simeq$ 100 M $\simeq$ 100 T_c. For this reason they are never present in equilibrium numbers; however, some are produced due to the rare collisions of particles with sufficient energy. This results in a present monopole to photon ratio[5-7]

$$n_M/n_\gamma \simeq 10^3 \ (m/T_{max})^3 \ \exp(-2m/T_{max}), \qquad (1)$$

where m is the mass of the monopole, and T_{max} is the highest temperature achieved after SSB. For reference, $\Omega_M \simeq 5 \times 10^{23}$ $(n_M/n_\gamma)(m/10^{16}$ GeV). In general $m/T_{max} \simeq O(100)$ so that $n_M/n_\gamma \simeq O \ (10^{-70})$--a negligible number of monopoles. However, the number produced is exponentially sensitive to m/T_{max}, so that a factor of 3-10 uncertainty introduces an enormous uncertainty in the predicted production. For the simplest model of new inflation (SU(5) with Coleman-E. Weinberg SSB) m/T_{max} can be calculated rather precisely (after making a number of reasonable assumptions). Preskill[6] finds that $m/T_{max} \simeq 230$ which results in less than one thermally produced monopole in the observable Universe today. Thermal production is likely to result in far too few monopoles to ever be detectable; however, one should bear in mind that the uncertainties are in a large, negative exponent.

Cosmology leaves the poor experimenter with two firm predictions: that there should be equal numbers of north and south poles; and that either far too few to detect, or far too many to be consistent with observation should have been produced. The detection of any superheavy monopoles would necessarily send theorists back to their chalkboards!

THEIR HISTORY FROM $t \simeq 10^{-34}$ SEC TO $t \simeq 3 \times 10^{17}$ SEC

Although monopoles don't interact with each other and annihilate, they do interact with the ambient particles (e.g., monopole + $e^- \rightleftarrows$ monopole + e^-), and thereby stay in kinetic equilibrium (KE $\simeq 3T/2$), until the epoch of $e^\pm$ annihilation (T $\simeq$

1/2 MeV, t $\simeq$ 10 sec). Thereafter, they effectively cease to interact with other particles in the Universe (except gravitationally).

At the time of $e^{\pm}$ annihilation monopoles should have an internal velocity dispersion $<v^2>^{1/2} \simeq 10$ cm s^{-1}(10^{16} GeV/m)$^{1/2}$. After this, their velocity dispersion redshifts away $\propto R(t)^{-1}$, and in the absence of gravitational interactions should today have a value of $\simeq 10^{-8}$ cm s^{-1}(10^{16}GeV/m)$^{1/2}$. Since they are collisionless their velocity dispersion is the only thing which can support them against gravitational collapse. Their Jeans length today $\lambda_J \simeq v_s t$ $\simeq 10^{10}$ cm $\simeq 10^{-8}$ LY is tiny, so they are gravitationally unstable on all interesting astrophysical scales.

After decoupling (T $\simeq$ 1 eV, t $\simeq 10^{13}$ sec) matter being free from radiation pressure support started to clump due to the gravitational (or Jeans) instability. Monopoles, too, should have clumped with the matter as structure (galaxies, etc.) formed. However, monopoles which are effectively collisionless particles could not have dissipated energy and further condensed into the more compact structures such as galactic disks, stars, etc., all of which require dissipation to form. One would naively expect to find monopoles in structures which do not require dissipation to form, the haloes of galaxies, clusters of galaxies, etc. [For the present I am ignoring the effect of astrophysical magnetic fields on the monopoles.]

In the Universe today the typical peculiar velocity (i.e., velocity relative to the Hubble flow) of a galaxy is 0(300-1000 km s^{-1}), the higher end being typical of galaxies which are members of rich clusters of galaxies. Within a galaxy like ours, the orbital velocity of a star is also 0(300 km s^{-1}). So even though monopoles initially have a negligible velocity dispersion, their velocity relative to us should be 0(300 km s^{-1}) $\simeq 10^{-3}$ c, whether they are orbiting in the halo of our galaxy or just uniformly distributed throughout the cosmos.

WHAT ARE THEY DOING TODAY?--ASTROPHYSICAL CONSTRAINTS

The three most conspicuous (and interesting properties) of a grand unified monopole are: (i) macroscopic mass (comparable to a bacterium); (ii) hefty magnetic charge h $\simeq$ 69e; (iii) ability to catalyze nucleon decay, perhaps at a strong interaction rate. I will now discuss each of these aspects of a monopole and the astrophysical constraints that follow.

Monopoles and the Mass Density of the Universe

Theoretical considerations/prejudices (particularly the

inflationary Universe) favor the very nearly flat ($\Omega \equiv \rho/\rho_{crit} \simeq 1$) cosmological model. Big bang nucleosynthesis strongly suggests that baryons alone cannot close the Universe; specifically, if $\Omega_b \gtrsim 0.2$, then deuterium is underproduced, and ^{4}He is overproduced.[8] Monopoles are certainly a candidate for providing closure density. In addition, luminous matter accounts for only a small fraction, $0(10^{-1})$, of the mass in haloes of spiral galaxies (as inferred by rotation curves), in binary galaxies, in small groups of galaxies, and in rich clusters of galaxies (as inferred by dynamics/the virial theorem). Here, too, monopoles are a candidate, especially since the formation of these structures does not involve dissipation.

Ω is not know with great precision; however, it is certainly $\lesssim 2$. If monopoles are distributed uniformly in the cosmos, then this restricts the monopole abundance and flux:

$$n_M \lesssim 2 \times 10^{-21}(10^{16} \text{ GeV/m}) \text{ cm}^{-3}, \qquad (2a)$$

$$F \lesssim 5 \times 10^{-15}(10^{16} \text{ GeV/m}) \ (v/10^{-3}) \text{ cm}^{-2} \text{ sr}^{-1} \text{ s}^{-1}. \qquad (2b)$$

For reference 10^{-12} cm^{-2} sr^{-1} s^{-1} = 1 m$^{-2}(\pi$ sr$)^{-1}$yr^{-1}. On the other hand, if monopoles are clustered in galaxies, then their local density can be $0(10^5)$ higher. Based upon the assumption that the mass within 30 kpc of the center of the galaxy is at most 10^{12} M$_\odot$ (as is indicated by galactic rotation curves), it follows that:

$$n_M \lesssim 10^{-16}(10^{16} \text{ GeV/m}) \text{ cm}^{-3}, \qquad (3a)$$

$$F \lesssim 3 \times 10^{-10}(10^{16} \text{ GeV/m}) \ (v/10^{-3}) \text{ cm}^{-2} \text{ sr}^{-1} \text{ s}^{-1}. \qquad (3b)$$

Since there is no way a monopole can disguise the fact that it has mass, there is no way that this bound on the average flux of monopoles in the galaxy can be circumvented.

Monopoles and Magnetic Fields

A monopole by virtue of its magnetic charge will be accelerated by magnetic fields, and in the process can gain KE. Of course, any KE gained must come from somewhere. Any gain in KE is exactly compensated for by a loss in field energy: $\Delta KE = -\Delta(B^2/8\pi) \times$ Vol. Consider a monopole which is initially at rest in a region of uniform magnetic field. It will be accelerated along the field and after moving a distance ℓ the monopole will have

$$KE = hB\ell \simeq 6 \times 10^7 \text{ GeV}(B/10^{-6}G) \ (\ell/pc), \tag{4a}$$

$$v = (2hB\ell/m)^{1/2} \simeq 10^{-4} \ c(B/10^{-6}G)^{1/2} \ (\ell/pc)^{1/2}(10^{16} \text{ GeV/m})^{1/2}. \tag{4b}$$

If the monopole is not initially at rest the story is a bit different. There are two limiting situations, and they are characterized by the relative sizes of the initial velocity of the monopole, v_o, and the velocity just calculated above, v_{mag}. First, if the monopole is moving slowly compared to v_{mag}, $v_o << v_{mag} \simeq (2hB\ell/m)^{1/2}$, then it will undergo a large deflection due to the magnetic field and its change in KE will be given by (4a). On the other hand, if $v_o >> v_{mag}$, then the monopole will only be slightly deflected by the magnetic field, and its change in energy will depend upon the direction of its motion relative to the magnetic field. In this situation the energy gained by a spatially isotropic distribution of monopoles, or a flux of equal numbers of north and south poles will vanish at first order in B--some poles will lose KE and some poles will gain KE. However, there is a net gain in KE at second order in B by the distribution of monopoles,

$$<\Delta KE> \simeq (hB\ell) \ (v_o/v_{mag})^2/4. \tag{5}$$

For the galactic magnetic field $B \simeq 3 \times 10^{-6}$ G, $\ell \simeq 300$ pc, and $v_{mag} \simeq 3 \times 10^{-3} c \ (10^{16} \text{ GeV/m})^{1/2}$. Since $v_o \simeq 10^{-3}$ c, monopoles less massive than about 10^{17} GeV will undergo large deflections when moving through the galactic field and their gain in KE is given by (4a). Because of this energy gain, monopoles less massive than 10^{17} GeV will be ejected from galaxies in a very short time, and thus are unlikely to cluster in the haloes of galaxies. In fact the second order gain in KE will "evaporate" monopoles as massive as $0(10^{20}\text{GeV})$ in a time less than the age of the galaxy. Although considerations of galaxy formation would suggest that monopoles should cluster in galactic haloes, galactic magnetic fields should prevent monopoles less massive than $0(10^{20}\text{GeV})$ from clumping in galactic haloes. [These conclusions are not valid if the magnetic field of the galaxy is itself produced by monopoles, a point to which I will return.]

The "no free-lunch principle" ($\Delta KE = -\Delta$ Magnetic Field Energy) and formulae (4a) and (5) can be used to place a limit on the average flux of monopoles in the galaxy.[9-11] If, as it is commonly believed, the origin of the galactic magnetic field is due to dynamo action, then the time required to generate/regenerate the field is of the order of a galactic rotation time $\simeq 0(10^8$ yr). Demanding that monopoles not drain the field energy in a time

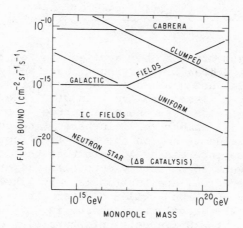

Fig. 1. Summary of the astrophysical bounds on the monopole flux
         as a function of monopole mass and Cabrera's flux
         (= 1/total exposure). "Clumped" and "uniform" refer to
         the bounds based upon the mass density contributed by
         monopoles. For the bound based upon monopole catalyzed
         nucleon decay in neutron stars (σv) = 10^{-27} cm^2 was
         assumed. (Note: 10^{-12}cm^{-2}sr^{-1}s^{-1}=1m^{-2}(π-sr)$^{-1}$yr^{-1}.)

shorter than this results in the following constraints:

m $\lesssim$ 10^{17} GeV:

$$F \lesssim 10^{-15} \text{ cm}^{-2} \text{ sr}^{-1} \text{ s}^{-1} (B/3 \times 10^{-6} \text{ G}) (3 \times 10^7 \text{ yr}/\tau) \times$$

$$(r/30 \text{ kpc})^{1/2} (300 \text{ pc}/\ell)^{1/2}, \tag{6a}$$

m $\gtrsim$ 10^{17} GeV:

$$F \lesssim 10^{-16} \text{ cm}^{-2} \text{ sr}^{-1} \text{ s}^{-1}(m/10^{16} \text{ GeV})(3 \times 10^7 \text{ yr}/\tau)(300 \text{ pc}/\rho\ell),$$

$$\tag{6b}$$

where v_o has been assumed to be 10^{-3} c, τ is the regeneration time
of the field, ℓ is the coherence length of the field, and r is the
size of the magnetic field region in the galaxy. [Purcell will
discuss bound (6a) in greater detail.] Constraint (6a) which
applies to 10^{16} GeV monopoles is very stringent (some 5 orders of
magnitude smaller than Cabrera's reported "flux") and is known as
the "Parker bound." For more massive monopoles ($\gtrsim$ 10^{17} GeV) the
"Parker bound" becomes less restrictive[9,11,12] (because the KE gain
is a second order effect); however, the mass density constraint
becomes more restrictive (cf. Fig. 1). These two bounds together

restrict the flux to be $\leq 10^{-13}$ cm^{-2} sr^{-1} s^{-1} (which is allowed for monopoles of mass $\simeq 3 \times 10^{19}$ GeV).

Analogous arguments can be applied to other astrophysical magnetic fields. Rephaeli and Turner[13] have analyzed intracluster (IC) magnetic fields and derived a flux bound of $O(10^{-18}$ cm^{-2} sr^{-1} s$^{-1})$ for monopoles less massive than $O(10^{18}$ GeV). Although the presence of such fields has been inferred from diffuse radio observations for a number of clusters (including Coma), the existence of IC fields is not on the same firm footing as galactic fields. It is also interesting to note that the IC magnetic fields are sufficiently weak so that only monopoles lighter than $O(10^{16}$ GeV) should be ejected, and thus it is very likely that monopoles more massive than 10^{16} GeV will cluster in rich clusters of galaxies, where the local mass density is $O(10^2$-$10^3)$ higher than the mean density of the Universe. Unfortunately, our galaxy is not a member of a rich cluster.

Drukier[14] has analyzed the survival of the magnetic fields of white dwarfs, and obtains a bound on the ratio of monopoles to nucleons inside a white dwarf, $n_M/n_N \leq 10^{-27}$. However, since it is unlikely that monopoles would have condensed with nucleons when the star was originally formed, this limit is not straightforward to interpret. For example, if one assumes that there were initially no monopoles present when the star formed (as I believe is most reasonable), then while the star was on the main sequence it would have accumulated $O(10^{41}$ cm^2-sr-s F) monopoles. (Assuming it captures every monopole which strikes its surface.) Drukier's bound $n_M/n_N \leq 10^{-27}$ then translates into F $\leq 10^{-11}$ cm^{-2} sr^{-1} s^{-1}-which is not nearly as restrictive as "Parker's bound".

Recently, Ritson[15] has analyzed the survival of stellar magnetic fields for a class of stars known as peculiar A stars. It is believed that the magnetic fields of these stars are not produced by dynamo action, but are "fossil fields", so that they cannot be regenerated. Such stars will stop monopoles less massive than about 5×10^{16} GeV which are incident upon them. Once inside, the monopoles are accelerated by internal magnetic fields, and in the process dissipate these fields. By demanding that the fields of these stars survive for a time $O(10^{10}$ yrs), Ritson derives a flux bound, F $\leq 10^{-19}$ cm^{-2} sr^{-1} s^{-1} for monopoles less massive than 5×10^{16} GeV. However, I don't believe that his conclusions are "water tight." For example, the monopoles could sink to the center of the star, "eat out" a region of magnetic field, and just sit, leaving the bulk of the magnetic field to survive. Ritson's limit relies critically upon the assumption that once inside the star, monopoles constantly move about inside dissipating the field as they do.

Monopole Catalyzed Nucleon Decay

Wilczek[16], Rubakov[17], and Callan[18] have all pointed out that grand unified monopoles can catalyze nucleon decay. Callan[18] and Rubakov[17] have argued that this process should proceed at a "typical strong interaction rate." I will not discuss the details here since Callan will speak on this topic. Kolb, Colgate, and Harvey[19]; Dimopoulos, Preskill, and Wilczek[20]; and Bais et al.[21] have used monopole-catalyzed nucleon decay inside of neutron stars to set the most stringent astrophysical bound on the monopole flux. The basic idea is simple. Monopoles less massive than $O(10^{20}$ GeV) which strike the surface of a neutron star lose sufficient energy while passing through the star to be stopped. Once captured, they catalyze nucleon decay within the neutron star, releasing ~ 1 GeV of energy per catalyzed decay. This energy is thermalized and radiated in the form of photons and neutrinos. The diffuse UV and X-ray background and surveys of X-ray point sources constrain the photon flux from neutron stars, and this in turn constrains the incident monopoles flux. Their limit is

$m \gtrsim 10^{17}$ GeV:

$$F \leq 10^{-22} \text{ cm}^{-2} \text{ sr}^{-1} \text{ s}^{-1} (\sigma v/10^{-27} \text{ cm}^2)^{-1} , \qquad (7a)$$

$m \lesssim 10^{17}$ GeV:

$$F \leq 10^{-21} \text{ cm}^{-2} \text{ sr}^{-2} \text{ s}^{-1} (10^{16} \text{ GeV/m}) (\sigma v/10^{-27} \text{ cm}^2)^{-1}, \qquad (7b)$$

where (σv) is the cross section times relative velocity divided by c for nucleon decay catalyzed by monopoles. The mass dependence results because the capture cross section is velocity-dependent, and monopole velocities for mass $\leq 10^{17}$ GeV are determined by galactic magnetic field acceleration [cf. (4b)]. It goes without saying that these bounds are discouragingly restrictive.

Monopoles moving more slowly than 3×10^{-5} c$(10^{16}$ GeV/m) and 10^{-3} c$(10^{16}$ GeV/m) can be stopped in the earth and Jupiter respectively due to Eddy current losses.[22] Once inside, they catalyze nucleon decay, resulting in the release of heat. The heat flow at the surface of these planets can be used to obtain bounds on the flux of monopoles at the earth and at Jupiter:[22]

$$F_{\oplus} \leq 7 \times 10^{-22} \text{ cm}^{-2} \text{ sr}^{-1} \text{ s}^{-1} (\sigma v/10^{-27} \text{ cm}^2)^{-1} f^{-1} , \qquad (8a)$$

$$F_{JUP} \leq 5 \times 10^{-19} \text{ cm}^{-2} \text{ sr}^{-1} \text{ s}^{-1} (\sigma v/10^{-27} \text{ cm}^2)^{-1} f^{-1} , \qquad (8b)$$

where for earth f is the fraction of the flux due to monopoles moving more slowly than 3×10^{-5} c(10^{16} GeV/m), and for Jupiter f is the fraction of the flux due to monopoles moving more slowly than 10^{-3} c(10^{16} GeV/m). If the local flux is primarily due to a cloud of monopoles orbiting the sun ($v_{orbit} \simeq 10^{-4}$ c), then for earth $f \simeq 10^{-3}$ (10^{16} GeV/m)^4. This implies a limit on the local flux of 10^{16} GeV monopoles: $F_{\oplus} \lesssim 10^{-18}$ cm^{-2} sr^{-1} s^{-1}, which is less restrictive than the neutron star limit, but applies directly to the local flux. In the case of Jupiter $f \simeq O(1)$ for 10^{16} GeV monopoles (whether or not there is an orbiting cloud of monopoles), resulting in the bound: $F_{JUP} \lesssim 10^{-18}$ cm^{-2} sr^{-1} s^{-1}, which is much less restrictive that the neutron star limit, but still rather interesting.

ODDS & ENDS

The "Harvard Effect"

With the exception of the limits based upon monopole catalysis of nucleon decay inside the earth and Jupiter,[22] all the astrophysical constraints discussed are bounds on the average flux of monopoles in the galaxy, not upon the local (at earth) flux. Of course, "the hunt" is taking place on earth. Dimopoulos et al.[23] have emphasized that we do not live in a typical place in the galaxy (we are order 10^5 times closer to the sun than we are to the next nearest star), and thus the average galactic flux may be a poor indicator of the local flux. Specifically, they suggested that the sun might gravitationally capture a cloud of monopoles which would orbit the sun and lead to a local flux which is enhanced by up to 8 orders-of-magnitude.

In order for a monopole to be captured into solar orbit it must first lose energy by passing through the sun or one of the planets (magnetic fields in the solar system are far too weak to be of any use). One can easily estimate the potential contribution to the monopole flux at earth due to each of the nine planets and the sun by: (i) calculating the number of monopoles which pass through the body of interest in a time interval τ due to an isotropic flux from outside the solar system and multiplying this by the efficiency ε for capture into solar orbit; (ii) dividing this number by volume occupied by the monopole orbits to obtain the number density of orbiting monopoles; (iii) multiplying this number density by ($v_{orbit}/4\pi$) $\simeq$ (10^{-4} c/4π) to obtain the potential contribution to the locally measured flux. The result is

$$F_{\oplus} \simeq 8800 \ (R_{10}^2 \ \tau_9 \ \varepsilon/V_{AU}) \ F_{Gal}, \tag{9}$$

where the radius of the object is $R_{10} *10^{10}$ cm, the orbital lifetime of a captured monopole is $\tau_9 * 10^9$ yrs, and V_{AU^3} is the volume occupied by the monopole orbits in units of AU^3. For the sun $F_\oplus/F_{Gal} \simeq 3 \times 10^5 \tau_9 \; \epsilon/V_{AU}$ and for Jupiter $F_\oplus/F_{Gal} \simeq 6 \times 10^3 \tau_9 \epsilon/V_{AU}$. Katherine Freese and I[24] have recently carefully estimated τ_9, ϵ, and V_{AU} and find that for the sun $\epsilon \simeq O(10^{-1})$ p, $V_{AU} \simeq O(100)$, and $\tau_9 \simeq O(10^{-1})$; while for Jupiter $\epsilon \simeq O(10^{-2})$, $V_{AU} \simeq O(100)$, and $\tau_9 \simeq O(1)$. [Orbiting monopoles see a time varying solar magnetic field due to the rotation of the sun, which causes them to gain or lose energy, thereby crashing into the sun or evaporating.] Bringing these numbers together we find that due to monopoles corralled by Jupiter $F_\oplus/F_{Gal} \simeq O(1.1-3)$, and due to the sun $F_\oplus/F_{Gal} \simeq O(50 \text{ p})$, both effects are <u>small</u>!

What is p? A monopole which is captured into orbit by passing through the sun has an additional problem--increasing its angular momentum so that its perihelion becomes greater than R_0, otherwise it will continue to pass through the sun on subsequent orbits and quickly spiral into the sun. The only plausible mechanism is for the monopole to get a radial kick from the solar magnetic field. The field strength required is $O(100 \text{ kG})$; p is the probability that a monopole receives the necessary kick. There is no evidence for fields of anywhere near this magnitude on the solar surface (highest field strengths observed are $\sim$ 5kG in sunspots); however, field strengths of up to 10^6 G well beneath the solar surface cannot be ruled out. Our analysis[24], which is largely kinematic in nature, strongly suggests that there is no large local enhancement in the flux of monopoles. In which case, bounds on the average galactic flux are applicable to the local flux as well.

Are Monopoles the Source of the Galactic Magnetic Field?

The two most stringent bounds on the monopole flux are the neutron star bound and "Parker's bound". The neutron star bound depends in a crucial way upon the rate for monopole catalyzed nucleon decay--a quantity which is not likely to be measured, and is difficult to compute. [Indeed, Wilczek[16] and Preskill[25] have argued that it should be suppressed by many powers of M_W.] That leaves the "Parker bound". Is it too vulnerable? Possibly, but probably not.

If the magnetic field of the galaxy is produced by magnetic plasma (i.e., local magnetic charge) oscillations,[9,11,26] then the magnetic field energy and KE of the monopoles just oscillate back and forth. Monopoles do not drain the energy of the field, they just borrow it every half cycle. The magnetic plasma frequency is

$$\omega_p = (4\pi h^2 n_M/m)^{1/2} , \tag{10}$$

where n_M is the average number density of monopoles plus

antimonopoles. Using the local galactic mass density as an upper bound, it follows that $\omega_p/2\pi \lesssim (3.5 \times 10^4 \text{ yr})^{-1}(10^{16} \text{ GeV/m})$. If these oscillations are not to damp out they must maintain spatial, as well as temporal coherence; otherwise, they will be subject to Landau damping. Given the spatial inhomogeneity within the galaxy, it is difficult to imagine how spatial coherence can be maintained. Temporal coherence requires that the phase velocity $\upsilon_{ph} = (\omega_p/2\pi)\ell$ (ℓ = spatial scale of the magnetic field) >> internal velocity dispersion of the monopoles $\simeq 10^{-3}$ c (virial velocity). For $\ell \simeq$ 300 pc, $\upsilon_{ph} \lesssim 3 \times 10^{-2}$ c(10^{16} GeV/m) and this condition can be satisfied for m $\lesssim 10^{17}$ GeV as long as the structure of the galactic field is not "fed down" to smaller scales where it can be Landau damped (e.g., by the shearing action of the electric plasma). However, one then has a galactic magnetic field which oscillates with a period of O(30,000 yrs). Such a short oscillation period exacerbates the problem of confining high-energy cosmic rays. Salpeter, Shapiro, and Wasserman[11] have argued that unless the period is $\gtrsim 3 \times 10^5$ yr grains will not be able to align fast enough to follow the oscillations. [The observed polarization of starlight (which is used to infer the structure of the galactic field) is caused by grains which align with the magnetic field.] Both of these considerations argue for longer oscillation periods (and smaller phase velocity)--there is little or no room to satisfy both υ_{ph} >> 10^{-3} c and $2\pi/\omega_p$ >> 3×10^5 yr. Although the existence of IC magnetic fields is less certain than galactic magnetic fields, if the bound based upon their survival is to be evaded by magnetic plasma oscillations, then υ_{ph} must be >> 3×10^{-2} for IC fields. This requires m << 10^{16} GeV (which seems rather unlikely).

Another difficulty for the magnetic plasma oscillation scenario is the observed spatial strurcture of the galactic magnetic field. As Purcell will discuss, the structure of the galactic field strongly indicates that $\vec{\nabla} \times \vec{B} \neq 0$, rather than the curl-free magnetic field one would expect to observe if the field were produced primarily by magnetic charges rather than by electric currents. Finally, large local magnetic charge density fluctuations are needed to produce the observed field strength: $\delta n/n_M \simeq 10^{-3}(m/10^{16} \text{ GeV})$ ($\delta n = n_M - n_{\overline{M}}$). How do these fluctuations initially get set up? While it is not possible to simply rule this possibility out, it does seem to be beset with a number of serious difficulties. Wasserman will discuss this scenario in more detail.

CONCLUDING REMARKS

Perhaps the most startling of all the startling properties of monopoles is their ability to catalyze nucleon decay. If the cross section associated with this process is indeed a strong interaction cross section, then we will be faced with a "Catch-22": the rate for this extraordinary process is so favorable that the bound from neutron stars restricts the flux to be hopelessly small, $F \lesssim 10^{-21}$ cm^{-2} sr^{-1} s^{-1} for 10^{16} GeV monopoles. If the cross section for this process is significantly smaller than $10^{-27}cm^2$ (e.g., it would zero if the GUT did not violate B conservation), then the most reliable and restrictive limit on the monopole flux is the "Parker bound," $F \lesssim 10^{-15}cm^{-2}$ sr^{-1} s^{-1} for 10^{16} GeV monopoles. The only way to evade this bound is to produce the galactic field by magnetic plasma oscillations. Although this possibility cannot be ruled out, the difficult obstacles this scenario faces make it seem rather implausible. I also think it is unlikely that the local flux of monopoles is significantly greater than the average flux in the galaxy, so the "Parker bound" should also apply to the flux of monopoles on earth. I am therefore led to the conclusion that the local monopole flux is at least as small as 10^{-15} cm^{-2} sr^{-1} s^{-1} and likely as small as 10^{-18} cm^{-2} sr^{-1} s^{-1} or 10^{-21} cm^{-2} sr^{-1} s^{-1}. However, the potential rewards for us all are so great that I wouldn't mind be proven wrong (even by 5 orders-of-magnitude!).

ACKNOWLEDGEMENTS

This research was supported in part by the DOE through DE AC02-80ER10773 A003 (at Chicago).

REFERENCES

1. G. 't Hooft, Nucl. Phys. B 79, 276 (1974); A. Polyakov, Pis'ma Zh. Eksp. Teor. Fiz. 20, 430 (1974) [JETP Lett. 20, 194 (1974)].
2. J. Preskill, Phys. Rev. Lett. 43, 1365 (1979).
3. Ya. B. Zel'dovich and M. Y. Kholopov, Phys. Lett. 79B, 239 (1979).
4. For a review of some of these scenarios, see P. Langacker, Phys. Rep. 72, 185 (1981).
5. M. S. Turner, Phys. Lett. 115B, 95 (1982).
6. J. Preskill, in the "Proceedings of the Nuffield Workshop on the Very Early Universe," G. Gibbons, S. W. Hawking, and S. Siklos, eds., (1983).
7. G. Lazarides, Q. Shafi, and W. P. Trower, Virginia Poly. Inst. preprint VPI-EPP-82-2 (1982).
8. K. Olive, D. N. Schramm, G. Steigman, M. S. Turner, and J. Yang, Astrophys. J. 246, (1981); J. Yang, M. S. Turner, G. Steigman, D. N. Schramm, and K. Olive, Univ. of Chicago preprint (1983).
9. M. S. Turner, E. N. Parker, and T. J. Bogdan, Phys. Rev. D 26, 1296 (1982).
10. E. N. Parker, Astrophys. J. 163, 225 (1971); S. Bludman and M. A. Ruderman, Phys. Rev. Lett. 36, 840 (1076); G. Lazarides, Q. Shafi, and T. Walsh, Phys. Lett. 100B, 21 (1981); E. M. Purcell, in these proceedings.

11. E. Salpeter, S. Shapiro, and I. Wasserman, Phys. Rev. Lett. 49, 1114 (1982).
12. J. Preskill in the "Proceedings of the Nuffield Workshop on the Very Early Universe," G. Gibbons, S. W. Hawking, and S. Siklos, eds., (1982).
13. Y. Rephaeli and M. S. Turner, Phys. Lett. B, in press (1982).
14. A. K. Drukier, Klinikum Rechts der Iser preprint (Munich, 1982).
15. D. M. Ritson, SLAC preprint 2977 (1982).
16. F. Wilczek, Phys. Rev. Lett. 48, 1146 (1982).
17. V. A. Rubakov, Pis'ma Zh. Eksp. Teor. Fiz. 33, 658 (1981) [JETP Lett. 33, 644 (1981)]; preprint P-0211 (1981).
18. C. G. Callan, Phys. Rev. D 25, 2141 (1982); "Dyon-Fermion Dynamics," (Princeton Univ. preprint, 1982).
19. E. W. Kolb, S. A. Colgate, and J. Harvey, Phys. Rev. Lett. 49, 1373 (1982).
20. S. Dimopoulos, J. Preskill, and F. Wilczek, Phys. Lett. 119B, 320 (1982).
21. F. A. Bais, J. Ellis, D. V. Nanopoulos, and K. A. Olive, CERN preprint TH 3383 (1982).
22. M. S. Turner, Nature, in press (1983).
23. S. Dimopoulos, S. L. Glashow, E. M. Purcell, and F. Wilczek, Nature 298, 824 (1982).
24. K. Freese and M. S. Turner, Phys. Lett. B, in press (1983).
25. J. Preskill, in preparation (1982).
26. This possibility has also been discussed by P. Eberhard, LBL preprint, (1982), and J. Arons and R. D. Blandford, Phys. Rev. Lett. 50, 544 (1983).

MONOPOLES AND THE GALACTIC MAGNETIC FIELD

E. M. Purcell

Physics Department
Harvard University
Cambridge, Massachusetts 02138

Polarization of starlight, discovered thirty years ago, was the earliest evidence for an interstellar magnetic field. It remains the only optical manifestation of a general galactic field. Radioastronomy has provided several other lines of evidence. These include observation of synchrotron radiation, of Faraday rotation of radiation from extragalactic sources, a few observations of Zeeman "splitting" in the 21 cm hyperfine line of H, and the measurement of Faraday rotation combined with dispersion in the radiation from pulsars. Here we shall summarize briefly what is known about the interstellar magnetic field, emphasizing aspects relevant to questions about GUT monopoles. Then we shall consider the motion of monopoles in such a field and discuss the limits on monopole flux implied by the existence of the field we observe.

Let's begin with a justly famous chart* compiled by Mathewson and Ford[3] from an immense number of observations, including many of their own, of starlight polarization. Dashes are stars whose light, when it reaches our telescope, proves to be linearly polarized. The polarization is not large, typically one or two percent. The orientation of the dash gives the position angle of the electric vector.

There is no doubt that the polarization results from the passage of the starlight through interstellar regions containing aspherical, partially aligned solid grains. It is quite certain, too, that the position angle of the linear polarization is

---

*We do not reproduce the chart here. The original is in the reference given. Reproductions are in many review articles, including Aannestad and Purcell[1] and Heiles.[2]

correlated with the direction of the magnetic field where the grains are. That must be true even if the original alignment of the grains was effected by a mechanism that did not involve the magnetic field at all. For it can be shown[4] that the axis of a spinning grain will precess, owing to the grain's electric charge and the resulting magnetic moment, around the direction of the local magnetic field B. And it will do so rapidly enough to obliterate any average alignment except that relative to B. The residual average alignment could manifest itself in polarization either parallel to or perpendicular to B. Other arguments make us confident that the polarization observed (electric vector) is in fact parallel to the local field direction. We may therefore yield to the nearly irresistable impression of "iron filings" to interpret Mathewson and Ford's map. It appears that the general field, insofar as a systematic large-scale pattern can be discerned, is mainly parallel to the galactic plane, and more or less perpendicular to the line of sight when we look toward ($\ell = 0°$) or away from ($\ell = 180°$) the galactic center. It is as if the field lines were stretched out along the spiral arms. Remember that in the visible, because of this very dust, we can see only two or three kiloparsecs into the galactic pancake.

As for the strength of the field, these observations provide only a lower bound, and that of a negative kind. No one has invented a way of aligning grains, so as to produce the polarization observed, in a field less than several microgauss.[5] Other lines of argument lead to estimates in the same range. It is consistent with the intensity of background synchrotron radiation from cosmic rays. Also, the resulting magnetic pressure $B^2/8\pi$ is about right to support the weight of the gas above the galactic plane. (Even the cold neutral interstellar gas sticks to the magnetic field; thanks to a few electrons it's conductivity, along the field direction, is roughly that of graphite!). If the magnetic field has a strength between 10^{-6} and 10^{-5} gauss, its energy density, like that of some other components of the interstellar medium, including cosmic rays and starlight, is 10^{-12} erg-cm^{-3} in order of magnitude.

We turn now to the most direct and quantitative evidence concerning the galactic field in the vicinity of the sun, the pulsar dispersion-rotation data. The observed dispersion in the fluctuating radio signal from a pulsar yields the integral along the line of sight of the free electron density n_e, while the Faraday rotation yields directly the integral of n_e times the longitudinal magnetic field component. Of course, to infer B from the ratio, we must assume that n_e and B are reasonably well correlated. The most recent analysis of which I am aware is that of Thomson and Nelson[6] which is based on 48 pulsars within 3 kiloparsecs of the sun. These authors attempt to separate in their models a "systematic" field configuration and a "random" field.

Their best-fitting model for a systematic field can be described as follows. A layer of horizontally directed field encloses the galactic mid-plane. The field strength is a Gaussian in the z coordinate:

$$B_{hor} = B_o \, e^{-z^2/h^2} \tag{1}$$

Their best value for B_o is 3.5 microgauss, and for h is 75 pc. In plan view, looking down on the galactic plane, their model looks something like Fig. 1, which is my oversimplified representation of it. A remarkable feature of the plan view is the field reversal along a line a few hundred pc inside the Sun's position. By the way, this field with manifestly non-zero curl could not, itself, be the field of a monopole distribution. For the method of fitting and important details, including estimates of uncertainty, the reader must consult the paper of Thomson and Nelson. Here we shall consider only a uniform slab of horizontal field, described by Eq. (1) with the constants as given, and extending uniformly over a horizontal region of the order of one kiloparsec in dimensions (see Fig. 2).

We note first that the areal energy density of the field just described is

$$\frac{1}{8\pi} \int_{-\infty}^{\infty} B_{hor}^2 \, dz = 1.4 \times 10^8 \text{ erg-cm}^{-2} \tag{2}$$

For our purposes the vertical width and the shape of the field distribution don't matter. All that counts is a constant we'll denote by W and define as

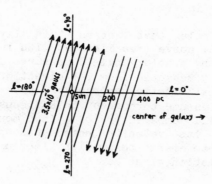

Fig. 1. Local field in the galactic plane according to Thomson and Nelson's analysis of pulsar rotation-dispersion data. (R.C. Thomson and A.H. Nelson M.M.R.A.S. 1980).

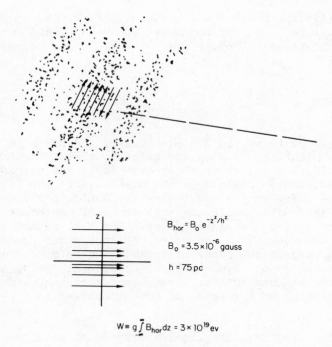

Fig. 2. Galactic field in plan view and in cross-section.

$$W = \int_{-\infty}^{\infty} g\, B_{hor}\, dz = 3 \times 10^{19} \text{ eV} \tag{3}$$

Here g is the monopole's magnetic charge. From now on all masses and energies will be in eV.

Consider a monopole that enters this layer, or slab, of magnetic field from above (see Fig. 3). Its initial direction of motion is specified by a polar angle θ measured from the z axis and an azimuthal angle ϕ measured from the direction of the field B in the slab. The initial kinetic energy of the particle, E_0, is $1/2mv^2$. We shall assume throughout this discussion that particle velocity remains small compared to c. The monopole moves down through the slab, its velocity components perpendicular to B remaining constant, to emerge below with kinetic energy E_1, which an elementary calculation gives as

$$E_1 = E_0 + W \tan\theta \cos\phi + \frac{W^2}{4E_0 \cos^2\theta} . \tag{4}$$

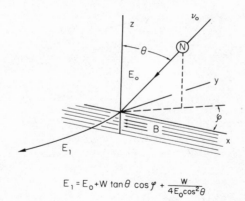

$$E_1 = E_0 + W \tan\theta \cos\varphi + \frac{W}{4E_0\cos^2\theta}$$

Fig. 3. A monopole passing through the galactic disk.

Denote by E_{esc} the kinetic energy the particle would need to escape from the Galaxy starting from this location. Of course this depends on the mass of the monopole. The escape velocity from the Sun's vicinity is about 300 km/sec. For the canonical GUT monopole of mass 10^{25} eV this corresponds to a kinetic energy of 5×10^{18} eV, about 1/6 of W. We now ask, for what range of initial conditions, E_0, θ, ϕ, will E_1 be _less_ than E_{esc}? Given a value of E_0, the cone of directions of incidence which result in $E_1 < E_0$ is bounded by the equality

$$\tan\theta \, \cos(\pi-\phi) = \frac{E_0 - E_{esc}}{W} + \frac{W}{4E_0}(1+\tan^2\theta) \qquad (5)$$

If $4E_0E_{esc} < W^2$, Eq. (5) has no solution. In that case every particle emerges from the slab with more than escape energy, regardless of its direction of incidence. In our example with $E_{esc} = W/6$, this occurs for all incident energies less than 1.5 W, which is 4.5×10^{19} eV.

On the other hand, if $E_0 > W^2/4E_{esc}$, there is a rather narrow cone of directions centered around $\phi = \pi$ and $\theta = \tan^{-1} 2E_0/W$ for which the particle, when it emerges from the slab, will have been slowed below escape velocity. When such a particle eventually returns to the slab, its angle of incidence will generally be different, and it may well emerge from its second transit of the slab with kinetic energy greater than E_{esc}. In the example just given, that almost always happens to a particle trapped by its first transit. A trajectory of that kind is shown in Fig. 4.

That only a rather energetic arriving particle can become a slow emerging particle is explained as follows. A slow particle

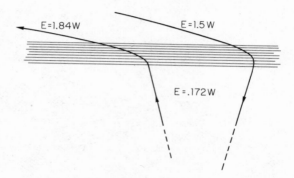

Fig. 4. A monopole which was slowed to less than escape velocity
         in its first passage through the disk.

leaving the slab has necessarily spent a long time within the slab,
where it was subjected constantly to a horizontal force gB. This
could have left it with a small horizontal velocity only if its
initial horizontal velocity was large and nearly opposite in
direction to gB. Even when this condition is met, the angular
window is narrow and the fraction of particles in an isotropic
initial distribution that are thus slowed down is small.   This
depends on E_{esc} being somewhat smaller than W as it was in our
example. To be somewhat more conservative, let us take E_{esc} = W/2
rather than W/6. With W = 3 × 10^{19} eV, as before, this is
approximately the energy of escape from the Galactic center for a
10^{25} eV monopole, or escape from solar vicinity for a particle of
mass 3 × 10^{25} eV. The critical initial energy for this case is
just W/2, which happens to be equal to E_{esc}. That gives our first
conclusion an especially paradoxical flavor: only a particle that
enters the slab with energy <u>greater</u> than E_{esc} can emerge with
energy <u>less</u> than E_{esc}. The largest angular window for capture
occurs for E_0 approximately equal to W. It extends in θ from 45°
to 71° and in φ from 150° to 210° (see Fig. 5). It includes less
than 5 percent of the incident trajectories in an isotropic
particle flux of that energy. For larger or smaller values of the
initial energy E_0 the fraction is even smaller.

     We can summarize these results in the statement that if the
escape energy is not more than W/2, practically <u>every</u> monopole that
enters the magnetic layer, whatever its initial kinetic energy,
will depart with <u>more</u> than escape energy, never to return. This
conclusion will not be significantly changed by the presences of
additional, more irregular, fields.   The Galaxy does not retain
monopoles -- it spits them out. It is a rare monopole indeed that
has passed through the galactic disk twice!

     This greatly simplifies the calculation of the effect upon the

For what θ, $\mathscr{S}$ in $E_1 < E_{esc}$?

$$\tan \theta \, \cos(\pi - \mathscr{S}) = \frac{E_0 - E_{esc}}{W} + \frac{W}{4E_0 \cos^2\theta}$$

no solution unless $4E_0 E_{esc} > W^2$

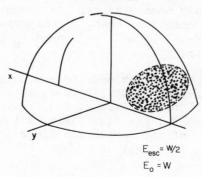

$E_{esc} = W/2$

$E_0 = W$

Fig. 5.   Shaded patch defines cone of directions of   incidence   for
          which   exit   velocity   is   less than escape velocity.   See
          text for assumptions.

galactic   magnetic   field   of   a   universal flux of monopoles.   The
passage through the disk of a monopole of initial kinetic energy E_0
extracts   an   amount   of energy given, in the mean, by the last term
in   Eq.   (4)   averaged   over   the   distribution   of   directions   of
incidence.   The apparent divergence at grazing incidence ($\theta = \pi/2$)
disappears if we take the finite   horizontal   extent   of   our   slab
roughly   into   account   by cutting off the integration at $\theta = \cos^{-1}$
(thickness of slab/width   of   slab).   The   mean   energy   gain   per
monopole incident in the isotropic flux is then

$$<E_1 - E_0> = \frac{W^2}{2 E_0} \, \ell n \; (\text{width/thickness}) \qquad\qquad (6)$$

If   the   slab   is   a   kiloparsec   in   extent,   not   an unreasonable
assumption, the logarithmic factor will not be very different   from
2 and our result becomes

$$<E_1 - E_0> \approx W^2/E_0 \qquad\qquad (7)$$

We   face   a   greater   uncertainty   in   the   choice   of E_0.   On the
assumption that the monopole velocity dispersion is small, v_0 will
be   typical   of   galactic   peculiar   velocities,   something   like
300 km-s^{-1}. Then for monopole masses   less   than 10^{26} eV, E_{esc}/W
remains   small   enough   to ensure that most monopoles make only one
transit,   with   a   mean   energy   gain   given   by   Eq.   (7).   For

$m = 10^{26}$ eV, with $E_0 = (\beta^2/2) \times 10^{26}$ eV $= 5 \times 10^{19}$ eV, the mean energy gain is 1.8×10^{19} eV.

We have already noted that the energy stored in the slab of horizontal field amounts to 10^{20} eV cm^{-2}. It follows that the passage of as many as 3 monopoles per cm^2 would drain half the field energy. We conclude that a universal flux of 10^{26} eV monopoles cannot exceed $3/t_r$ cm^{-2} s^{-1} where t_r is the time constant for field regeneration.

This flux limit is proportional to monopole mass for given v_0. Being proportional also to $\int B^2 dz/[\int B\ dz]^2$, it is independent of the field strength B_0, as long as W is at least comparable to E_{esc}. If this last condition holds, the flux limit is inversely proportional to the scale height h of the horizontal field. Expressing the flux in cm^{-2} ster^{-1} sec^{-1}, the limit is

$$\text{flux} < \frac{1}{\pi} \frac{3}{t_r} \left(\frac{m}{10^{26}\ \text{eV}}\right) \left(\frac{75\ \text{pc}}{h}\right)\quad \text{cm}^{-2}\ \text{ster}^{-1}\ \text{sec}^{-1} \qquad (8)$$

Notice that the neutrality of the monopole flux is irrelevant. A flux consisting of equal numbers of N and S poles extracts just as much energy from the field as an equal particle flux of N poles only. The difference is only that N and S poles will be shot out of the Galaxy in different directions. A neutral incident flux will be converted to separate outgoing streams of non-zero magnetic current, which, owing to the non-zero $\nabla \times \underline{E}$ that entails, develop sheaths of electric current -- and so on! But in order to establish our limit on the monopole flux we were not obliged to describe the scene that would evolve if the galactic field cannot repair itself!

We have here yet another version of the Parker limit. The only merit I claim for it is that it uses observed properties of a portion of the galactic disk. Indeed, of a significant portion, for even if the magnetic field is quite different a few kiloparsecs away, the destruction of the local field out to one or two kiloparsecs would still proceed as described.

If we follow Parker's original suggestion[7] of 10^8 years for the lower limit on t_r, we obtain from Eq. (8) a flux limit of 3×10^{-16} cm^{-2} ster^{-1} sec^{-1} for monopoles of mass 10^{26} eV, and one tenth of that for the canonical mass of 10^{25} eV. These are somewhat more severe than the limits usually quoted.

REFERENCES

1. P. A. Aannestad, E. M. Purcell, Ann. Rev. Astron. Astrophys. 11, 309 (1973).
2. C. Heiles, Ann. Rev. Astron. Astrophys. 14, 1 (1976).
3. D. S. Mathewson, V. L. Ford, Mem. Astron. Soc. 74, 139 (1970).
4. A. Z. Dolginov, I. G. Mytrophanov, Ap. Space Sci. 43, 291 (1976); P. G. Martin, M.N.R.A.S. 153, 279 (1971).
5. E. M. Purcell, L. Spitzer, Ap. J. 167, 31 (1971); E. M. Purcell, Ap. J. 231, 404 (1979).
6. R. C. Thomson, A. H. Nelson, M.N.R.A.S. 191, 863 (1980).
7. E. N. Parker, Ap. J. 160, 383 (1970).

THE PLASMA PHYSICS OF MAGNETIC MONOPOLES IN THE GALAXY

Ira Wasserman

Center for Radiophysics and Space Research
Cornell University
Ithaca, New York 14853

The possible discovery of a Dirac magnetic monopole[1] has rekindled interest in astrophysical bounds on the cosmic monopole abundance.[2,3,4] Particularly intriguing is the idea that the Galaxy may possess a magnetic monopole halo sufficiently massive to stabilize the Galactic disk against bar-mode deformations.[5] Typical parameters for such a stabilizing halo are given in Table 1. The local monopole flux, $n_M v_H$, from a Galactic monopole halo would lead to an event rate

$$\Gamma_D \approx n_M v_H A_D \approx 1.8 \times 10^{-2} \ A_D(cm^2) \ m_{16}^{-1} \ yr^{-1} \tag{1}$$

in a monopole detector of cross-sectional area A_D. As Cabrera has pointed out,[1] $\Gamma_D \approx (2.8 \ m_{16} \ yr)^{-1}$ for $A_D = 20 \ cm^2$, in substantial agreement with the occurrence[6] of a single candidate event in $\sim$1yr of observation if $m_{16} \sim 1$, as is generally the case[6] for GUT's.

The existence of a Galactic magnetic monopole halo may, however, be irreconcilable with observed properties of the Galactic magnetic field.[7,8] It has long been known that the acceleration of magnetic monopoles in an _assumed_ _static_ Galactic field shorts out the field far more rapidly than any plausible interstellar dynamo can regenerate it.[9] The efficiency of this "short circuit" strongly suggests that the back action of the monopole plasma must be taken into account in any self-consistent calculation of the damping of Galactic magnetic fields embedded in a monopole halo. In this review we summarize recent work[3,4] on the plasma physics of magnetic monopoles in the Galaxy, and the constraints on monopole properties and abundances implied by these calculations.

Table 1. Basic Halo Properties

Parameter	Symbol	Relation to Other Quantities	Numerical Value**
Mass	M_H	*	$2 \times 10^{11} M_\odot$
Radius	R_H	*	20 kpc
Velocity Dispersion	v_H	$v_H = \left(\dfrac{GM_H}{R_H} \right)^{1/2}$	$210 \cdot \dfrac{km}{s}$
Crossing Time	t_H	$t_H = \left(\dfrac{R_H^3}{GM_H} \right)^{1/2}$	10^8 yrs.
Particle Density	n_M	$n_M = \dfrac{3M_H}{4 \pi R_H^3 m_M}$	$2.6 \times 10^{-17} \, m_{16}^{-1} \, cm^{-3}$
Particle Number	N_M	$N_M = \dfrac{M_H}{m_M}$	$2.4 \times 10^{52} \, m_{16}^{-1}$

* The total halo mass within a radius equal to the radius of the Galactic disk must be comparable to the total disk mass to guarantee stability against bar-mode deformations.

** $m_{16} \equiv$ [Monopole mass (m_M)] / [10^{16} GeV].

Two dimensionless parameters are extremely useful for describing a hypothetical magnetic monopole halo.[3] The first is

$$\gamma_1 = \frac{Gm_M^2}{g^2} = 2 \times 10^{-8} \, m_{16}^2 \tag{2}$$

which is the ratio of the gravitational to magnetic force between two monopoles of magnetic charge $g = e/2\alpha$ and mass m_M. The second is

$$\gamma_2 = \frac{B^2 R_H^3}{GM_H^2/R_H} \approx 8 \times 10^{-4} \tag{3}$$

which is the ratio of magnetic to gravitational energy in the halo for a field strength $B \approx 2 \times 10^{-6}$ Gauss (as in the interstellar medium[7,8]). The halo mass (M_H) and radius (R_H) are given in Table 1. The halo is presumed to have zero net magnetic charge overall. The (small) parameters γ_1 and γ_2 impose strict constraints on any large scale magnetic charge density fluctuations $\delta N_M / N_M$. For a gravitationally bound halo we require

$$\frac{\delta N_M}{N_M} < \gamma_1^{1/2} = 1.4 \times 10^{-4} \, m_{16} \, , \tag{4}$$

while if, in addition, the density fluctuations are not to produce a magnetostatic field stronger than $\sim 2 \times 10^{-6}$ Gauss, we require

$$\frac{\delta N_M}{N_M} < (\gamma_1 \gamma_2)^{1/2} \approx 4 \times 10^{-6} \, m_{16} \, . \tag{5}$$

From Eq. (5) we see that linear perturbations in a Galactic monopole plasma can have a significant impact on the Galactic magnetic field.

We shall separately discuss the two topologically distinct cases of longitudinal magnetic fields, which are generated by magnetic charge density fluctuations and satisfy $\vec{\nabla} \times \vec{B} = 0$, $\vec{\nabla} \cdot \vec{B} \neq 0$, and transverse magnetic fields, which are generated by electric currents in the Galactic disk plasma and satisfy $\vec{\nabla} \cdot \vec{B} = 0$, $\vec{\nabla} \times \vec{B} \neq 0$. For each of these cases we should ideally understand both the effects of the monopole halo on the sources of the magnetic field (which may be the monopoles themselves) and the effects of the magnetic field on the monopole halo, and iterate. In practice, only linear calculations are currently available although some direct simulations[10] are now in progress.

Consider first longitudinal magnetic fields. Since in this case the Galactic field is due to large scale magnetic charge density fluctuations we require $\delta N_M / N_M \sim (\gamma_1 \gamma_2)^{1/2} \approx 4 \times 10^{-6} m_{16}$. The "natural" scale for these fluctuations is $L \sim R_H \approx 20$ kpc, but fluctuations down to the magnetic Debye length

$$\lambda_D \approx R_H \gamma_1^{1/2} \approx 3 \text{ pc} \times m_{16} \ll R_H \tag{6}$$

are not significantly damped.[11] On length scales $L \gg \lambda_D$, longitudinal magnetic fields undergo oscillations at the plasma frequency

$$\omega_M = \sqrt{2} \left(\frac{4\pi g^2 n_M}{m_M} \right)^{1/2} . \tag{7}$$

(The factor of $\sqrt{2}$ arises because monopoles and antimonopoles are equally massive.) These plasma oscillations are Landau damped but only insignificantly since $\omega_M L > v_H$ for $L >> \lambda_D$, which means that very few monopoles (or perhaps none) comove with the longitudinal waves, thereby suffering <u>coherent</u> magnetostatic accelerations and heating. Thus field energy and monopole bulk kinetic energy are periodically interchanged in these large scale plasma oscillations with essentially no net energy loss.

A longitudinal Galactic magnetic field generated by magnetic charge density fluctuations would therefore be characterized by conservative oscillations in amplitude and direction on the "oscillation timescale"

$$t_{osc} \equiv \frac{2\pi}{\omega_M} = \frac{2\pi}{\sqrt{3}} t_H \gamma_1^{1/2}$$

$$\tag{8}$$

$$\approx 5 \times 10^4 \, m_{16} \, \text{yrs.}$$

Unfortunately, no direct observational information exists on the temporal behavior of the Galactic magnetic field.[7,8] The most promising "fossils" for the purpose of tracing the magnetic history of the Galaxy may be interstellar dust grains, which are believed to be aligned with the Galactic field as a result of measurements of the linear polarization of Galactic starlight.[8] Grain alignment is not yet understood[8,12] although it is generally accepted that the <u>minimum</u> alignment timescale is

$$t_{align} > 2.4 \times 10^5 \, \text{yrs} \tag{9}$$

or perhaps longer.[13] Requiring that $t_{align} \leq 1/2 \, t_{osc}$ we get

$$m_M > 10^{17} \, \text{GeV} \tag{10}$$

and, from Eq. (1),

$$\Gamma_D < 1.8 \times 10^{-3} \, A_D(\text{cm}^2) \, \text{yr}^{-1}. \tag{11}$$

The flux limit in Eq. (11) corresponds to $\Gamma_D < (28 \, \text{yr})^{-1}$ for $A_D = 20 \, \text{cm}^2$, which conflicts (but not seriously) with the occurrence of one candidate event in ~ 1 yr.[1]

A more serious challenge to the possibility that the Galactic magnetic field is generated by a hypothetical monopole halo is posed by the observed topology of the Galactic field. The general Galactic field is parallel to and winds around the Galactic disk, with fluctuations $\delta B/B \sim 1$ on length scales ~ 100 pc.[14] (The field also appears to be confined to the disk but, since the disk plasma and interstellar grains have scale heights ~ 100 pc, it is safe to say that little or nothing is known about the magnetic field far above or below the Galactic plane.) Small-scale fluctuations on length scales ~ 100 pc could be due to intrinsic magnetic charge density fluctuations as long as $\lambda_D < 100$ pc or, equivalently [cf. Eq. (6)], $m_M < 10^{18}$ GeV. However, there is no compelling reason for a halo-generated field to be parallel to and wind around the disk plane. Although it is conceivable that the magnetic field in the plane of the Galaxy connects to a larger scale halo field that is dragged and wound up by the rotating disk plasma, this possibility seems unlikely since the Galactic rotation period t_{rot} $\sim t_H \gg t_{osc}$ unless $m_M > 2 \times 10^{19}$ GeV.

Next we consider transverse magnetic fields generated by electric currents in the disk plasma. In the absence of magnetic monopoles the dynamo theory of the Galactic magnetic field,[15] which is widely (but not universally[16]) accepted, maintains that cyclonic turbulence in the interstellar medium regenerates the field on a characteristic timescale $t_{reg} \sim t_{rot} \sim t_H \approx 10^8$ yrs. Magnetic monopoles are vigorously accelerated by the interstellar field, sucking up the entire magnetic field energy of the Galaxy in a time $\sim \gamma_1^{1/2} t_H \sim t_{osc} \approx 5 \times 10^4 m_M$ yrs if they are present at the full halo density n_M (Table 1).[6] This argument can be translated into the well-known Parker bound[2,3,9] on the magnetic monopole flux in the Galaxy.

However, as monopoles are accelerated by the Galactic field, strong magnetic currents $\vec{J}_M$ are generated, which are non-vanishing even when the Galactic field has fallen to zero. These magnetic currents generate electric fields via the modified Faraday induction law

$$\vec{\nabla} \times \vec{E} + \frac{1}{c}\frac{\partial \vec{B}}{\partial t} = -\frac{4\pi}{c}\vec{J}_M \,. \tag{12}$$

These electric fields, in turn, can generate electric currents in the disk plasma which regenerate the Galactic field. Thus one expects transverse magnetic field oscillations, possibly decaying in magnitude because of Landau damping, with a characteristic oscillation time $t_{osc} = 2\pi/\omega_M \sim \gamma_1^{1/2} t_H$. This is roughly what one finds from the linearized plasma theory of sinusoidal perturbations on an assumed uniform background state. Qualitatively similar (but more complex) behavior is found from analogous calculations that attempt to model the Galactic dynamo as well.[4]

More accurate calculations should incorporate realistic
geometries, non-linear effects, and self-gravity. A
translationally invariant background state, while convenient
mathematically in linear plasma theory, is not an adequate model
for the Galaxy. In the next simplest geometry (which is closer to
reality for the Galaxy) one could assume azimuthal symmetry around
and translational invariance along the z-axis, so that [in
cylindrical coordinates (r,θ,z)] $\vec{B} = \hat{e}_\theta B(r)$. In this case monopoles
and antimonopoles both tend to spiral outwards, with opposite
senses of rotation about the symmetry axis. Although azimuthal
monopole motions, which give rise to magnetic currents $\vec{J}_M =
\hat{e}_\theta J_M(r)$, couple back to $\vec{B}$ through Faraday's law (leading to
oscillations at a frequency ω_M), radial monopole kinetic energy
does not couple to $\vec{B}$ and is therefore not recoverable. This
(non-linear) energy drain does not appear for uniform background
states, and could be important. In still more realistic
geometries, one should account for the fact that a Galactic
magnetic monopole halo occupies a far larger volume in space than
does the interstellar plasma, which is confined to the thin
Galactic disk of height h $\sim$100 pc $\sim$5 × $10^{-3}R_H$. Magnetic monopoles
cross the disk in a time

$$t_{cross} \sim \frac{2h}{v_H} \approx 10^6 \text{ yrs} \approx 20\ m_{16}^{-1}\ t_{osc} \tag{13}$$

so there is a constant drain of monopole energy and magnetic
currents from the disk to the halo. Halo magnetic currents can,
through Faraday's law, have some regenerative effect on the
magnetic field in the disk, but probably less than in the
translationally invariant case.

Direct simulations of a magnetized disk Galaxy surrounded by a
collisionless monopole halo can be performed with existing N-body
codes originally designed to study the dynamics of self-gravitating
collisionless systems.[10] With appropriate choices of γ_1 and γ_2, and
model circuits to simulate Galactic electric currents, one can
carry out calculations for realistic magnetic field strengths and
geometries even though N << N_M. However, for small γ_1 and $N \sim 10^3$,
two-body relaxation occurs rapidly (generally on timescales <t_H) so
the simulated system is highly collisional. Thus the calculations
now under way choose relatively large values of γ_1 and, hence, m_{16}.
The simulations currently in progress at Cornell[10] assume $\gamma_1 = 1$
or, equivalently, $m_M = 7 × 10^{19}$ GeV.

The final word is clearly not yet in on the expected behavior
of transverse magnetic fields in a monopole halo. Nevertheless it
is useful to try to anticipate the results, and to derive
constraints on the monopole mass and flux imposed by astronomical
observations. Since large temporal fluctuations in the Galactic
field are expected on timescales $\sim t_{osc}$, grain alignment requires

$$m_M > 10^{17} \text{ GeV} \tag{14}$$

and therefore

$$\Gamma_D < 1.8 \times 10^{-3} A_D (\text{cm}^2) \text{ yr}^{-1} \tag{15}$$

for transverse as well as longitudinal fields. The large scale structure of the Galactic field, which is almost certainly due to the differential rotation of the Galaxy in transverse-field "dynamo" models, very likely imposes a much more stringent lower bound on the monopole mass. Since Galactic rotation can only be important if $t_{rot}/t_{osc} \sim t_H/t_{osc} \lesssim 1$, we require

$$m_M > 2 \times 10^{19} \text{ GeV} \tag{16}$$

and therefore

$$\Gamma_D < 9 \times 10^{-6} A_D (\text{cm}^2) \text{ yr}^{-1}. \tag{17}$$

This (unfortunately) somewhat qualitative topological constraint is clearly quite restrictive, leading to very pessimistic limits on m_M and Γ_D.

From Eqs. (11), (15) and (17) it is clear that the continued detection of cosmic magnetic monopoles at a rate $\Gamma_D \sim 1$ yr^{-1} would pose formidable problems for astrophysical theories of the Galactic magnetic field. This would be true even if the density of magnetic monopoles in the Galactic monopole halo is a factor $f < 1$ smaller than the critical density n_M, given in Table 1, which would be required to stabilize the Galactic disk. Although the lower bounds on m_M given in Eqs. (10), (14) and (16) scale as $f^{1/2}$ the ratio n_M/m_M remains fixed so, for a given halo speed v_H (which is independent of mass or concentration) the upper bounds on Γ_D given in Eqs. (11), (15) and (17) also scale as $f^{1/2}$, and are therefore reduced. The idea[17] that the source of observed magnetic monopoles is local would become increasingly attractive in the wake of continued monopole detections, and deserves further detailed quantitative study.

It is a pleasure to thank R. I. Epstein and C. J. Pethick for their hospitality at NORDITA, where this review was written, and to acknowledge the collaboration of R. Farouki, E. E. Salpeter and S. L. Shapiro on the less speculative aspects of this work. This research was supported in part by National Science Foundation Grant No. AST 81-16370 to Cornell University.

REFERENCES

1. B. Cabrera, Phys. Rev. Lett. 48, 1378 (1982).
2. M. S. Turner, E. N. Parker and T. J. Bogdan, Phys. Rev. D 26,
    1296 (1982).
3. E. E. Salpeter, S. L. Shapiro and I. Wasserman, Phys. Rev.
    Lett. 49, 1114 (1980).
4. J. Arons and R. D. Blandford, preprint (1982).
5. J. P. Ostriker and P. J. E. Peebles, Ap. J. 186, 467 (1973).
6. G. t'Hooft, Nucl. Phys. B 79, 276 (1974), and B 105, 538
    (1976). A. Polyakov, JETP Lett. 20, 194 (1974).
7. C. Heiles, Ann. Rev. Astron. Ap. 14, (1976).
8. L. Spitzer, "Physical Processes in the Interstellar Medium,"
    Wiley, New York, (1978).
9. E. N. Parker, Ap. J. 160, 383 (1970).
10. R. Farouki, E. E. Salpeter, S. L. Shapiro and I. Wasserman, in
    preparation.
11. See for example, S. Ichimaru, "Basic Principles of Plasma
    Physics," Benjamin, Reading (1973).
12. P. Aannestad and E. M. Purcell, Ann. Rev. Astron. Ap. 11, 309
    (1973).
13. E. M. Purcell has suggested in the course of this conference
    that the Barnett effect may cause paramagnetic dust grains
    to follow the oscillating field, thus leading to a very
    short effective alignment time. No detailed calculation of
    this possibly very important effect is as yet available.
14. J. R. Jokipii and E. N. Parker, Ap. J. 155, 799 (1969); J. R.
    Jokipii and I. Lerche, Ap. J. 157, 1137 (1969); J. R.
    Jokipii, I. Lerche and R. A. Schammer, Ap. J. (Letters)
    157, L119 (1969).
15. E. N. Parker, Ap. J. 163, 255 (1971).
16. For an interesting exchange on this issue see J. H.
    Piddington, Cosmic Electrodynamics 3, 60 (1971) and 3, 129
    (1972), and E. N. Parker, Ap. and Sp. Sci. 22, 279 (1973).
17. S. Dimopoulos, S. L. Glashow, E. M. Purcell and F. Wilczek,
    Nature 298, 824 (1982).

MONOPOLONIUM

Christopher T. Hill

Fermi National Accelerator Laboratory*
P.O. Box 500
Batavia, Illinois   60510

If the Universe is populated by magnetic monopoles it becomes conceivable that a monopole-antimonopole boundstate, i.e. monopolonium, can be formed in the laboratory or may have been formed naturally in the Universe at large. Such objects are purely classical and have an interesting physical evolution in time, dependent upon their masses, their initial radii and their core structure.[1] For GUT monopoles with masses of the order of 10^{16} GeV, the lifetimes of monopolonium systems range from days, for an initial diameter of about a fermi, up to many times the Universe' lifetime with diameter of about a tenth of an angstrom or more. While behaving as a classical system as long as $r > r_{co}$, they will radiate characteristic dipole radiation up to high energies, $M_{mo} \sim 10^{16}$ GeV. Thus a monopolonium system provides a window on the physics of elementary processes up to the extremely high energy scale characterized by it's mass.

Moreover, in the early Universe we argue that a sizeable and potentially detectable abundance of ultra long-lived monopolonium may have been formed. Remarkably, we find that this process would have occurred during the relatively late period of helium synthesis and depends only upon the assumption of the existence of an acceptable abundance of ordinary heavy monopoles at that time. For SU(5) monopoles we find that in a typical cosmologically averaged cubic light year containing on average 10^{32} monopoles, there will be today about 10^{15} monopolonia and roughly 400 decays per year. In galaxies and clusters these abundances and rate densities may be significantly larger. There may also exist mechanisms to significantly enhance the formation and we view the above results as conservative lower limits. The objects of larger diameter are spinning down producing radio frequency radiation from which we may place lower bounds on the masses of GUT monopoles. The cataclysmic decay events may produce visible cosmic-ray and high-energy gamma-ray events in large scale earthbound or orbiting detectors. Indeed, monopolonium may be easier to find than monopoles themselves.

Assume for the sake of discussion that we have an SU(5) monopole separated a distance r from an anti-monopole. For SU(5) we assume

$$M_X \approx 5 \times 10^{14} \text{ GeV}$$
$$\alpha_{GUT} \approx 1/40 \quad (1)$$
$$M_m \approx \alpha_{GUT}^{-1} M_X = 2 \times 10^{16} \text{ GeV}.$$

The effective Rydberg for the monopolonium system at large separation ($r >> 10^{-13}$ cm) is

$$R = \overline{M} g_m^4 / 2N^2 \approx 293 \, M_m \text{ GeV} \quad (2)$$
$$\overline{M} = M_m/2 \quad \text{(reduced mass)}$$

where g_m is the magnetic charge and N the "monopole number":

$$\vec{B} = \frac{g_m \hat{r}}{r^2}; \quad g_m e = \frac{N\hbar c}{2}; \quad g_{Dirac} = 3.28 \times 10^{-8} \text{ esu.} \quad (3)$$

For SU(5) monopolonium the above Rydberg is valid only at distances larger than a few fermi. As r becomes comparable to (Λ_{QCD}^{-1}) the SU(3) color chromomagnetic field turns on. The chromomagnetic field terminates at a distance scale of .2 fm<r<1fm due to the confinement effects of QCD, believed generally to be a shielding by color-magnetic monopole-like fluctuations in the ordinary QCD vacuum.[2]

For $r = 1/M_W$, the U(1) group of electromagnetism decomposes into the U(1) and diagonal generator of SU(2) of the full Weinberg-Salam electroweak model. For all scales less than a fermi the various operant coupling constants are evolving with energy by the usual logarithmic renormalization effects. These renormalization effects lead to a net evolution of the effective magnetic charge, g_m.[3]

Remarkably, however, the evolution of g_m is very small over the full range of the desert, even though these various heirarchical effects are setting in and the individual coupling constants are evolving considerably in this range. With λ a threshold parameter of $O(1)$, the magnetic coupling constant is

$$E \lesssim 1 \text{ GeV} \qquad g_m^2 = 1/4e^2$$

$$1 \text{ GeV} \lesssim E \lesssim \lambda M_{Z^0} \qquad g_m^2 = 1/4e^2 + 1/3g_3^2 \qquad \text{(4a,b)}$$

$$\lambda M_{Z^0} \lesssim E \lesssim \lambda M_X \qquad g_m^2 = 1/4g_1^2 + 1/4g_2^2 + 1/3g_3^2$$

$$E \approx M_X \qquad \approx \qquad g_m^2 = \alpha_{GUT}^{-1} \qquad \text{(4c,d)}$$

where E is the characteristic energy scale $= 1/r$. Numerically we see that $g_m^2(r \gg 1\text{fm}) = 137/4 = 34.25$, while $g_m^2(r = M_X^{-1}) = 40$. The various heirarchy effects lead to only a net 15% change in g_m^2 over the full range of the desert, and we shall ignore these in our analysis of monopolonium energetics. However, we will have to include these effects in our discussion below of gamma, Z-boson, and hadron production via gluon jets.

Assume now that the monopole anti-monopole pair is in a circular orbit about the center of mass. We have

$$g_m^2/r^2 = \bar{M}\omega^2 r \qquad \text{(5)}$$

where ω is the angular frequency and r the diameter. The energy is

$$E = \frac{1}{2} \bar{M}\omega^2 r^2 - g_m^2/r = -\frac{1}{2} g_m^2/r \qquad \text{(6)}$$

The system will lose energy by classical dipole radiation and the Larmor power formula is indeed valid for monopoles as well as electric charges. Thus

$$\frac{dE}{dt} = -2\left(\frac{2}{3}\right) g_m^2 \left(\omega^2 \frac{r}{2}\right)^2 c^{-3} = -\frac{64}{3} E^4/(g_m^2 M_m^2 c^3). \qquad \text{(7)}$$

Integrating Eq. (7) and using Eq. (6) we obtain the lifetime

$$\tau \approx M_m^2 c^3 r_0^3/(8g_m^4) . \qquad \text{(8)}$$

In Table 1 we give numerical values of the lifetime vs. classical diameter, energy, principal quantum number and v/c. Remarkably, a system of a GUT monopole with $r = 10^{-13}$ cm lives about 43 days while with r = 1/10 angstrom, about a tenth the size of a hydrogen atom, we obtain 10^{11} years! This latter result raises the spectre of relic monopolonium produced in the very early Universe surviving up to the present and decaying today.

The classical decay of the system may be viewed quantum mechanically as a cascade of jumps through sequentially decreasing principal quantum numbers. The energy is given by the virial theorem and by the Bohr formula[4]

$$E = -\frac{1}{2} g_m^2/r = -R/n^2 \tag{9}$$

We see in Table 1 that the principal quantum number of the instantaneous orbit is O(40) as v/c-->1. Simultaneously the

Table 1. Monopolonium Properties

Classical Diameter (cm)	Lifetime (sec)	Binding Energy (GeV)	Transition Energy (eV)	Principal Quantum Number	V/C
10^{-8}	3.71×10^{22}	3.35×10^{-5}	1.61×10^{-7}	4.17×10^{11}	4.10×10^{-11}
10^{-9}	3.71×10^{18}	3.35×10^{-4}	5.09×10^{-6}	1.32×10^{11}	1.30×10^{-10}
10^{-10}	3.71×10^{15}	3.35×10^{-3}	1.61×10^{-4}	4.17×10^{10}	4.10×10^{-10}
10^{-11}	3.71×10^{12}	3.35×10^{-2}	5.09×10^{-3}	1.32×10^{10}	1.30×10^{-9}
10^{-12}	3.71×10^{9}	3.35×10^{-1}	1.61×10^{-1}	4.17×10^{9}	4.10×10^{-8}
10^{-13}	3.71×10^{6}	3.35	5.09	1.32×10^{9}	1.30×10^{-8}
10^{-14}	3.71×10^{3}	3.35×10^{1}	1.61×10^{2}	4.17×10^{8}	4.10×10^{-8}
10^{-15}	3.71	3.35×10^{2}	5.09×10^{3}	1.32×10^{8}	1.30×10^{-7}
10^{-16}	3.71×10^{-3}	3.35×10^{3}	1.61×10^{5}	4.17×10^{7}	4.10×10^{-7}
10^{-18}	3.71×10^{-9}	3.35×10^{5}	1.61×10^{8}	4.17×10^{6}	4.10×10^{-6}
10^{-20}	3.71×10^{-15}	3.35×10^{7}	1.61×10^{11}	4.17×10^{5}	4.10×10^{-5}
10^{-22}	3.71×10^{-21}	3.35×10^{9}	1.61×10^{14}	4.17×10^{4}	4.10×10^{-4}
10^{-24}	3.71×10^{-27}	3.35×10^{11}	1.61×10^{17}	4.17×10^{3}	4.10×10^{-3}
10^{-26}	3.71×10^{-33}	3.35×10^{13}	1.61×10^{20}	4.17×10^{2}	4.10×10^{-2}
10^{-28}	3.71×10^{-39}	3.35×10^{15}	1.61×10^{23}	4.17×10^{1}	4.10×10^{-1}

orbital diameter approaches the core size of a the GUT monopole, $r \longrightarrow 1/M_X$.

The instantaneous transition energy is given by

$$E' = R \left(\frac{1}{n^2} - \frac{1}{(n+1)^2} \right) \approx 2R \frac{1}{n^3} \tag{10}$$

The system decays by emitting photons until E' becomes greater than the pion threshold, when $r \longrightarrow 10^{-18}$ cm and the lifetime remaining is 10^{-9} sec, and $n = 4.2 \times 10^6$. Here the system radiates both photons and gluons by classical dipole transitions.

As the system passes through the principal quantum number, $n = 4 \times 10^5$, the Z^0 threshold opens up as the U(1) of electromagnetism decomposes into the U(1) and diagonal generator of SU(2)$_{el}$. Above all thresholds we note that the three normalized probabilities are simply expressed in terms of the running coupling constants

$$P_{gluon} = (4/3g_3^2) \frac{1}{D}$$

$$P_\gamma = \left(\frac{g_2^2}{g_1^2} + \frac{g_1^2}{g_2^2} \right) \frac{1}{(g_1^2+g_2^2)} \frac{1}{D} \tag{11}$$

$$P_{Z^0} = \frac{2}{(g_1^2+g_2^2)} \frac{1}{D}$$

$$D = 1/g_1^2 + 1/g_2^2 + 4/3g_3^2$$

where p_i is the probability to emit species i. These are plotted in Fig. 1.

We therefore have the number of quanta of species i produced in an energy window E to E + dE:

$$\frac{dn_i}{dE} = \frac{2^{1/3}R^{1/3}}{3} p_i(E)(E^{-4/3}) \tag{12}$$

We find by a numerical integration that from a scale of 1 GeV up to M_X that the total number of direct photons is 4×10^6 while there are 2.3×10^5 Z-bosons and 1.3×10^5 gluons produced. In Fig. 2 we plot the three multiplicity distributions.

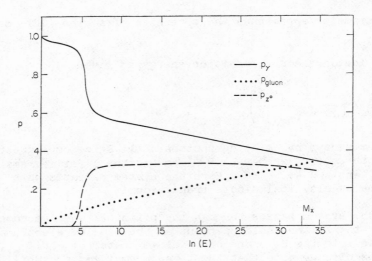

Fig. 1. Normalized probabilities, p_γ, p_{Z^0}, p_{gluon}.

The fragmentation of gluons into high multiplicity jets of hadrons and secondary decay products, including photons from pi decays, substantially modifies the spectrum. Most of the relatively soft photons will be secondaries in this range. We first estimate the total yield of hadrons. In QCD the multiplicity of charged hadrons produced in a gluon jet of energy E is expected to be given in leading log QCD[5]

$$N_h(E) = a \ \exp \ (b\sqrt{\ln(E/\Lambda)}) \ + n_0 \tag{13}$$

where a and n_0 are uncalculated and b is determined

$$b = 4(C_a/b_0)^{1/2}; \ b_0 = 11 - \frac{2}{3} n_f; \ C_a = 3 \tag{14}$$

Phenomenologically, the Petra data including quark jets is well fit by b = 2.7 ± .28, fully consistent with the above result, with a = .027 and n_0 = 2.[6] The total hadron yield is thus determined by convoluting the gluon distribution with the fragmentation multiplicity

$$N_h = \int_{E_0}^{M_x} N_h(E) \ \frac{2^{1/3} R^{1/3}}{3} \ E^{-4/3} \ P_{gluon}(E) \ dE \tag{15}$$

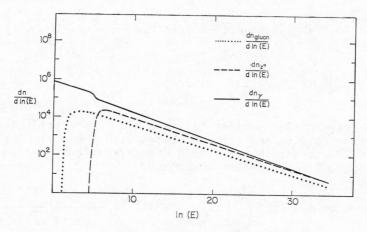

ln (E)

Fig. 2.   γ, Z^0, gluon d ln(n)/d ln E vs. ln E.

where E_0 is of order 10 GeV.

We find a total yield of 10^7 hadrons for leading log QCD. (The naive parton model predicts a ln(E) multiplicity in a jet which is already inconsistent with the low energy data and we thus exclude it.)

We may further estimate the spectrum of hadrons and secondary photons, though here we are on somewhat thinner ice. The exact x-distribution for fragmentation of a gluon jet is not known, and only a few properties, such as the total multiplicity and more recent observations of a peak at very low x have been determined.[5] Indeed, it is not clear how much can be determined theoretically. For our purposes, the important features are to realize the correct multiplicity, assure that the first moment of the distribution be normalized properly to unity, and try to guess the correct large-x behavior, which we take to be $(1-x)^2$. We will build the multiplicity into the low-x behavior of the distribution. For leading log QCD multiplicity we find that the following distribution works reasonably well:

$$\frac{dN_h}{dx} = N(b) \exp (b\sqrt{-\ln x}) \frac{(1-x)^2}{x\sqrt{\ln-x}} \tag{16}$$

where $N(b)$ is determined by the condition that the first moment of the distribution is normalized to unity (energy conservation). We obtain

$$N(b) = \frac{1}{2}\left[e^{b^2/4}I(b) - \sqrt{2}e^{b^2/8}I(b/\sqrt{2}) + \frac{1}{\sqrt{3}}e^{b^2/12}I(b/\sqrt{3})\right]$$

$$I(b) = \frac{\sqrt{\pi}}{2}\left(1 + erf\left(\frac{b}{2}\right)\right) \tag{17}$$

The fragmentation distributions are converted into hadron energy distributions and are convoluted with the gluon distribution from Eq. (21) to obtain the hadron energy spectrum, e.g., for the leading log QCD distribution we obtain

$$E\frac{dN_h}{dE} = \frac{(2R)^{1/3}}{3}N(b)\int_{E_0}^{M_x}\left(1 - \frac{E}{E'}\right)^2(E')^{-4/3}\left(\frac{\exp(b\sqrt{\ln(E'/E)})}{\sqrt{\ln(E'/E)}}\right)$$

$$\cdot P_{gluon}(E')dE' \tag{18}$$

We choose as the upper limit of this convolution the energy scale corresponding to the point at which the cores of the monopoles are overlapping. The hadrons we count do not include the neutrals that end up as photons. Here we may again appeal to the Petra data in which the naive expectation that about 30% of the total distribution converts quickly to photons is born out. The results are plotted in Fig. 3 along with the gluon distribution. We note that at accelerator energies the baryon yield in jets is anomalously higher than one would have expected from naive hadronization ideas and constitutes about 10% of the spectrum. This may continue up to the energies under consideration here and, if so, we may have a novel mechanism for producing ultra high energy cosmic rays by the decays of relic monopolonium. One can readily see that the differential energy spectrum of hadrons falls as a power of energy where the exponent lies between -1 and -2. This is a unique spectrum in that the multiplicity is dominated by low energy objects while most of the energy is carried away by a few hadrons. Such a spectrum is an efficient one for producing the very high-energy cosmic rays (e.g., the "ankle" in the CR spectrum). A production spectrum with exponent less than -2 has both the multiplicity and energy concentrated in low energy particles and the total energy of the source must be enormous to provide a few very energetic hadrons.

Reference to Table 1 shows that at a principal quantum number of n = 40 the classical diameter of the monopolonium system has shrunk to $r = 10^{-28}$ cm = $1/M_x$ and the cores of the monopoles themselves are now overlapping. Simultaneously, v/c-->1 and our

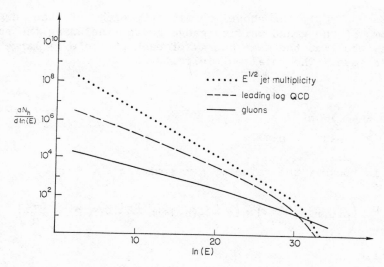

Fig. 3. Charged hadron spectrum (a) leading log QCD, (b) $E^{1/2}$
          multiplicity, and (c) gluon spectrum. γ-distribution $\backsim$1/3
          hadron distribution.

classical approximations are invalid. At this stage we still have
about 75% of the system's total energy to liberate. Here we expect
to produce a burst of particles of all types contained in the bare
unified gauge theory.

A simple approximation to the physics of the burst is to
assume that the system's total energy is uniformly distributed
throughout a local region of diameter $\sim 2/M_x$. Particle
multiplicities are then determined by a universal amplitude and by
phase space alone. Similar statistical models are successful in
hadroproduction at high energy.[7]

The universal amplitude, A, will be seen to have dimensions of
area and is expected to be related to the volume of the system

$$A \backsim \left(\frac{4}{3}\pi R^3\right)^{2/3} \backsim 2.6R^2 \backsim \frac{10.4}{M_x^2} \tag{19}$$

We obtain the partial width into n identical quanta:[8]

$$\Gamma^n \approx \frac{1}{E_0{}^4} \left(\frac{AE_0{}^2}{4\pi}\right)^n \; [n!(n-1)!(n-2)!]^{-1} \tag{20}$$

We can apply the above result to each of the degrees of freedom of the grand unified gauge group, including the superheavy bosons, provided the mean energy of each particle is large compared to it's mass. The total multiplicity is

$$\bar{n} = \sum_{n=1}^{\infty} n\Gamma^n / \sum_{n=1}^{\infty} \Gamma^n \tag{21}$$

and $\bar{n}$ is roughly the value of n maximizing Γ^n. From

$$\Gamma^n \left(\exp\, n\ln\left(\frac{AE_0{}^2}{4\pi}\right) - n\ln\frac{n}{e} - (n-1)\ln\frac{n-1}{e} - (n-2)\ln\frac{n-2}{e} \right) \tag{22}$$

we obtain

$$\bar{n} \approx \left(\frac{AE_0{}^2}{4\pi}\right)^{1/3} e \approx 25\,. \tag{23}$$

Eq. (23) is, we emphasize, the total number of fundamental degrees of freedom of equal statistical weight that are produced. Here the energy is uniformly shared by the outgoing particles. The average energy per particle is $E = 1.2 \times 10^{15}$ GeV which makes our neglect of superheavy masses reasonable.

In SU(5) we have 24 gauge and 24 Higgs bosons. In counting the number of X- and Y- bosons, we must take a total of 12×2 degrees of freedom from the gauge bosons and 12×1 (longitudinal) degrees of freedom from the Higgs. We further have a total of 24×3 degrees of freedom altogether. Thus the fraction of X- and Y- bosons produced is $12 \times (2+1)/(24 \times 3) = 1/2$. Thus $25/2 = 12$ X- and Y-bosons are expected. Table 2 presents the approximate yields and fractions in the burst phase of the various SU(5) gauge and Higgs bosons.

The decays of the superheavy gauge and Higgs bosons as well as the fragmentation of the gluons will produce very high energy hadron jets as well has leptons. With an average energy of $O(10^{15})$ GeV, the expected multiplicity per jet is $= 10^4$ from the leading log QCD and, though it somewhat increases the multiplicity at very high energies, it is a negligible correction to the hadron spectrum of Fig. 3.

The particles produced in the burst will decay into leptons, quarks, and the lighter gauge and Higgs bosons. Of the 25 degrees of freedom initially excited, roughly $25 \times 2 \times 8/(2 \times 24 + 24)$, or

Table 2. Fractions and Approximate Yields

Species	Fraction	Approx. Yield
$X, \bar{X}$	1/4	6
$Y, \bar{Y}$	1/4	6
W^+, W^- Z^0 }	1/8	3
gluons	2/9	6
γ	1/36	1
color 8 Higgs }	1/9	3
weak Higgs }	1/72	0

≈6 are gluons. The remaining 20 objects will typically decay into two body final states. Ignoring gauge bosons, we expect typically 25% of these to be leptons and 75% quarks. Thus we get roughly 10 leptons and 30 quark jets in addition to the 5 original gluon jets. These jets should be distributed more or less isotropically in space and might be cleanest along the z-axis of the system where the $1-\cos^2(\theta)$ Larmor distribution is zero. (Of course p_T effects and multiple scattering will give a nonzero background here).

As stated above, objects of an initial size of order a tenth of an angstrom or more have lifetimes equal to or exceeding that of the Universe. This suggests the possibility that such monopolonia were formed in the early Universe and may have survived up to the present. Some fraction of these will be decaying presently and the high   multiplicity   of   final   fragments   may be observable. Alternatively, the larger objects are presently spinning down and should be producing a diffuse radio background. From this we can place joint limits on the masses and closure fractions of arbitrary monopoles   as   this   part of the annihilation is completely insensitive to the GUT assumption.

We see that the binding energy of monopolonia with sizes between 1/10 to 10 angstrom ranges from = 340 keV to 3.4 keV. Thus, this is the relevant temperature scale for the formation and corresponds to the Universe age of from 10 to 10^4 seconds. First we shall assume a uniform distribution of monopoles and antimonopoles with a common density r_M. At a temperature $T < M_m$ we will have a Maxwell-Boltzmann distribution

$$r_m \left(\frac{M_m}{2\pi T}\right)^{3/2} \exp\left(\frac{-M_m v^2}{2T}\right) d^3v d^3r \tag{24}$$

Consider a monopole and antimonopole pair, each described by the distribution of Eq. (24). The equilibrium fraction of pairs that are bound with binding energy in the range E_b to $E_b + dE_b$ is given by:

$$df = \left(\frac{M_m}{2\pi T}\right) \int d^3\vec{v}_1 d^3\vec{v}_2 d^3\vec{r}_1 d^3\vec{r}_2 \ \exp\left\{\frac{-1/2 M v_1^2 - 1/2 M v_2^2 + g_m^2/|\vec{r}_1 - \vec{r}_2|}{T}\right\}$$

$$\cdot \ \delta(E_b - (1/2)M(\vec{v}_1 - \vec{v}_2)^2 + g_m^2/|\vec{r}_1 - \vec{r}_2|) dE_b r_m r_m$$

$$= \left(\frac{2M_m}{2\pi T}\right)^{3/2} \int d^3V d^3R \ \exp\left(-\frac{1}{2}(2m_m)V^2\frac{1}{T}\right) \ dr_{m\bar{m}} \tag{25}$$

$$dr_{m\bar{m}} = \frac{dE_b}{|E_b|} \ \frac{\pi^3}{2} \left(\frac{1}{2\pi T|E_b|}\right)^{3/2} r_m r_{\bar{m}} \ e^{|E_b|/T} g_m^6$$

This may be viewed as formation by multibody monopole collisions in thermal equilibrium. In a comoving volume we have N_M monopoles. We may write

$$r_m = aT^3(r_m/r_\gamma) \qquad a = \pi^2/30 \tag{26}$$

Thus the number of boundstates in the comoving volume with binding energy E_b to $E_b + dE_b$ becomes

$$dN_{m\bar{m}} = \frac{dE_b}{|E_b|} \ \frac{\pi^5}{60} \left(\frac{T}{2\pi|E_b|}\right)^{3/2} e^{|E_b|/T} g_m^6 \ \frac{r_m}{r_\gamma} N_m \tag{27}$$

The ratio r_M/r_γ is cosmologically invariant for the low formation rates discussed here.

We see that the binding energy enters as an exponential and thus the most probable states by Eq. (27) have infinite binding energy. This is the well known disease of the Coulomb gas and it is readily interpreted. In general, as the members of a pair get very close together, the Boltzmann factor in Eq. (25) is diverging. But such a pair is also no longer in thermal equilibrium and is dominated by the local mutual Coulomb force. Only when the energy is within an order of magnitude of the temperature is the process expected to be reliably described as an equilibrium one. Thus we should only apply Eq. (27) for $E_b = -\eta T$, where η is a parameter of $O(1)$. Thus we have for the differential number of objects instantaneously bound with energy E to E + dE

$$dN_{m\bar{m}} = \frac{dE}{E} \left(\frac{\pi^5}{60}\right)\left(\frac{1}{2\pi\eta}\right)^{3/2} e^\eta g_m^6 \frac{r_m}{r_\gamma} N_m \tag{28}$$

where N_m is the number of monopoles in a comoving volume.

As the system cools these objects remain behind as boundstates (like flotsam left behind by a receding tide). We've compared a computation of formation by radiative capture in which a pair collide and accelerate sufficiently to Larmor radiate their energy away and become bound.[9] Although this is actually different physics than that embodied in Eq. (28), it gives nonetheless essentially the same result.[1]

The formula of Eq. (28) possesses a pleasant scaling behavior. Reference to the lifetime formula of Eq. (8) shows that we may write

$$\frac{3dE_b}{|E_b|} = \frac{d\tau}{\tau} \tag{29}$$

Thus, the decay rate of monopolonia in a comoving volume follows immediately:

$$\frac{dN_{m\bar{m}}}{d\tau} = \frac{1}{\tau} \left(\frac{\pi^5}{180}\right) \left(\frac{1}{2\pi\eta}\right)^{3/2} e^\eta g_m^6 \frac{r_m}{r_\gamma} N_m \tag{30}$$

We see that the decay rate increases as the age of the comoving volume decreases. Today, a typical cubic light-year would contain roughly 3×10^{32} GUT monopoles if we saturate the closure density. From Eq. (30) we obtain 350 decays of monopolonia per year per cubic light-year, assuming the conservative $r_M/r_\gamma = 10^{-24}$, and 3.5×10^{12} saturating the helium abundance limit of $r_M/r_\gamma = 10^{-19}$. We may further estimate the total fraction of monopolonia by converting from lifetime to diameter through the convenient scaling law

$$\frac{3dr}{r} = \frac{d\tau}{\tau} \tag{31}$$

Thus, we may integrate from a radius of r = 1/10 Angstrom up to an upper limit, R, for which we expect the formation to terminate

$$(N_{m\bar{m}}/N_m) = g_m^6 \left(\frac{\pi^5}{60}\right)\left(\frac{1}{2\pi\eta}\right)^{3/2} e^\eta \left(\frac{r_m}{r_\gamma}\right) \ln\left(\frac{R}{r_0}\right) \qquad (32)$$

We thus see that the specific choice of R is effectively irrelevant as we are only logarithmically sensitive to it. In practice we would expect R to correspond to a value for which a monopolonium is readily ionized by a magnetic field or other traumatizing event. (We note that at 1/10 Å it requires a B-field of order 10^{10} Gauss to ionize). Putting in numbers yields about 10^{14} (or 10^{19} assuming the larger monopole to photon ratio) GUT monopolonia per cubic light year, or a fraction of 10^{-18} monopolonia to monopoles.

The discussion presented here can be amplified considerably. We should in principle carry out a detailed balance formation calculation including the effects of ionization in the reverse direction (though below the binding energy temperature the photons exponentially tail away and we expect our formation process to dominate). We should allow for "pumping" of deeply bound, short-lived monopolonium up into long-lived states by photons below the full ionization threshold. Such a mechanism may enhance the abundance of monopolonium. Furthermore, if monopoles are gravitationally clumped at early times, the formation would increase substantially. We thus regard the quoted abundances and decay rate densities as conservative lower limits.

We will not give here a detailed description of the observability of relic monopolonium. However we will remark that we can place nontrivial limits on the monopole masses and closure fractions from the above considerations and that we believe that the decay products of these systems might be detectable in a variety of experimental configurations. A systematic discussion of the observational implications, constraints, signatures and other general considerations is in preparation.[1]

ACKNOWLEDGEMENTS

I have greatly benefited from discussions with Prof. J. D. Bjorken with whom many of these ideas were initially developed. Also, Prof. D. Schramm has contributed in developing the cosmological arguments. Also W. Bardeen, J. Rosner, C. Quigg, J. Ostriker, R. Carrigan, and E. Kolb gave useful suggestions and discussions.

REFERENCES

1. C. T. Hill, Fermilab-Pub-82/70-THY; C. T. Hill and D. Schramm,
    in preparation.
2. C. P. Dokos, T. N. Tomaras, Phys. Rev. $\underline{D21}$, 2940 (1980);
    M. Daniel, G. Lazarides, Q. Shafi, Nucl. Phys. $\underline{B170}$, 151
    (1980).
3. S. Coleman, 1981 Int'l School of Subnuclear Physics, "Ettore
    Majorana."
4. J. D. Jackson, "Classical Electrodynamics," J. Wiley & Sons,
    N. Y., (1963).
5. A. H. Mueller, Phys. Lett. $\underline{104B}$, 161 (1981); A. Bassetto,
    M. Ciafaloni and G. Marchensini, Nucl. Phys. $\underline{B163}$, 477
    (1982); A. H. Mueller, Columbia University Preprint,
    Oct. 1982.
6. See e.g., K. H. Mess, B. H. Wiik, Desy 82-011 (1982) for a
    review.
7. D. Horn, F. Zachariasen, "Hadron Physics at Very High
    Energies," W. A. Benjamin, Reading, Mass, (1973).
8. J. D. Bjorken, S. Brodsky, Phys. Rev. $\underline{D1}$, 1416 (1970).
9. J. Preskill, Phys. Rev. Lett. $\underline{19}$, 1365 (1979).

Note: After completion of this work we became aware of the
interesting paper of D. Dicus, D. Page and V. Teplitz, Phys. Rev.
D $\underline{26}$, 1306 (1982), which gives the most comprehensive analysis of
monopole annihilation and which is quite relevant to the
cosmological discussion herein.

*Operated by Universities Research Association, Inc., under
contract with the U.S. Department of Energy.

STATUS OF STANFORD SUPERCONDUCTIVE MONOPOLE DETECTORS

Blas Cabrera

Physics Department
Stanford University
Stanford, California 94305

INTRODUCTION

The theoretical similarities between flux quantization in superconductors and Dirac magnetic monopoles make superconductive systems natural detectors for these elusive particles. More recently, grand unification theories have been shown to predict the existence of stable supermassive magnetically charged particles possessing the Dirac unit of magnetic charge. These particles would be nonrelativistic, weakly ionizing, and extremely penetrating; and thus may have eluded previous searches. In this paper we describe three generations of superconductive detectors designed to look for a cosmic-ray flux of such particles.

Superconductive technologies, many developed at Stanford University over the last decade, have led naturally to very sensitive detectors for magnetically charged particles. Much of the work utilizes ultra low magnetic field regions achieved using expandable superconducting shields.[1] Application of this ultra low field technology has allowed full utilization of modern SQUID (Superconducting Quantum Interference Device) sensitivities. In the second section, we summarize the elementary theory of magnetic charges and in the third section, a very similar derivation for flux quantization is presented together with experimental data clearly demonstrating flux quantization in a five centimeter diameter superconducting ring. The exact coupling of a magnetic charge to a superconducting ring is computed in the fourth section. Then in the fifth secion, the design and operation of the original single axis detector with its 10 cm^2 isotropic sensing area is summarized, and in the sixth section, the initial operation of our new three loop detector with a seven times greater loop sensing area is presented. In the closing section, we describe our design for the next generation of larger superconductive detectors with sensing areas approaching 1 m^2.

ELEMENTARY THEORY FOR MAGNETIC MONOPOLES

Before discussing the coupling of a magnetically charged particle to a superconductor, let us briefly consider the elementary theory for magnetic monopoles.[2] We begin by considering the generalized Maxwell equations

$$\vec{\nabla} \times \vec{B} - \frac{1}{c} \frac{\partial \vec{E}}{\partial t} = \frac{4\pi}{c} \vec{j}_e$$

$$\vec{\nabla} \cdot \vec{E} = 4\pi \rho_e$$

$$\vec{\nabla} \times \vec{E} + \frac{1}{c} \frac{\partial \vec{B}}{\partial t} = - \frac{4\pi}{c} \vec{j}_m \tag{1}$$

$$\vec{\nabla} \cdot \vec{B} = 4\pi \rho_m .$$

The magnetic charge density ρ_m and the magnetic current density $\vec{j}_m$ are normally set to zero from lack of experimental evidence for magnetically charged particles. The minus sign in the $\vec{j}_m$ term insures that the continuity equation for magnetic charges and currents has the same form as for electric charges and currents. The symmetrization of Maxwell's equations in this way is not a strong reason for believing in the possible existence of magnetic charges. One can define new electric and magnetic fields as linear combinations of the present electric and magnetic fields

$$\vec{E}' = \vec{E} \cos \theta + \vec{B} \sin \theta$$

$$\vec{B}' = - \vec{E} \sin \theta + \vec{B} \cos \theta , \tag{2}$$

where θ is an arbitrary angle and then

$$\rho'_e = \rho_e \cos \theta$$

$$\rho'_m = -\rho_e \sin \theta . \tag{3}$$

Thus, every charged particle would carry both electric and magnetic charge in a universal ratio $\rho'_m/\rho'_e = -\tan\theta$. If we pick $\theta = \pi/4$, Maxwell's equations become symmetric but no new physics has been added. We must thus find a more fundamental reason for suggesting the existence of magnetic charges.

In 1931, Dirac[3] found such a reason by asking whether the existence of magnetically charged particles could be made consistent with quantum mechanics. Dirac considered a single electron in the field of a magnetic charge and found that for the electron wave function to remain single-valued, a quantization condition must exist between the elementary electric and magnetic charges

$$eg = 1/2 \; \hbar c \; . \tag{4}$$

Thus, if magnetic charges existed, they would explain the experimentally observed quantization of all electric charges. At that time no other theoretical explanation for this observed quantization existed.

We can understand the Dirac quantization by considering the Schrödinger current density for a single electron in the field of a magnetic charge

$$\vec{j} = \frac{\hbar e}{2im} (\psi^* \vec{\nabla} \psi - \psi \vec{\nabla} \psi^*) - \frac{e^2}{mc} \psi^* \psi \vec{A} . \tag{5}$$

Expressing $\psi = |\psi| e^{i\phi}$ as a magnitude and phase, we find

$$\psi^* \vec{\nabla} \psi - \psi \vec{\nabla} \psi^* = 2i \psi^* \psi \vec{\nabla} \phi \; . \tag{6}$$

A line integral of $\vec{\nabla}\phi$ will depend only on the end points. Thus, upon integrating around a closed path, the phase change $\Delta\phi$ must equal $n(2\pi)$, where n is any integer, in order to obtain a single-valued wave function ψ as required physically. Then for any closed path Γ, not through the monopole

$$\frac{mc}{e^2} \oint_\Gamma \frac{\vec{j} \cdot d\vec{\ell}}{\psi^* \psi} = n \; \frac{hc}{e} - \int_{S_\Gamma} \vec{B} \cdot d\vec{s} \; . \tag{7}$$

If we take two surfaces S_Γ and S'_Γ each bounded by the path Γ and with the two together enclosing the monopole, then the current density term in Eq. 7 is identical for both. Subtracting the two equations, we obtain

$$k \left(\frac{hc}{e}\right) = \oint_{S_\Gamma - S_\Gamma'} \vec{B} \cdot d\vec{s} = 4\pi g \, , \tag{8}$$

where $k = n-n'$ is an integer and the closed surface integral equals the total flux emanating from the pole. Thus, Eq. 4 is obtained.

For k not zero we have introduced a discontinuity in the vector potential which produced an ambiguity in the phase change of ψ around the path Γ. The phase change obtained when considering the surface S_Γ differs by exactly $k(2\pi)$ from that obtained considering S_Γ'. But as Dirac stressed, only phase changes modulo 2π are physically observable. Thus, only magnetic charges possessing an integer Dirac charge are consistent with all physical observables in quantum mechanics remaining unambiguous.

The Dirac condition does not exactly symmetrize electric and magnetic charges because the elementary magnetic charge is predicted to be much stronger than the elementary electric charge. From Eq. 4

$$g = \left(\frac{\hbar c}{e^2}\right) \frac{e}{2} = \frac{e}{2\alpha} \, , \tag{9}$$

137/2 times greater than the electric charge e. Therefore, two magnetic charges a certain distance apart feel a force which is $(137/2)^2$ greater than that between two electric charges the same distance apart. The coupling constant, $g^2/\hbar c = \alpha \, (g/e)^2 \approx 34$, would thus be stronger than that of any known force.

Note also from Eq. 8 that the flux emanating from a Dirac charge is

$$4\pi g = \frac{hc}{e} \, . \tag{10}$$

As we shall see, this value is exactly twice the flux quantum of superconductivity!

The Dirac suggestion motivated many experimental searches for particles possessing magnetic charge, but no convincing candidates were found.[4]

Over the last decade, work on unification theories has unexpectedly yielded strong renewed interest in monopoles. In 1974, 't Hooft and independently Polyakov[5] showed that in true unification theories (those based on simple or semi-simple compact groups) magnetically charged particles are necessarily present. These include the standard SU(5) Grand Unification model. The modern theory predicts the same long-range field as the Dirac solution; now, however, the near field is also specified leading to a calculable mass. The Dirac theory was not able to predict a mass. The standard SU(5) model predicts a monopole mass of 10^{16} GeV/c^2 horrendously heavier than had been considered in previous searches. A detailed discussion of these theories can be found elsewhere in this volume in excellent papers by expert authors.[6]

Supermassive magnetically charged particles would possess qualitatively different properties.[7] These include necessarily nonrelativistic velocities from which follow weak ionization and extreme penetration through matter. Thus such particles may have escaped detection in previous searches.

To understand why a superconducting ring makes a natural detector for such slow monopoles, we next discuss flux quantization in superconductors.

FLUX QUANTIZATION

In this section, a simple theoretical derivation for flux quantization is presented. It includes all that we need to understand the coupling of a hypothetical Dirac magnetic charge to a superconducting ring. In 1959, Gorkov[8] demonstrated theoretically that the Ginzburg-Landau (GL) equations for superconductivity are an exact consequence of the Bardeen-Cooper-Schreiffer[9] (BCS) theory in the limit of the temperature T approaching the transition temperature T_c of the superconductor. Global properties such as flux quantization are straightforward consequences of the GL equations but often become impossibly difficult to derive from the full microscopic BCS theory. It has been theoretically shown that the global properties derived from GL are valid even well below T_c as observed experimentally.[24]

The GL supercurrent density equation

$$\vec{j} = \frac{\hbar e}{2im'} (\psi^* \vec{\nabla} \psi - \psi \vec{\nabla} \psi^*) - \frac{(e')^2}{m'c} \psi^* \psi \vec{A} , \qquad (11)$$

looks identical with the nonrelativistic single particle Schrödinger current density (Eq. 5), but the interpretation is somewhat different. The particles participating in the supercurrent are the Cooper pairs and have twice the mass m' = 2m and charge e' = 2e of the electron. In addition, ψ, called the order parameter, now represents a coherent many-body state of

Cooper pairs with $\psi^{*}\psi = 1/2\ n_s$ representing the local pair density, thus half the superelectron density n_s. As with Eq. 5, we set $\psi = |\psi|e^{i\phi}$ and again find, through Eq. 6, that upon integrating around a closed path, the phase change $\Delta\phi$ must equal $n(2\pi)$ in order to obtain a singled-valued order parameter ψ. This requirement is exactly the origin of flux quantization and is experimentally observable whenever the coherence of the order parameter extends around a closed path entirely within the superconductor.

Making these substitutions into Eq. 11, we obtain

$$\vec{j} = \frac{\hbar e n_s}{2m}\vec{\nabla}\phi - \frac{e^2 n_s}{mc}\vec{A} \ . \tag{12}$$

Before taking the line integral around a closed path, it can easily be shown that the current density $\vec{j}$ vanishes inside of a thick superconductor by taking the curl of both sides twice (note that n_s remains constant to a high degree throughout the superconductor)

$$\vec{\nabla}\times(\vec{\nabla}\times\vec{j}) = -\frac{e^2 n_s}{mc}\vec{\nabla}\times\vec{B} \ . \tag{13}$$

Using Maxwell's equation $\vec{\nabla}\times\vec{B} = 4\pi/c\ \vec{j}$ and taking $\vec{\nabla}\cdot\vec{j} = 0$ for a time independent current

$$\nabla^2\vec{j} - \frac{1}{\lambda^2}\vec{j} = 0 \ , \tag{14}$$

where

$$\lambda = \sqrt{\frac{mc^2}{4\pi e^2\ n_s}} \ , \tag{15}$$

is the London penetration depth, typically 300-500 angstroms. Solutions for the supercurrent density $\vec{j}$ in Eq. 14 fall exponentially to zero over a characteristic length λ into the superconductor.

Now integrating Eq. 12 around a closed path, we obtain

$$\frac{4\pi\lambda^2}{c}\oint_{\Gamma}\vec{j}\cdot d\vec{l} = n\frac{hc}{2e} - \int_{s_{\Gamma}}\vec{B}\cdot d\vec{s} \ , \tag{16}$$

where s_{Γ} is the area bounded by the path Γ. Then for a superconducting ring thick compared to λ we can always find a path

Γ along which $\vec{j}$ vanishes. Thus the magnetic flux through the ring must be an integer number of

$$\phi_0 \equiv \frac{hc}{2e} = 2.07 \times 10^{-7} \text{ gauss-cm}^2 , \qquad (17)$$

the flux quantum of superconductivity.

Flux quantization was first observed in 1961, at Stanford by Fairbank and Deaver and independently in Germany by Doll and Nabauer.[10] Using the ultra low field technology and modern SQUID sensitivities, S. Felch, J. T. Anderson, and myself[11] have recently demonstrated flux quantization in a single turn 5 cm diameter ring made of 0.005 cm diameter niobium wire. The data in Fig. 1 show the ring biased within several milliKelvin of its superconducting transition temperature T_c. The magnetic flux was measured with a second superconducting loop of higher transition temperature closely coupled to the ring and connected to a SQUID sensor. The potential barriers separating the various flux quantum states are reduced in height as the temperature approaches T_c until finally, the kT thermal energy associated with the normal electrons is sufficient to occasionally kick the ring from one quantum state to the next. If the temperature is too low, e.g., far left of the figure, no transitions are seen; whereas, if the temperature is too close to T_c, the transitions between states occur too rapidly to follow with the 1 Hz bandwidth used. There is a range 2 to 3 milliKelvin below T_c, where the transition rate is slow enough so that each of the quantum states are clearly shown. The kT thermal excitations carry the ring through no more than 6 or 7 quantum states because the energy associated with one flux quantum $\phi_0^2/2L$ is smaller than kT by a factor of nearly 10. Our measurements on this 5 cm diameter ring are the first in this regime and represent the largest area for which flux quantization has been directly observed.

As discussed in the next section, the coupling of a Dirac magnetic charge to this ring would change the quantum state by 2. Thus, direct observation of flux quantization in a superconducting ring necessarily demonstrates sufficient resolution for the detection of monopoles in that system.

MONOPOLE COUPLING TO A SUPERCONDUCTING RING

Rather than presenting the detailed calculation,[12] here we obtain the correct result for the current change induced by the passage of a magnetic charge through a superconducting ring using a simple magnetic field line picture. Consider a magnetic charge g moving along the symmetry axis of a superconducting ring made with

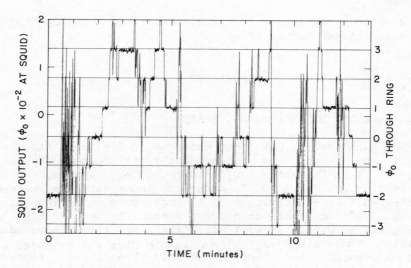

Fig. 1. Thermally induced transitions between quantum states
clearly demonstrate flux quantization. The passage of a
single Dirac charge through this 5 cm diameter
superconducting ring would induce exactly a 2 ϕ_0 change.

wire many λ in diameter. From the third section, we know that the
magnetic field inside the wire is always negligibly small. Thus no
magnetic field lines may pass through the wire and as the charge
passes through the ring, the field lines are brushed back.
However, as the particle proceeds to infinity after passing through
the ring, all of the magnetic field lines emanating from the charge
cannot return through the ring. The answer to this apparent
dilemma is shown schematically in Fig. 2. Each flux line is
continuously deformed as the charge g approaches the wire and each
separates into two disjoint lines--one left surrounding the wire
and the other remaining connected to the charge. At the point
where the crossover occurs, the magnetic field is zero.

For a trajectory along the ring's symmetry axis, the magnetic
field lines can be found analytically. They always lie along the
surface of tubes of constant flux which are now figures of rotation
about the z axis. Thus, for the charge g, a distance z' away from
the plane of the ring with radius b, the flux $\phi(r,z)$, through a
circle of radius r a distance z away from the ring, is given by

$$\phi(r,z) = \phi_0 \left[1 + \frac{z'-z}{\sqrt{(t'-t)^2+r^2}} \right] - \phi_0 \ \frac{M}{L} \left[1 + \frac{z'}{\sqrt{(t')^2+b^2}} \right] , \quad (18)$$

where L is the self inductance of the ring and M the mutual

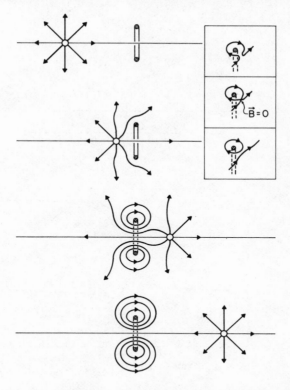

Fig. 2.   Schematic representation of magnetic field lines   as
          monopole passes through superconducting ring.

inductance between the ring and the circle. The first term on   the
right-hand side is the flux from the monopole and the second, the
flux from the induced supercurrent in the ring. M (in henries) can
be expressed in terms of the complete elliptic integrals E and K as

$$M = \mu_0 \ (rb)^{1/2} \left[\left(\frac{2}{k} - k \right) K \ (k) - \left(\frac{2}{k} \right) E \ (k) \right] \ , \qquad (19)$$

where

$$k^2 = \frac{4rb}{(r+b)^2 + z^2} \ . \qquad (20)$$

Using   these   equations,   a video simulation of the changing field
lines as a monopole passes through a superconducting ring has   been
made by Gordon Smith at Stanford.

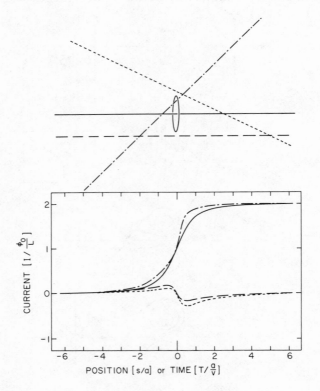

Fig. 3. The induced current in a superconducting ring of radius
"a" is plotted against time (or position) for each of four
different monopole trajectories. The particle travels at
constant velocity v.

     Regardless of the exact magnetic field structure, it is clear
from Fig. 2 that every field line emanating from the pole must have
left a closed loop around the ring wire as the particle moved
through.   Thus, the net change of flux through the ring must
exactly equal the total flux from the pole--two ϕ_0 for a monopole
of unit Dirac charge. The same result is obtained for all particle
trajectories which pass through the ring, whereas no net flux
change   results   from trajectories which miss the ring.   To
illustrate, in Fig. 3 the induced supercurrent resulting from each
of four different trajectories has been computed. Note that the
rise times are similar and are of order the ring radius divided by
the particle velocity. For trajectories which miss the ring only
transient currents are obtained with a maximum peak to peak
excursion of one ϕ_0.

     Thus if a magnetic charge g passes through the ring, the
number of flux quanta threading the ring will change by two;

whereas, if the particle does not pass through, the flux will remain unchanged. Finally if the magnetic charge passes through the bulk superconductor, such as the wire of the ring, it would leave a trapped doubly quantized vortex, and some intermediate total current would persist.

A superconductive system based on these properties is sensitive only to magnetic charges and thus makes a natural detector. The passage of any known particle possessing electric charge or a magnetic dipole would cause very small transient signals but no dc shifts. A cosmic flux of magnetically charged particles such as the predicted supermassive GUT monopoles can be detected by monitoring the current in a superconducting loop or scanning the surface of a superconducting sheet for spontaneously appearing doubly quantized vortices. Several such devices are described below.[13]

PROTOTYPE SINGLE LOOP DETECTOR

To test the feasibility of such devices an existing instrument, originally built as a magnetometer for measuring the ultra low field superconducting shields, was used in a prototype search for a particle flux of supermassive monopoles.[12] This instrument has been operated as a monopole detector for a total of 382 days. It consists of a four turn 5 cm diameter loop made of 0.005 cm diameter niobium wire. The coil is positioned with its axis vertical and is connected to the superconducting input coil of a SQUID. The passage of a single Dirac charge through the loop would result in an $8 \phi_0$ change ($2 \phi_0$ couple to each of the four turns) in the flux through the superconducting circuit, comprised of the detection loop and the SQUID input coil. As shown in Fig. 4, the SQUID and the loop are mounted inside an ultra low field shield in an ambient field of 5×10^{-8} gauss.

The presence of the superconducting shield complicates the expected detector response. In addition to the trajectory coupling directly to the loop, doubly quantized supercurrent vortices appear in the walls of the shield at the points where the particle enters and exits (see Fig. 4). Thus a field change is produced in the shield which also couples to the loop even if the trajectory misses the loop. Thus the possible values for the total current induced in the loop are no longer sharp.

The distribution of all possible total induced current values for an isotropic distribution of straight line particle trajectories is shown in Fig. 6a. The density of sensing area D_σ (averaged over 4π solid angle) is plotted as a function of the total induced current. All of those trajectories which intersect the loop induce current changes between $7.5 \phi_0$ and $8 \phi_0$ and the

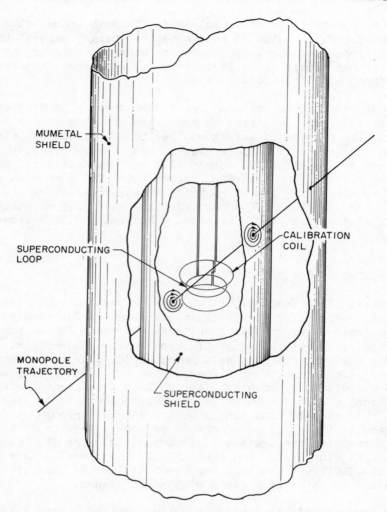

Fig. 4.   Schematic of prototype monopole detector.

integral of that portion of the distribution is 10.1 cm^2. This area
corresponds exactly to the cross section of a 5.08 cm diameter loop
averaged over 4π solid angle (1/2 of the loop area). In addition,
for those trajectories which intersect the shield but not the loop,
induced currents between 0 and $0.5\ \phi_0$ would be seen. The integral
under this portion is infinite, but the noise level of the detector
sets a natural lower cutoff. For our prototype detector with its
noise level of $0.2\ \phi_0$, we obtain an additional sensing area of
145 cm^2. Thus for every event between 7.5 and 8 ϕ_0, one would
expect 14 events between 0.2 and 0.5 ϕ_0.

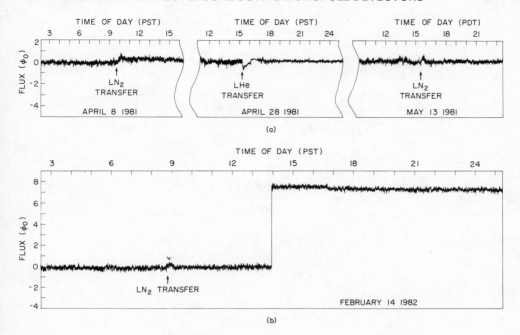

Fig. 5. Data of (a) typical stability during one month, and (b)
the candidate event.

Several intervals throughout a continuous one month time
period are shown in Fig. 5a, where no adjustments of the dc level
has been made. Typical disturbances caused by daily liquid
nitrogen and weekly liquid helium transfers are evident. A single
large event was recorded (Fig. 5b). It is consistent in magnitude
with the passage of a single Dirac charge within a combined
uncertainty of ±5% (resulting from the calibration uncertainty and
the distribution D_σ in Fig. 6a). It is the largest event of any
kind in the record.

Several histograms have been obtained from a detailed analysis
of the first 151 days. The detailed analysis of the remaining 231
days is not complete, but no large events were seen.   In Fig. 6b
are plotted the 27 spontaneous events exceeding a threshold of 0.2
ϕ_0 which remain after excluding known disturbances (Fig. 7a) such
as transfers of liquid helium and nitrogen. An event is defined as
an offset with stable levels for at least one hour before and
after. Only six events were recorded during the 70% of the running
time when the laboratory was unoccupied.   Since many more small
events were seen during working hours it is clear that the detector
is sensitive to disturbances. As we shall discuss below, this is
consistent with its known sensitivity to mechanical disturbances.

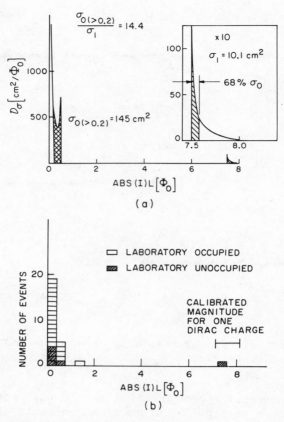

Fig. 6. (a) Calculated density of sensing area versus signal
        mangitude for the prototype detector. (b) Data histogram
        of all spontaneous events during first 151 days.

However, even if we assmume that the five small events and the
large event occurring during nonwork hours are real, the absence of
an additional 7 small events suggests that the actual rate of large
events should be several times lower. Thus a better upper bound on
the particle flux is obtained by considering the lack of small
events.

    The most likely sources for spurious signals are associated
with the mechanical sensitivity of the apparatus. Mechanically
induced offsets have been intentionally generated. These could be
caused by shifts of the four turn loop wire geometry, which would
produce inductance changes and inversely proportional current
changes (the flux remains constant). Alternatively, trapped
supercurrent vortices in or near the SQUID could move from

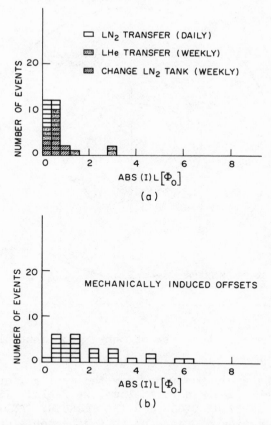

Fig. 7.   Data histograms (a) of all events during first 151 days
which directly correlated with dewar maintenance and (b)
of all mechanically induced events from test raps.

mechanically induced local heating. In any case, sharp raps with a
screwdriver handle against the detector asssembly cause offsets.
The magnitude distribution of 31 such induced events is shown in
Fig. 7b.

Although we have not been able to produce such events which
appear as clean as the candidate event, magnitudes close to the
candidate event magnitude have been generated. If we further
assume that mechanical stresses frozen into the detector upon
initial cooling from room temperature could spontaneously release
or be triggered by a small initial external disturbance, it would
be possible to imagine a large spurious signal.   The difficulty
with easily accepting this scenario as a plausible explanation for
the candidate event is that the data from this detector has been
accumulated over five separate runs, each commencing with a

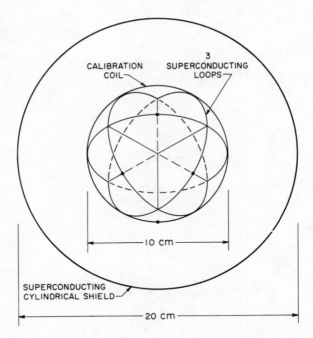

Fig. 8. Schematic top view of three loop monopole detector.

cooldown of the apparatus from room temperature, and no other such
signals were ever seen. Nevertheless, we feel IT IS NOT POSSIBLE
TO RULE OUT A SPURIOUS CAUSE FOR THE FEBRUARY CANDIDATE EVENT. We
are planning a series of tests on this detector aimed at reducing
the uncertainty in our calibration and further clarifying the
origin of the mechanically induced offsets.   These tests will
include repeated thermal cycling of the apparatus.

     Our recent primary effort has gone into building a new
detector with a larger sensing area and with sufficient cross
checks to rule out spurious causes for any new events.   It is
described in the next section.

PROGRESS ON NEW THREE LOOP DETECTOR

     In the spring of 1982, a group at Stanford, including
M. Taber, J. Bourg, S. Felch, R. Gardner, R. King and me began work
on a larger detector. We have completed construction of a new
three loop detector.   This new detector is based on the same
principles as the prototype system described in the fifth section,
but has improved mechanical stability, a greater sensing area, and
much better spurious signal discrimination. As shown schematically

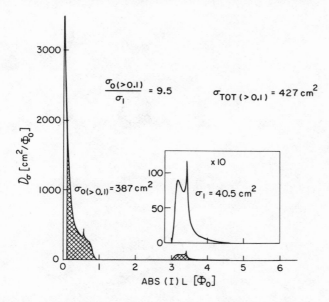

Fig. 9. Density of sensing area versus signal magnitude for one
        loop of three loop detector.

in Fig. 8, it consists of three mutually orthogonal superconducting
loops, each 10.2 cm in diameter (twice that of the original
device). In addition to a sevenfold increase (70.9 cm²) in direct
average isotropic sensing area for trajectories passing through at
least one loop, 57% of those trajectories also intersect at least
two loops, providing valuable coincidence information.   The
sensitivity of this new instrument to near miss trajectories which
traverse the shield but do not intersect the three loops provides
an additional effective sensing area for single Dirac charges 53
times greater (540 cm²) than that of the prototype detector for
signals of magnitude 0.1 ϕ_0 or larger. This near miss sensitivity
can provide lower flux bounds if no candidate events are seen.
However, a near miss candidate would not be as definitive as one
corresponding to a trajectory passing through at least one loop,
because of the smaller signal-to-noise levels.

    We are completing a detailed analysis of the distribution
functions for this new detector.[14] Our preliminary results are
summarized in Figs. 9-11. First consider only one of the three
detector loops.   In Fig. 9, the density of sensing area D_σ in cm²
per ϕ_0 has been plotted against the signal size in ϕ_0. Thus, for
example, the total sensing area for trajectories which intersect
the loop is given by the area between 3 and 4 ϕ_0 and is equal to
40.5 cm²--exactly the area of an 81 cm² circle averaged over 4π
solid angle. For our three axis detector, the sensing areas

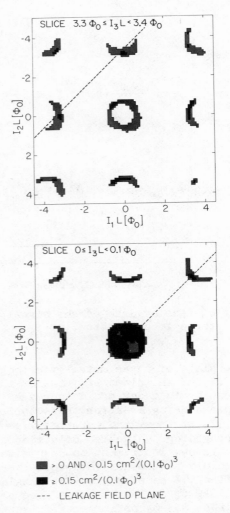

Fig. 10. Two slices through the three-dimensional distribution
         function for the density of sensing area in the new three
         loop detector. The top slice corresponds to signal sizes
         in loop 3 between 3.3 and 3.4 ϕ_0 and the bottom slice
         between 0 and 0.1 ϕ_0. The diagonal line in each slice is
         the locus of all possible signals resulting from leakage
         fields.

overlap by design and thus we must calculate a three dimensional
density of sensing area function $D_\sigma = D_\sigma(\phi_1, \phi_2, \phi_3)$ where each
coordinate corresponds to the signal size in each of the three
loops. Thus each point in this space corresponds to a three loop
event.

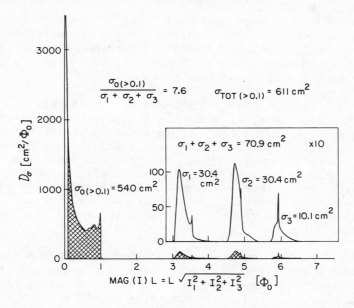

Fig. 11. Density of sensing area versus three loop signal magnitudes (square root of the sum of the squares of each loop signal).

In our preliminary Monte Carlo calculation the volume of this phase space was broken into cubes 0.1 ϕ_0 on an edge and 100,000 trajectories were computed. Two slices through this three dimensional density of sensing area function D_σ are shown in Fig. 10. The top slice is for signals in loop 3 between 3.3 and 3.4 ϕ_0 and the bottom for signals in loop 3 between 0 and 0.1 ϕ_0. All trajectories correspond to the signals in the shaded regions, with the darker shading representing $(0.1 \phi_0)^3$ cubes, each contributing more than 0.15 cm^2 to the sensing area averaged over 4π solid angle. In Fig. 10 (top), the shaded areas in the corners correspond to trajectories that intersect all three loops, those in the middle of the sides to trajectories that intersect exactly two loops, and the central area to trajectories that intersect loop 3 but miss loops 1 and 2. The central region of Fig. 10 (bottom) corresponds to trajectories that do not intersect any loop but do intersect the superconducting cylinder walls and produced a "three-signal-magnitude" (the square root of the sum of the squares of the three signal sizes) greater than 0.1 ϕ_0, which we have taken as our achievable noise level.

The data from this three dimensional function can be folded into one dimension by plotting the density of sensing area as a function of the three-signal-magnitude. This has been done in

Fig. 11 where the three disjoint peaks for increasing magnitude
correspond to trajectories passing through exactly one, two and
three loops. The total integrated average sensing area for
three-signal-magnitudes greater than 0.1 ϕ_0 is 611 cm^2, a factor of
60 greater than for the prototype detector.

These preliminary results indicate that the volume of this
phase space occupied by all possible monopole trajectories passing
through at least one detector loop is about 1% of the volume of a
sphere with radius just equal to that of the largest allowed event
magnitude (about 6.5 ϕ_0). Thus, for a candidate event to be
consistent with a possible monopole trajectory, it must lie within
the allowed volume. This requirement is particulary stringent for
trajectories that intersect at least one loop, but does not
effectively discriminate the smaller signal levels of near miss
trajectories.

In addition, two known sources of spurious signals lie within
well defined planes in this phase space. These are magnetic field
changes at the three loops caused either by leakage fields in
through the top of the superconducting shield or by the pendulum
motion of the loop assembly in the shield field gradient. These
planes can be experimentally determined to better than 1%
resolution. Thus the most convincing candidate event would
correspond to a trajectory intersecting at least one loop and lying
away from these planes in the phase space. Data from an external
fluxgate magnetometer and an accelerometer, both mounted on the
outside of the detector, will remove most, if not all, signals from
the planes of confusion produced by the leakage fields and the
shield magnetic gradient.

To monitor the detector, we have designed a computer based
data acquisition system. The output signals from the fluxgate and
the accelerometer, together with the outputs from the three
independent SQUID loop current sensors, are continuously sampled at
200 readings per second per channel. These data are temporarily
stored into a computer ring buffer (about 30 seconds long) and
digitally filtered to 0.1 hz bandwidth (one point every 5 seconds)
in real time. The filtered readings are then stored on magnetic
disc to form a permanent continuous record. In addition, these
filtered data are constantly compared to detect changes in the
SQUID sensor levels above a 0.1 ϕ_0 threshold. Such an event then
triggers the permanent storage of the ring buffer contents around
the time of the event. The high bandwidth information allows
convincing discrimination against mechanically induced spurious
events of any kind. However, the 10 millisecond resolution is NOT
capable of observing supercurrent rise times of order 0.1 to 1
microsecond expected from the passage of a Dirac charge with
velocity between 10^{-3} and 10^{-4} times the speed of light. This data
acquisition system has been assembled and is expected to become
fully operational by early 1983.

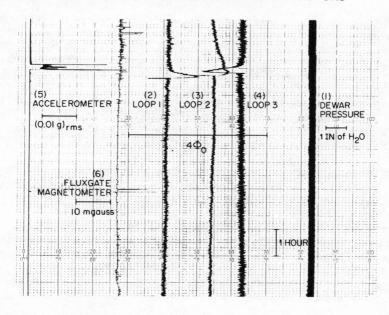

Fig. 12.   Recent typical data taken at 0.1 hz bandwidth   on   backup
          six   pen   strip   chart   recorder.   It   includes a liquid
          nitrogen transfer cycle.

As a backup to the computer, the detector is also continuously
monitored   on   a   six   channel   strip   chart   recorder   at   0.1 hz
bandwidth.   As of December, 1983, we have operated the   system   for
nearly   five   months.   Figure 12 shows ten hours of typical recent
strip chart data.   We have accumulated 97 days of data primarily on
the strip chart recorder with SQUID noise levels of 1 ϕ_0 or better.
The signal from any Dirac magnetic charge passing through at   least
one   loop   would have been detectable.   NO SUCH SIGNAL HAS YET BEEN
SEEN.   The present noise level does not allow   clear   detection   of
near   miss   signals.   However,   using only the direct loop sensing
area, the dominant contribution to our experimental   particle   flux
bound now comes from our new three axis detector.

As of December 6, 1982, the limit   on   the   particle   flux   of
magnetic   charges   passing   through   the   earth's   surface   at   any
velocity and of any mass is less than 8.7×10^{-11} cm^{-2} s^{-1} sr^{-1}.
This limit is derived combining 382 days of data from the prototype
detector with its 10.1 cm^2 of averge loop sensing area and 97   days
from the new three axis detector with its 70.9 cm^2 loop area.

Cosmological theories based on GUTs[15] lead to impossibly   high
or   unobservably   low predictions for monopole particle flux limits
with the results being exponentially model dependent.   Thus we turn

to astrophysical arguments for several observational limits. Here we assume a particle mass of 10^{16} GeV/c^2 to obtain representative numbers. Then an absolute upper bound for the galactic monopole particle flux of 4×10^{-10} cm^{-2}s^{-1}sr^{-1} is obtained from the limits on the local galactic dark mass.[12] A much smaller upper bound of 10^{-15} cm^{-2}s^{-1}sr^{-1} is obtained assuming an isotropic flux from arguments based on the existence of the 3 microgauss galactic magnetic field.[16] However, possible models incorporating monopole plasma oscillations may allow a much larger particle flux.[17] All of the above bounds suggest particle velocities near 10^{-3} c.

An enhanced monopole density, gravitationally bound to our own solar system,[18] would allow much smaller average galactic flux levels and lead to lower particle velocities near 10^{-4} c.

Perhaps the most important question regarding direct experimental detection is: can conventional ionization devices with their much larger sensing area detect the passage of single Dirac charges with velocities of order 10^{-4} to 10^{-3} c?[19] New calculations based on fundamental quantum mechanical arguments suggest that helium gas devices would provide such a sensitivity.[20]

Finally, it has been shown that the supermassive monopoles arising from grand unification theories will catalyze nucleon decay processes.[21] If the cross section for such events is of order the hadron cross section, as has been suggested,[22] then all attempts at direct detection of these monopoles may be doomed to failure. Arguments based on X-ray flux limits from galactic neutron stars[23] and which assume a strong cross section lead to an upper bound for a magnetic particle flux of about 10^{-22} cm^{-2} s^{-1} sr^{-1}--about one per year through the entire land area of Rhode Island!

LARGER SENSING AREA SUPERCONDUCTIVE DETECTORS

Superconductive devices are the only known monopole detectors which are sensitive to arbitrarily low particle velocities. Their response is exactly calculable using simple fundamental concepts. Thus an important question is how large can the sensing area of such devices be made using existing technologies?

First with respect to our new three axis system, we are now working to reduce the detector noise levels below 0.1 ϕ_0 and have obtained short periods of such operation. The ability to reliably detect near miss events will increase the sensing area from 70.9 cm^2 to over 600 cm^2. However, candidates in the near miss category are less convincing. Larger diameter loops with their larger self inductances provide ever decreasing signal-to-noise ratios and for diameters larger than about 10 cm cannot be effectively used as monopole detectors.

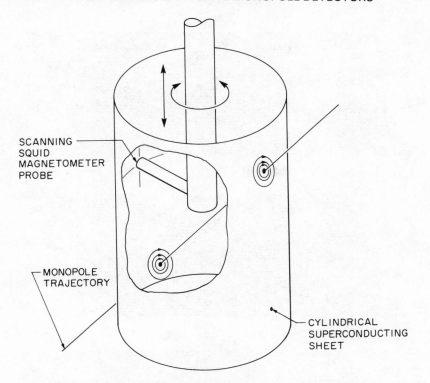

SCANNING
SQUID
MAGNETOMETER
PROBE

MONOPOLE
TRAJECTORY

CYLINDRICAL
SUPERCONDUCTING
SHEET

Fig. 13. Schematic of proposed large area scanning detectors.

To overcome this limitation, we are designing a new generation
of superconductive detector which will use up to several square
meters of a thin superconducting sheet in the form of a cylinder
for recording magnetically charged particle tracks. It is
analogous to a photographic emulsion which records electrically
charged particle tracks. A magnetic charge, traversing the
cylinder twice in most cases, would record signatures consisting of
doubly quantized trapped flux vortices in the walls (Fig. 13). As
long as the sheet remains superconducting, the strong flux pinning
due to lattice and surface defects will prevent any motion. The
ambient trapped flux pattern, about one quantized vortex per cm^2 in
a 10^{-7} gauss field, will be periodically recorded using a small
scanning coil coupled to a SQUID. Two simulated scans, over 10 cm
by 10 cm area, are shown in Figs. 14a and 14b. A typical SQUID
noise level has been added. In Fig. 14b, a new doubly quantized
vortex has appeared and is most clearly seen when 14a is subtracted
from 14b as shown in 14c.

With our new three loop detector, or later with the larger
sensing area scanning detectors, either we will begin to see more

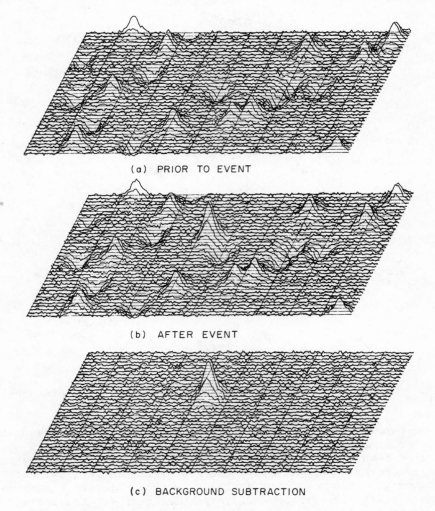

(a) PRIOR TO EVENT

(b) AFTER EVENT

(c) BACKGROUND SUBTRACTION

Fig. 14. Simulated data from scanning detectors.

events which would be very convincing, or we will set an upper limit approaching $10^{-12}\,\mathrm{cm}^{-2}\mathrm{s}^{-1}\mathrm{sr}^{-1}$ on a magnetic particle flux--100 times smaller than our present value. Either way, this is a very exciting time for our group.

ACKNOWLDEGEMENTS

    The work described in this paper has been funded in part by NSF grant DMR 80-26007 and DOE contract DE-AM03-76SF00-326.

REFERENCES

1. B. Cabrera, Ph.D. Thesis, Stanford University, 1975; B. Cabrera and F. van Kann, Acta Astronautica 5, 125 (1978); B. Cabrera, in "Third Workshop on Grand Unification," P. H. Frampton, S. L. Glashow, and H. van Dam, eds., Birkhauser, Boston (1982).

2. For complete review, see: P. Goddard and D. I. Olive, Rep. Prog. Phys. 41, 1357 (1978).

3. P.A.M. Dirac, Proc. Roy. Soc. A 133, 60 (1931); Phys. Rev. 74, 817 (1948).

4. For review of experiments, see: A. S. Goldhaber and J. Smith, Rep. Prog. Phys. 38, 731 (1975); B. Cabrera and W. P. Trower, "Foundations of Physics," Feb. (1983); G. Giacomelli in these proceedings.

5. G. 't Hooft, Nucl. Phys. 79B, 276 (1974) and Nucl. Phys. 105B, 538 (1976); A. M. Polyakov, JETP Letts. 20, 194 (1974).

6. See papers by A. S. Goldhaber and J. Ellis in these proceedings.

7. J. P. Preskill, Phys. Rev. Lett. 19, 1365 (1979); G. Lazarides, Q. Shafi and T. F. Walsh, Phys. Lett. 100B, 21 (1981).

8. L. P. Gorkov, Sov. Phys. JETP 9, 1364 (1959).

9. See for example: A. L. Fetter and J. D. Walecka, "Quantum Theory of Many-Particle Systems," McGraw-Hill, San Francisco (1971), Chap. 13.

10. B. S. Deaver and W. M. Fairbank, Phys. Rev. Lett. 7, 43 (1961); R. Doll and M. Nabauer, Phys. Rev. Lett. 7, 51 (1961).

11. B. Cabrera, S. Felch and J. T. Anderson, "Precision Measurement and Fundamental Constants II," B. N. Taylor and W. D. Phillips, eds., Nat. Bur. Stand. (U.S.), Spec. Publ. No. 617, in press.

12. B. Cabrera, Phys. Rev. Lett. 48, 1378 (1982).

13. See also past work on superconductive detectors for static matter searches: L. W. Alvarez, Lawrence Radiation Laboratory Physics Note 470, 1963 (unpublished); P. Eberhard, Lawrence Radiation Laboratory Physics Note 506, 1964 (unpublished); L. J. Tassie, Nuovo Cimento 38, 1935 (1965); L. Vant-Hull, Phys. Rev. 173, 1412 (1968); P. Eberhard, D. Ross, L. Alvarez and R. Watt, Phys. Rev. D 4, 3260 (1971). For other recent work on superconductive detectors, see papers by C. C. Tsuei and D. Cline in these proceedings.

14. B. Cabrera, R. Gardner and R. King, in preparation.

15. For review, see: A. H. Guth in these proceedings.

16. M. S. Turner, E. N. Parker and T. J. Bogdan, Phys. Rev. D 26, 1296 (1982); also see papers by M. S. Turner and E. Purcell in these proceedings.

17. E. E. Salpeter, S. L. Shapiro and I. Wasserman, Phys. Rev.
     Lett. 49, 1114 (1982); also I. Wasserman in these
     proceedings.
18. S. Dimopoulos, S. L. Glashow, E. M. Purcell and F. Wilczek,
     Nature 298, 824 (1982).
19. S. P. Ahlen and K. Kinoshita, Phys. Rev. D 26, 2347 (1982); S.
     P. Ahlen in these proceedings.
20. S. D. Drell, N. M. Kroll, M. T. Mueller, S. J. Parke and
     M. A. Ruderman, SLAC-PUB-3012.
21. C. Dokos and T. Tomaras, Phys. Rev. D 21, 2940 (1980); A.
     Blaer, N. Christ and J. Tang, Phys. Rev. Lett. 47, 364
     (1981); F. Wilczek, Phys. Rev. Lett. 48, 1146 (1982).
22. V. Rubakov, JETP Lett. 33, 644 (1981); Nucl. Phys. B 203, 311
     (1982); C. Callan, Phys. Rev. D 26, 2058 (1982).
23. E. W. Kolb, S. Colgate and J. A. Harvey, Phys. Rev. Lett. 49,
     1373 (1982); S. Dimopoulos, J. Preskill and F. Wilczek,
     Phys. Lett., in press; J. Ellis in these proceedings.
24. F. Bloch, Phys. Rev. B 2, 109 (1979).

A PREVIEW OF CURRENT STUDIES ON MAGNETIC

MONOPOLE DETECTION USING INDUCTION COILS

C. C. Tsuei

IBM Thomas J. Watson Research Center
Yorktown Heights, New York 10598

INTRODUCTION

A single candidate event for magnetic monopoles reported recently by Cabrera[1] has revived interest in detecting these elusive, magnetically singly-charged particles. Only five months after the publication of Cabrera's paper, there are about ten or more research groups engaging in or contemplating experiments on induction-coil detection of monopoles. Since the estimated flux of monopoles is extremely small, such experiments should be well-designed so that, among others, the following three requirements can be met. 1) The sensing areas of the detector should be large so that, in the absence of any monopole event, and in a reasonable length of time, one can establish an upper limit on monopole flux comparable with or lower than those set by other means of detection such as the ionization technique.[2] 2) If a positive monopole event is observed, it should be unequivocally discernible from any spurious signals. 3) As much information as possible should be extracted from such an event. These conditions are particularly difficult to satisfy in experiments on monopole detection in view of the fact that theoretical estimates on the characteristic properties and abundance of monopoles vary greatly from theory to theory.[3,4]

A preview of current activities in induction-coil detection at this early stage of the search can serve as a forum for exchanging information and ideas and hopefully may be useful in the optimization of the techniques for the present and future monopole detectors. For this reason, I was asked to take on the task of summarizing the basic features of several existing schemes for induction-coil detection with the exception of Cabrera's which is described in detail in another article in these proceedings.

201

This article is based on unpublished sources provided by the following persons whom I gratefully acknowledge: H. J. Frisch, J. Incandella (University of Chicago), R. Carrigan, M. Kuchnir, J. Lach (Fermilab), W. P. Kirk, H. Armbruster (Texas A & M University), A. D. Caplin, C. N. Guy, M. Hardiman, J. G. Park (Imperial College), D. E. Morris, P. H. Eberhard (Lawrence Berkeley Laboratory), M. J. Price (CERN, Switzerland), T. Datta (University of South Carolina), B. Deaver (University of Virginia), C. Tesche, J. Chi, P. Chaudhari (IBM T. J. Watson Research Center).

As a result of the Faraday electromagnetic induction, the passage of a Dirac magnetic monopole ($g = hc/e$) through a metallic loop with self-inductance L and resistance R will result in a change in the circulating current I by ΔI (see Fig. 1) and

$$\Delta I = 2\Phi_0/L \tag{1}$$

where $\Phi_0 = hc/2e = 2.07 \times 10^{-7}$ G-cm^2 is the flux quantum. The energy deposited in the coil by a monopole is

$$\Delta E = \frac{(2\Phi_0)^2}{2L} = \frac{2\Phi_0{}^2}{L} \tag{2}$$

For a square coil (of the size of 100×100 cm^2) made of 5 mil wires, the inductance is about 7 μH and the maximum current induced in the coil ΔI is about 5×10^{-10} A. The change in energy ΔE is about 10^{-24} J. These are indeed very small signals. Any viable

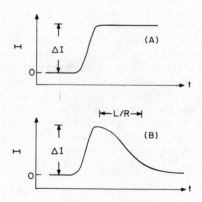

Fig. 1. The monopole induced current I, as a function of time t, in an induction coil with self-inductance L and resistance R, the maximum rise in current, ΔI is persistent if the coil is superconducting (A), and is transient over a period of L/R if the coil is normal (B).

techniques of induction detection should provide resolutions of at least one tenth of these quantities. The induced current ΔI is persistent as long as the coil is in the superconducting state (Fig. 1A) and ΔI is transient over a period of L/R if the coil is in the normal state (Fig. 1B). The rise time for ΔI is, in principle, determined by the monopole velocity and trajectory. In practice, it is limited by the material properties such as R, L of the coil and also by the parameters associated with the detection technique such as noise and the response time of the sensing devices. In this article, techniques of using normal-state and superconducting induction coil for monopole detection will be described. Emphasis will be on problems arising from the requirement of large-area detection and the discrimination against spurious signals.

MONOPOLE DETECTION USING NON-SUPERCONDUCTIVE COILS

     The primary motivations of employing a non-superconductive detection are to have a large area detection at relatively low cost and to eliminate the possibility of flux jumping as a possible explanation of any observed events. In this kind of experiment, the usual requirements of effective shielding against external time varying magnetic fields and mechanical isolation are important. The crucial problem here is to make sure that the monopole signal is not overwhelmed by the Johnson noise in the detection coils. A non-superconductive induction monopole detector has been proposed by D. E. Morris[5] at Lawrence Berkeley Laboratory using an FET amplifier and by M. J. Price[6] at CERN Switzerland, using appropriate band-pass and w-filters. A signal-to-noise ratio (S/N) of about 10 has been claimed. As an example, the detecting system proposed by Morris will be described briefly. As shown in Fig. 2, the detection coil (about 10^4 turns of 0.15 mm diameter copper wire) has an outside diameter of 15 cm and a coil thickness of 3 cm and is sitting inside of an OFHC Cu shield container which provides an effective attenuation of the external magnetic noises. The container and the pick-up coil are maintained at liquid helium temperature to reduce the Johnson noise ($4 k_B RT$). The coil resistance R at 4.2 K is about 100 Ω. The emf induced by a monopole passing through the coil is of the order of 10^{-6} volts and is detected as a current of about 10^{-7} A by several FETs (cooled at 150 K) connected in parallel. The decay time constant L/R for the coil is about 0.3 sec and is long enough for the FETs to work as a current sensitive device. The output of the FETs is further amplified and filtered. The noise voltage for two FETs (U311) in parallel is about 7×10^{-10} V/Hz$^{1/2}$. This would give a S/N of about 20 for a bandwidth of 32 kHz. After taking into account the Johnson noise arising from the coil resistance (R~100Ω at 4.2 K) and noise from other parts of the electronic circuits, the total S/N ratio for the detector is estimated to be 15. It should be

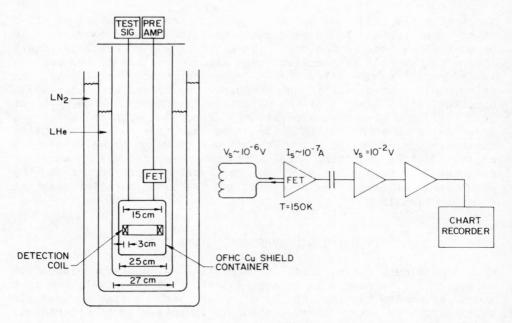

Fig. 2. A non-superconducting induction detector for magnetic
         monopoles using an FET amplifier as proposed by
         D. E. Morris.

noted that this S/N probably represents an upper limit for this
particular design of the non-superconductive detector and the
feasibility of such a detector has not been demonstrated
experimentally yet.

MONOPOLE DETECTION USING SUPERCONDUCTIVE COILS

About four years after the quantization of the flux (strictly
fluxoid) in superconductors was experimentally established,
L. J. Tassie suggested[7] that hollow superconducting cylinders can
be used to detect Dirac magnetic monopoles. Making use of a
combination of the symmetrized Maxwell equation and London fluxoid
quantization condition, he showed the passage of a magnetic
monopole through a superconducting loop resulted in a magnetic flux
change of $2\Phi_0$. It was also pointed out that the superconductive
induction technique is valid for detecting non-ionizing monopoles.
It should be pointed out that this was all before the advent of the
supermassive, weakly ionizing monopoles as required in the Grand
Unification theories.

As we have pointed out in the beginning of this article, the monopole signals for an even modestly large detection coil is very small and an ultra sensitive magnetic flux monitor is needed. In terms of the capability of measuring a minute change in the flux threaded through a large superconductive coil, a SQUID is probably the most sensitive device available for this need. A SQUID is a superconducting quantum interferometer based on the Josephson effects and the flux quantization in a superconductive ring containing one or two Josephson weak links.[8] The energy resolution of a commercial SQUID (e.g., SHE Hybrid rf-SQUID) is about 10^{-28} J in 1 Hz bandwidth, giving an upper limit for S/N ratio of about 10^4 in the case of the superconductive square loop (100 × 100 cm^2 in size) mentioned before. For monopole detection, this impressively large S/N ratio, however, should be taken with proper perspective. The link between the Dirac magnetic charge and flux quantization in superconductors also makes the SQUID sensitivity a liability in discriminating against spurious signals.

In what follows, some characteristics of several current superconductive monopole detectors will be described. The basic experimental arrangement (Fig. 3) is identical for all these detection systems. The difference lies in the design of the magnetic shields and the detection coils and will be discussed later. As shown in Fig. 3, a superconductive pick-up coil with self-inductance L_d is connected to the SQUID input coil which has a self-inductance of L_i. The SQUID output is amplified and filtered and is fed into a strip chart recorder and/or a computer. Both the SQUID and the detection coil are kept at 4.2 K and are enclosed in a superconductive magnetic shield.

Magnetic Shielding

A major difference between the current experiments on

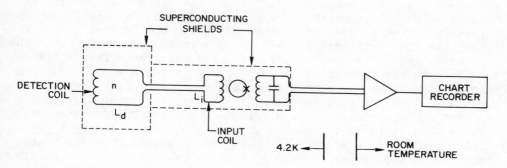

Fig. 3. A schematic diagram showing the experimental arrangement for a superconductive induction detector for magnetic monopoles.

relatively large area detection and the Cabrera's original work is that the detection coil sits in a relatively high magnetic field environment (namely, an ambient field, B of about 0.01 to 0.5 G instead of 10^{-8} G or less as in Cabrera's case). The reason why ultra-low magnetic field is not practical for large-area detection can be seen from the following argument: If we demand a S/N ratio of at least 10, then the flux change in the pick-up coil due to fluctuations in ambient field, ΔB, should satisfy the following inequality:

$$\Delta\Phi_{noise} = \Delta B \times A \leq \frac{1}{10} (2\Phi_0) = 4 \times 10^{-8} \text{ G-cm}^2 \tag{3}$$

where A is the coil area and is 10^4 cm^2 for the square loop discussed before. Therefore, ΔB should be about 4×10^{-12} G or less. To achieve this, an ambient field of 10^{-12} G should be maintained, since ΔB is roughly the same order of magnitude of the field B itself unless B can be stabilized by certain means. Because a field of 10^{-10} G or less is extremely difficult, if not impossible to achieve, and to be kept for a long period of time (say one year) over a large area, stabilizing the ambient field appears to be a viable alternative to the production of an ultra-low field environment. This is done with the aid of superconducting magnetic shields. The principle and techniques of superconducting shielding[9,10] have been discussed in the literature. Basically, it makes use of the following properties of superconductors: 1) the zero resistance and the Meissner effect, 2) the conservation and quantization of magnetic flux trapped in superconductors of multi-connected configurations, and 3) flux-pinning[11] by structural defects such as dislocations, grain boundaries, impurities, and vacancies, etc. To provide an effective shielding against small time-varying fields for frequencies up to 10^{10} Hz, only the infinite conductivity of a superconductor is needed. To reduce the ambient dc fields, the Meissner effect is crucial as long as the shield is significantly thicker than the London penetration depth and the external field is lower than the critical field of the shielding material. In principle, the flux quantization and the Meissner effect allow us to eliminate the magnetic field completely inside a superconducting shield. However, a complete Meissner effect has never been realized because of flux-trapping by various flux-pinning centers in real superconductors. To obtain an ultra-low field environment such as that achieved in Cabrera's experiment, flux-pinning in the shield should be reduced as much as possible. In the "large detection area" experiments, strong flux-pinning is highly desirable for stabilizing the ambient field. For this reason, type II superconductors such as Nb-Ti alloys, and A15 materials should be used for shielding. A theoretical discussion on using type II superconductors as shielding materials has been reported by Perepelkin and Minenko.[12]

With the considerations just discussed, a double shell, rigid superconducting shield system has been designed. As an example, the Pb shield arrangement used in a prototype detector at IBM is shown in Fig. 4. The outer Pb shield, mounted over a hollow glass cylinder, is 7 inches in diameter and 24 inches in height. This shield provides a shielding factor of about 10^5 against fluctuations in the ambient field. The second shield in which the detection coils are enclosed gives a further shielding of about the same order of magnitude. When the double shield is cooled slowly in an ambient field of about 10^{-2} G, most of the field is trapped in the shield and the trapped flux lines are strongly pinned in the shield to give enough long-term stability for monopole detection. Similar double, but completely closed shields have also been used successfully in the experiments at Texas A & M University,[13] the University of Chicago, and Fermilab.

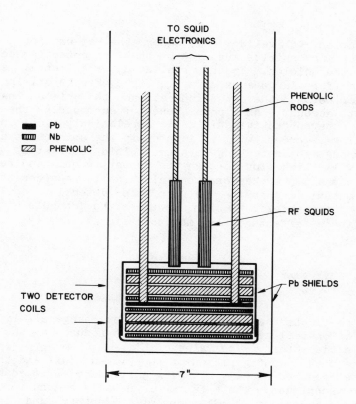

Fig. 4. A schematic cross-sectional view of the IBM prototype monopole detector. The two parallel, 6th derivative gradiometer detectors (10×10 cm^2 in size) are 1 1/4" apart.

It is pointed out by D. E. Morris[5] that a composite shield made of a non-superconductive metal with high conductivity such as Cu and Al coated with a superconducting film can have damping effects[15] which slow down the motion of the trapped flux line and hence increase the stability against flux jump. The time lag due to the damping effects also help to discriminate against spurious signals caused by flux jumps in the shields.

Detection Coils

When an n-turn detection coil (self-inductance $L_d \propto n^2$) is connected to the input coil (L_i) of a SQUID in a magnetometer configuration as shown in Fig. 3, the monopole signal (V_S) as a voltage output of the SQUID can be expressed by the following equation:

$$V_s \propto \frac{n \times (2\Phi_0)}{L_i + L_d} \tag{4}$$

This equation is valid only under the assumption that no superconducting shields are used or that the detection coil is far away from the shields. In any realistic experimental situation, especially for large area detection, the effects arising from the interaction between the detection coil and the shields has to be taken into account. As a consequence of surrounding the detection coil with superconductive shields in a relatively close proximity, a screening current will be induced in the shields as a monopole passes through the detecting system. This persistent screening current will link flux through the detection coil and reduces the monopole signal. Furthermore, one also should consider the effect of the mutual inductance between the detection coil and the superconducting shield, M which increases the monopole signal. To include these effects on the monopole signal, Eq. (4) is modified as follows:

$$V_s \propto \frac{n \times (2\Phi_0) \times F}{L_i + L_d'} \tag{5}$$

where F represents the fraction of the flux change $2\Phi_0$ linked into the detection coil after subtracting that induced by the screening current in the shield. The value of $L_d' = L_d - M$ only depends upon the geometrical configuration of the pick-up coil and the shield, and can be measured readily by using a standard ac impedance bridge or other techniques. The F factor is usually a much more dominant effect and depends on the monopole trajectory and the shield geometry. In certain simple cases, the F factor has been calculated. For instance, H. Frisch[14] has shown that for a shield and detection coil configuration such as that in Fig. 5(a), the

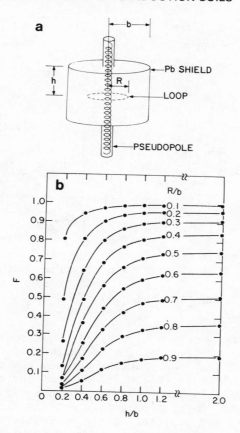

Fig. 5. (a) A superconductive shield and detector coil configuration. The "pseudopole" is a calibration solenoid (after H. Frisch). (b) The factor F, calculated from Eq. (6) for the case of z = 0, as a function of R; b and h.

fraction of the thin solenoid (so-called "pseudopole" which simulates a monopole trajectory along the cylindrical axis) flux detected by the loop is

$$F(R,b,h) = 1 - \left[\frac{R^2}{b^2} + \frac{2R}{b} \sum_{n=1}^{\infty} \frac{\cosh(K_n Z)}{\cosh(K_n h)} \frac{J_1(K_n R)}{X1_n[J_2(X1_n)]} \right] \quad (6)$$

where Z indicates the vertical position of the detector coil as measured from the center of the shield, $X1_n$ is nth root of J_1, the

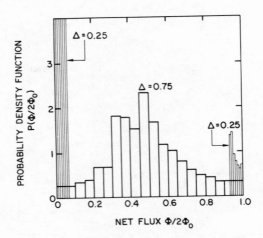

Fig. 6.   Probability density function, P $(\Phi/2\Phi_0)$ as a function of
          the net flux Φ in units of $2\Phi_0$ for a single coil monopole
          detector in an infinitely long cylindrical superconductive
          shield (after Park and Guy).

order 1 Bessel function, and $K_n = X_{1_n}/b$. Equation (6) is plotted in
Fig. 5(b) as a function of the dimensions of the shield and the
coil. It is clear that the coupling between the pick-up coil and
the shield is minimal (F > 0.95) only if the detector to shield
diameter ratio (R/b) is less than 0.2 and the aspect ratio for the
cylindrical shield (h/b) should be about one (i.e., h/b $\geq$ 1). The
requirement of R/b $\leq$ 0.2 points to the ineffectiveness of using a
detection coil as a magnetometer for a given shield size and to the
need of a more effective coil design.   Along a similar vein,
J. G. Park and C. N. Guy[16] have calculated the probability density
function, P (F), as a function of the net flux threaded through the
detection coil, Φ, in units of $2\Phi_0$ (i.e., $\Phi = 2F\Phi_0$) for a single
coil monopole detector in an infinitely long cylindrical shield.
Based on an isotropic distribution of incident monopoles, the
results of this calculation are shown in Fig. 6 for the cases of
the coil to shield diameter ratio Δ = 0.25 and 0.75. It is clear
that from this figure that $2\Phi_0$ is no longer a clear monopole
signature for the case of Δ = 0.75. In other words, when the coil
diameter is comparable with that of the superconducting shield, the
magnitude of the signal cannot be used to discriminate against
spurious signals.[17] To remedy this situation, Park and his
co-workers suggest that a compensating coil L_0 be inserted into the
detection circuit to convert it into an "asymmetric static pair"
(see upper figure in Fig. 7). The two coils (L_0 and L) have the
same area-turns product, couple equally to a uniform field but very
differently to incident monopoles.   Results for P($\Phi/2\Phi_0$) after
conversion for Δ = 0.75 are essentially reduced to those of the

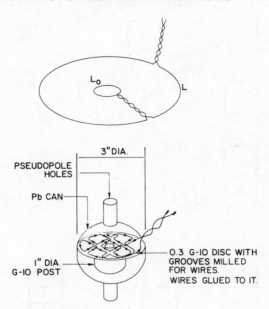

Fig. 7.   Asymmetric astatic pair (Park and Guy) and the "9-loop
          macramé (Frisch).

case Δ = 0.25 shown in Fig. 6. A solution proposed by the team of
University Chicago and Fermilab is shown in Fig. 7 (lower figure).
The detector, called a "9-loop macramé" was tested for several
weeks and was found to be insensitive to external fields. The
twisted nature of the loops in the detection coil cancel the
effects arising from the screening current in the shield. At IBM
Watson Research Center, a hierarchy of coplanar gradiometer coils[18]
has been developed to make full use of the area within a shield.
The pick-up coil consists of coplanar superconducting loops, wound
in opposite directions (+ and -), and connected in series. The
process of generating such arrays is illustrated in Fig. 8.   A
detector coil made of two loops of opposite senses (so-called
figure-8 configuration) can null out effectively the uniform-field
component of the external field and is therefore a planar first
derivative gradiometer (N = 1) with a spatial resolution of about
the size of the individual loops. The extension of this process to
higher order gradiometers as shown in Fig. 8 is quite
straightforward.   It is interesting to note that the induced
currents flowing between two adjacent loops with the same sign will
be cancelled with each other and consequently the wire between them
is not needed. This is an important feature of the coil geometry
which allows large area detection while providing low sensitivity
to external magnetic field and low self-inductance (i.e.,
relatively large signal) and remaining sensitive to local flux

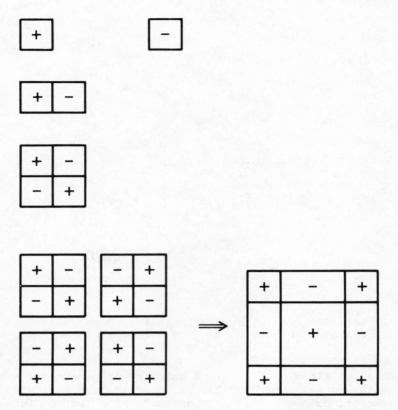

Fig. 8.   An illustration showing the process of generating a
          hierarchy of planar high-order gradiometers.

changes such as that caused by a monopole passing through any one
of the cells.   Examples of the planar gradiometers generated by
this process are shown in Fig. 9.   Details about the
characteristics and performance of such high-order planar
gradiometer detectors will be published elsewhere.[18]

Preliminary Results on the
Performance of a Prototype Monopole Detector

        As a preparation for large-area detection, several modest-size
(detector area ~100 cm^2) prototype detectors (either in the form of
a SQUID magnetometer or a SQUID gradiometer) have been constructed
and operated successfully in a relatively high magnetic field (~10
to 500 mG) environment. The overall signal-to-noise ratio obtained
in these experiments is about 10 to 20 at a bandwidth of 1 Hz. In
the following, some preliminary results on the IBM detector will be
briefly described.   A 6th derivative gradiometer configuration

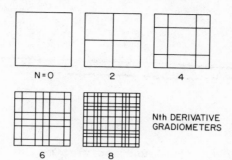

Fig. 9. Nth order derivative planar, square gradiometers generated
         by the process illustrated in Fig. 8.

(N = 6 in Fig. 9) was used for the detector coils which were wire
wound using 5 mil Nb wires and were glued to grooves on phenolic
plates. The outer length of the square coils was d = 10 cm, and
the dimensions of the basic unit cells were $\ell \times \ell$ = 1.25 cm $\times$
1.25 cm. The inductance of these coils was found to be 3.4 µH,
allowing an effective matching with the SQUID input-coil inductance
(L_i = 2 µH). As far as the undesirable coil-shield coupling is
concerned, the use of a high-order gradiometer coil reduces the
coil to shield diameter ratio Δ from d/D to an equivalent ratio of
ℓ/D (a gain of $2^{N/2}$ which translates into a gain in detector area
of about 2^N for a given diameter D of the shield).

    A monopole detector using a high-order gradiometer coil,
operated in a well-shielded low field environment can and did
detect flux jumps which mimic a monopole event. Flux jumps can be
caused by any thermal, mechanical or electrical disturbances in the
superconducting shields surrounding the detector coils, the SQUID
input coil and the SQUID itself. A flux change of the order of
$10^{-3}\Phi_0$ in the SQUID loop (Fig. 3) is sufficient to produce a
"monopole-like" step in the output of the SQUID. To safeguard
against such spurious signals, coincidence detection between at
least two independent detectors is definitely required. Two such
coincident SQUID gradiometer (N = 6) detectors were used in the
preliminary experiment (see Fig. 4). The Nb and Pb plates between
the detector coils were to assure the independence of the coils. A
trace of the SQUID output as a function time is shown in Fig. 10,
giving an S/N ratio of about 10 at 1 Hz bandwidth. As an attempt
to intercept all possible monopole trajectories, a coincident
arrangement for two planar, and four cylindrical high-order
gradiomenter coils is shown in Fig. 11. This design takes
advantage of the fact that the mutual inductance M between
orthogonal coils is zero and that M between the cylindrical
detectors is also very small.

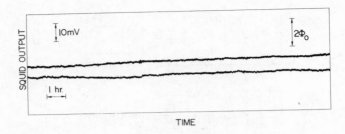

Fig. 10. SQUID output as a function of time for the two parallel
         gradiometer detector coils shown in Fig. 4.

Multiplexing

The main advantages of using a scheme of multiplexing for
monopole detection can be listed as follows:

1. It permits large-area detection without suffering from a
   serious deterioration in the S/N ratio.

2. The SQUIDs used for interrogating the detector coils can be
   turned off when not in use. This significantly reduces the
   chances of observing a flux jump at the input coil or at the
   SQUID loop.

3. More ambitiously, this technique, in addition to coincidence
   detection, allows a determination of the monopole trajectory
   and even velocity.

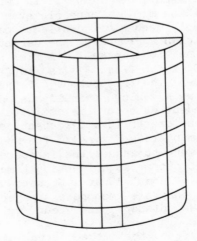

Fig. 11. A proposed cylindrical monopole detector using high-order
         gradiometer coils.

4. Possibly, it also can reduce the number of SQUIDs needed for a
    given area of detection and hence can cut the cost of the
    measurement.

       A proposed multiplexing scheme is schematically shown in
Fig. 12. The detection coils in the form of a long series of
twisted superconducting loops are coupled with the input coil of
the rf SQUID through a flux transformer. The normally closed
superconducting switches dictate the sequence of interrogation.
These switches can be opened by means of resistance or laser
heating. The action of switching ideally should induce no flux
jumps in the detection system. The mutual inductance between two
adjacent coils can be minimized by a staggered arrangement as shown
in Fig. 12. Two orthogonal sets of such detectors on a planar
substrate can provide spatial information about a monopole event.
Three or more such planar detecting systems operating in parallel,
in principle, is capable of determining monopole trajectories.

       B. Deaver[19] (University of Virginia) has considered a monopole
detector consisting of a two dimensional array of small
superconducting loops on a wafer. Each of these loops might be a
SQUID, or the pick-up loop of a SQUID, so that the passage of a
monopole is recorded by the persistent current induced in the loop.
The SQUID can provide both a pulse for timing and a measure of the
magnitude of the flux change. Two parallel wafers separated by a
few centimeters will make possible a velocity determination by
determining the time of flight between the two detecting loops and
positions of the two loops that were triggered. A third parallel
wafer would serve to verify the path. From knowledge of the
direction of travel and the sense of the induced current in the
loop, the sign of the monopole can be obtained. For details of
multiplexing, the reader is referred to the author of this design.

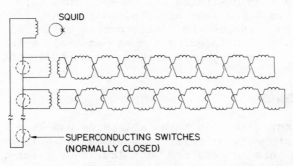

Fig. 12. A schematic diagram for a multiplexing scheme for
          monopole detection.

SUMMARY

Cabrera's monopole candidate event has generated interest in research on induction-coil detection of monopoles. These new studies address themselves to problems arising from the requirements of large-area detection and reliable discrimination against spurious signals. Techniques of using normal-state metallic coils have been proposed. The key problem associated with this approach is to reduce the Johnson noise due to the coil resistance. Theoretical estimates indicate that a S/N ratio of about 10 to 15 can be achieved if the detection coil is in a well-shielded, low temperature (4.2 K) environment. The technique of using a SQUID to monitor the change of magnetic flux threaded through a ·superconductive coil, in principle, can provide a S/N ratio of about 10^4 for a detector coil with a size of $10^4 cm^2$. The quantum interference effects from which this superior flux sensitivity is derived also makes this detection method vulnerable to spurious signals due to flux jumps. More work, theoretical and experimental, is needed on flux motion and flux jump in conjunction with SQUID detection of monopoles. The importance of magnetic shielding, and the coil-shield interaction to the performance of large-area detectors has been recognized. The use of multiple, rigid superconducting shields to stabilize the ambient magnetic field (10 to 500 mG) represents a viable alternative to Cabrera's impressive ultra-low field ($\sim10^{-8}$G) approach. In terms of reducing the probability of a flux jump, the advantage of starting with a relatively low ambient field (say 10^{-5} to 10^{-6} G), however, cannot be overemphasized. Low field of this magnitude can be achieved over a large region, preferably by μ-metal shielding. Various designs of the superconductive detector of the superconductive detector coils in the form of coplanar gradiometers have been tested in several prototype detectors (detection area ~100 cm^2) to reduce the effects of screening currents in the superconducting shields and to make full use of the available detection area in a shield. A signal-to-noise ratio of about 10 to 30 has been experimentaly demonstrated in these prototype detectors. To safeguard against spurious signals, coincidence detection between at least two independent detectors is essential. The topics on detection using a muliplexing scheme has been briefly touched in this article. Clearly, more work is required in this area.

ACKNOWLEDGEMENTS

The author wishes to thank all those who have kindly provided him with information prior to publication. In particular, he is grateful to his co-workers at IBM, C. Tesche, J. Chi and P. Chaudhari for many useful discussions.

REFERENCES

1. Blas Cabrera, Phys. Rev. Lett. 48, 1378 (1982).
2. G. Giacomelli, "Experimental Status of Monopoles" in these
    proceedings.
3. Edwin E. Salpeter, Stuart L. Shapiro, and Ira Wasserman, Phys.
    Rev. Lett. 49, 1114 (1982), and references therein.
4. S. Dimopoulos, S. L. Glashow, E. M. Purcell and F. Wilczek,
    Nature 298, 824 (1982).
5. D. E. Morris, a preprint on "A Non-Superconducting Induction
    Detector for Magnetic Monopoles Using an FET Amplifier."
6. M. J. Price, a preprint on "The Detection of Cosmic Magnetic
    Monopoles Using a Room Temperature Coil."
7. L. J. Tassie, Nuovo Cimento 38, 1935 (1965).
8. T. Van Duzer and C. W. Turner, "Principles of Superconductive
    Devices and Circuits" (Elsevier, 1981).
9. W. O. Hamiton, Revue de Phys. Appl. 5, 41 (1970).
10. S. I. Bondarenko, S. S. Vinogradov, G. A. Gogadze, S. S.
    Perepelkin, and V. L. Sheremet, Sov. Phys. Tech. Phys. 19,
    824 (1974); S. I. Bondarenko, V. I. Sheremet, S. S.
    Vinogradov, and V. V. Ryabovol, Sov. Phys. Tech. Phys. 20,
    73 (1975).
11. R. P. Huebener, "Magnetic Flux Structures in Superconductors"
    (Springer-Verlag, 1979).
12. S. S. Perepelkin and E. V. Minenko, Sov. Phys. Tech. Phys. 21,
    1422 (1976).
13. W. P. Kirk and H. Armbruster, a preprint on "Monopole
    Search--A Superconducting Induction Technique at Texas
    A & M University."
14. H. J. Frisch, a preprint on "Superconducting Magnetic Monopole
    Detector Development."
15. R. B. Harrison, J. P. Pendrys, and L. S. Wright, J. Low Temp.
    Phys. 18, 113 (1975).
16. J. G. Park and C. N. Guy, a preprint on "Inductive Monopole
    Detectors of Coil-Shield Area Ratio ~1: The Signal Size
    Distribution."
17. In a preprint entitled, "Unambiguous Detection of Monopoles,"
    Timir Datta points out that the original multiply quantized
    vortex $(2\phi_0)$ created by a monopole in the shield is
    energetically unstable and will decay into 2 stable
    vortices each enclosing one fluxon ϕ_0. This effect could
    change the results of the above calculations. In real
    superconductors, the flux-pinning force is probably strong
    enough to make this effect insignificant.
18. C. D. Tesche, C. C. Chi, C. C. Tsuei, and P. Chaudhari, a
    preprint on "An Inductive Monopole Detector Employing
    Planar High Order Superconducting Gradiometer Coils."
19. B. Deaver (private communication).

ACOUSTIC DETECTION OF MONOPOLES

B. C. Barish

Lauritsen Laboratory
California Institute of Technology
Pasadena, California 91125

INTRODUCTION

Physicists have long been interested in the existence (or non-existence) of magnetic monopoles. Historical interest stems from the symmetry between electric and magnetic fields in Maxwell's equations, and from the lack of abundant free magnetic charges compared to electric charges.

In 1931, Dirac showed that the existence of free magnetic charges (Dirac monopole) could provide reason for quantization of electric charge ($n\hbar/c$).[1,2] This motivated renewed interest in searching for monopoles, however subsequent searches for such monopoles were negative.

More recently, 't Hooft and Polyakov[3,4] showed that monopoles exist as solutions in many non-abelian gauge theories. The possibility of these GUT monopoles provided stimulus for much recent theoretical and experimental interest in the subject. The fact that Grand Unified Monopoles are extremely heavy ($M \sim 10^{16}$ GeV) makes most previous searches irrelevant. The fluxes in cosmic rays are expected to be extremely small, and experimental techniques required for detection are different.

Experimentally, there is a major challenge to develop techniques capable of detecting these heavy monopoles. The work described in this paper is responsive to the possibility that very large area detectors ($\geq 10^4 m^2$) may be required. One possibility to consider for such large areas is the detection of the acoustic signal resulting from eddy current losses in a conductor.

FLUX OF GUT MONOPOLES

Estimates of the fluxes of GUT monopoles in cosmic rays require understanding of the production mechanism, acceleration ($W = g_0 Bl = 2$ MeV/Gauss), gravitational dynamics, annihilations, trapping, etc. These have not been worked out in any detail due to the various uncertainties in the calculations. The best that can be done with our present knowledge is to place some <u>constraints</u> on the presence of GUT monopoles in cosmic rays.

The constraints come from two sources: (1) limits on the total mass of the Universe, and (2) the existence of Galactic magnetic fields.

Mass of the Universe

The most straightforward astrophysical limit comes from assuming that magnetic monopoles account for most of the mass of the Universe.[5]

The mass density contained in galaxies ρ_G accounts for about .02 ρ_c, the critical density to close the Universe. This implies that the number of monopoles,

$$n_m \leq \frac{10 \times [(\text{Galactic Mass/Monopole Mass})]}{(\text{Dist between Galaxies})^3}$$

$$n_M \sim 4 \times 10^{-20} \text{ cm}^{-3}$$

For comparison, the number of nucleons is

$$n_n \sim 4 \times 10^{-6} \text{ cm}^{-3}$$

and,

$$n_M/m_N \leq 10^{-14} \frac{\text{monopoles}}{\text{nucleon}}$$

From this, one obtains a flux limit

$$F \leq 5.4 \times 10^{-18} \text{cm}^{-2} \text{sr}^{-1} \text{sec}^{-1} \text{m}_{19}^{-1} \left[\frac{v}{10^{-3}c} \right]$$

A more optimistic flux limit is obtained using the fact that monopoles could cluster like mass in our galaxy ($10^{12} \odot$ within 30 kpc). This implies a flux limit of

$$F_c \leq 3 \times 10^{-13} \text{cm}^{-2} \text{sr}^{-1} \text{sec}^{-1} m_{19}^{-1} \left[\frac{v}{10^{-3}c} \right]$$

Survival of Galactic Field

If galactic fields are due to persistent currents, $(\nabla \times B_{gal} \neq 0)$, monopoles move along field lines and gain kinetic energy at the expense of the field. In order for the fields to survive, the field energy cannot be dissipated more rapidly than currents can be regenerated by dynamo action ($t_{re} \simeq 10^8$ yr). This requirement places a limit on the fluxes, usually called the Parker Bound.[6]

$$F \leq 10^{-16} \text{cm}^{-2} \text{sr}^{-1} \text{sec}^{-1} \qquad \text{(Parker Bound)}$$

This bound has been reexamined by Turner, Parker, and Bogdan using the monopole mass, velocity distributions, etc. They obtain a less restrictive bound for $M_M > 10^{16}$ GeV. Combined with the flux bounds from the mass-density of the Universe

$$F_{max} \leq 10^{-12} \text{cm}^{-2} \text{sr}^{-1} \text{sec}^{-1}$$

for $M_M \sim 10^{19}$ GeV, $\beta \sim 3 \times 10^{-3}$, and clustering.

Experimental Status

Cabrera has reported one possible event from an experiment using superconducting rings connected to SQUID magnetometers.[7] This technique is sensitive to a monopole of any β. Including the newest data, this one event corresponds to a limit,

$$F \leq 1.2 \times 10^{-10} \text{cm}^{-2} \text{sr}^{-1} \text{sec}^{-1}$$

or a rate of $F \sim 6 \times 10^{-11} \text{cm}^{-2} \text{sr}^{-1} \text{sec}^{-1}$. This event, if confirmed, may correspond to a rate much higher than the astrophysical bounds.

This has led to various speculations to explain a rate higher than these bounds. Two possible explanations have been discussed: (1) Monopoles are the source of the Galactic magnetic field. If monopoles are the source of the field, kinetic energy could be transferred to the mcnopoles by magnetic plasma oscillations.[6] If so, previous limits would not apply, and (2) local sources of monopoles. This possibility results from the idea that monopoles could be captured in the field of the sun. In this case, the flux could be enhanced, however the velocity would be typically slower ($\beta \sim 10^{-4}$, instead of $\beta \sim 10^{-3}$ for galactic monopoles).

Obviously, the Cabrera event needs confirmation and vigorous work is proceeding in that direction. Local sources would also explain negative results from scintillator experiments. Scintillator experiments with at least 10 times the sensitivity have reported negative results, however scintillators probably would not respond to such slow monopoles.

The experimental situation appears to have narrowed down to two directions. First, a number of experiments are underway using superconducting rings, etc., that should be capable of confirming unambiguously a flux of monopoles of any β near the Cabrera rate. The second direction is to build large area detectors which have capabilities for detection of galactic monopoles ($\beta \sim 10^{-3}$). The use of plastic scintillators for these large area detectors is the most obvious solution. Work is proceeding both empirically using various scintillator arrays and theoretically in understanding the response of scintillator to slow monopoles.[8]

Looking forward to the possibility of building very large arrays, the alternatives to plastic scintillator need to be carefully considered. The work underway on acoustic detection could lead to an alternative for a very large area array.

MECHANISMS FOR ENERGY LOSS

A monopole passing through matter will lose energy by ionization loss. The amount of energy loss for charged particles (or monopoles) is well understood for relativistic particles. However, the ionization loss for very slow particles is much less understood. Discrepancies in the various calculations for slow monopoles ($\beta \sim 10^{-3}$) have plagued the interpretation of the present generation of scintillator experiments and evaluation of scintillator for a future large array. Much work is presently being done to clarify this situation.

A second mechanism for energy loss exists for monopoles passing through conductors. The energy loss by this mechanism has been calculated by Martem'yanov and Khakimor[9] and more recently by Ahlen and Kinoshita.[10] Ahlen obtains

$$\frac{dE}{dx} = \frac{4\pi^2 N_e e^2 g^2}{m_e c^2} \frac{\beta c}{v_F} \simeq 1 \text{ GeV/cm}$$

for aluminum and $\beta \sim 3 \times 10^{-3}$. This is considerably larger than ionization loss, but also varies as β making detection more difficult for very slow monopoles.

PROTOTYPE DETECTOR FOR THERMAL-ACOUSTIC SIGNALS

Design

Introduction. At Caltech, we have built a prototype detector to study empirically the problems of detecting monopoles by acoustic techniques. The present detector is meant to be a test device to study sensitivity and not a design for a "real" monopole detector. This prototype monopole detector consists of two aluminum disks which are 144 cm in diameter and 10 cm thick. Six transducers are coupled to each disk to detect ultrasonic pulses. These transducers are sensitive from 8 to 12 MHz and were developed specifically for this project. Two scintillator arrays are mounted on the disks, one between the disks and one above them, to provide triggering information and a cosmic-ray veto (see Fig. 1).

Three properties are required of the monopole in order for it to fit our detector scheme. It must be very massive, around 10^{16} GeV; it must be slow, with a characteristic velocity of $10^{-3}c$; and its magnetic charge must be a multiple of $137e/2$. This monopole would be very penetrating with dE/dx due to ionization in nonconductors smaller than minimum ionizing. The case when dE/dx is not so small is due to eddy currents when the monopole travels through a conductor. Here the electromagnetic coupling to the free electrons is much stronger, and dE/dx has been calculated to be about 1 GeV/cm and varies linearly with β.[11,12] This energy is then transferred to the aluminum lattice in the form of a thermoacoustic pulse. The wavelength of this pulse is determined by the properties of the aluminum disk once the pulse has traveled a distance that is long in comparison to the monopole to electron impact parameter.[13] This wavelength has been calculated to be .6 mm, which corresponds to a frequency of 10 MHz in aluminum. The pulse amplitude at the transducer should be about $10^{-2}dyne/cm^2$, which translates to a pulse of ~.05 μV out of the transducer. This estimate allows for a factor of three attenuation due to dispersion and self absorption in the aluminum and interface losses. The transducer electromechanical coupling coefficient has been taken to be 0.7 (see Table 1).

Apparatus. Aluminum was chosen for the detector because it transmits ultrasonic pulses with minimal attenuation and is also a good conductor. To insure good sound transmission the aluminum alloy chosen was 2124 T851, which is one of the hardest alloys available. The alloy has been worked to minimize grain size and therefore the sound dispersion.

At six equally spaced locations around the edge, the disk is polished flat. At three places 120 degrees apart a pair of transducers are mounted. At the other three, transducers to put in test pulses are coupled to measure the sensitivity of the receiving transducers as a function of time.

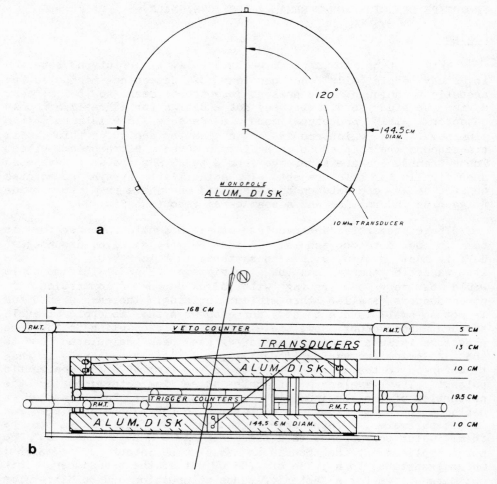

Fig. 1   (a) The CALTECH prototype monopole detector uses aluminum
         disks to detect the monopole acoustically. Two disks are
         used which are 144 cm in diameter and 10 cm thick.   The
         disks   are   separted   by ≈ 20 cm.   10 MHz piezoelectric
         transducers are coupled to   the aluminum   to detect   the
         acoustic pulse.   These transducers are mounted in pairs on
         the edge of the disk at three locations spaced 120 degrees
         apart.   Sending transducers are mounted opposite each
         receiving pair to provide calibration. Both the surface
         of the disk at the mounting site, and the aluminum face of
         the transducer are polished. The transducers are coupled
         to the disk with phenyl salicylate.

         (b) Two scintillator planes are included in the prototype.
         One is mounted between the disks and acts as a monopole
         trigger.   The other is mounted above the array and is used
         as a cosmic-ray veto.

Table 1. Physical Constants for 2124 Aluminum

Longitudinal Sound Velocity	v_l	6.3	$10^5\, cm/s$
Long. Acoustic Impedance	Z_l	17.3	$10^6\, kg/m^2 s$
Conductivity	σ_c	2.2	$10^5\, mho/cm$
Specific Heat	C_m	.215	$cal/g°K$
Thermal Expansion Coeff.	β	2.32	$10^{-5}/°K$
Young's Modulus	E	1.02	$10^{11} N/m^2$
Poisson's Ratio	σ	.355	
Density	δ	2.77	g/cm^3
Wavelength of 10 MHz Longitudinal Waves	λ_{10MHz}	.6	mm

To maximize the acoustic transmission from the disk to the piezoelectric ceramic in the transducer it was necessary to redesign the transducers that are commercially available. The standard 10 MHz transducer has a ruby face plate bonded to the ceramic piezoelectric material with a soft backing that helps dampen ringing. The transducer that is now in use has an aluminum face plate that is polished on both sides to improve sound transmission and is 0.6 mm thick. These changes improved the sensitivity of the system by a factor of five over the standard design transducers.

Electronics. The impedance characteristics of the transducer made the design of the preamplifier a difficult problem. The transducer output impedance can be characterized by a resistor and capacitor in parallel, with the resistor value less than 50Ω. To match this low impedance while maintaining both low noise and high frequency characteristics the preamp has three separate stages. The input stage consists of four bipolar transistors in parallel that effectively match the transducer while keeping the noise at the 5 microvolt level. The next stage is a tuned RF amplifier. The preamp is followed by two more amplifiers so that the entire gain is 10^4.

Trigger logic. The detector consists of two scintillator planes and two aluminum disks. The uppermost scintillator is used to veto cosmic rays. The second scintillator is a monopole trigger which starts a gated clock. The clock is fanned into three channels of an octal scalar with separate stops. Pulses from each transducer pair are used as stops for one of the channels. Because it is not known whether a monopole can be detected in scintillator, an alternate trigger method has been devised. In this method, the transducer pair closest to the monopole track acts as the trigger and the measured distances recorded by the other two channels are offset by the distance from the track to the closest (triggering) transducers.

Efficiency for Thermal Acoustic Pulse, and Characteristics

Signal from monopole. We have discussed the energy loss of a magnetic monopole passing through matter and know that it is still controversial. However, if the energy loss is known, the subsequent process of how the energy goes into heat and sound is understood.

At the starting point, we don't know how important heat conduction is, so we'd better consider it and have the equation

$$\frac{\partial T(x,t)}{\partial t} = \frac{K_H}{C_P} \nabla^2 T(x,t) + \frac{1}{C_P} Q(x,t) \tag{1}$$

where $T(x,t)$ is the temperature distribution at time t, K_H is the heat conductivity, C_p is the specific heat, and $Q(x,t)$ is defined such that $Q(x,t)dVdt$ is the energy deposited in volume element dV during a time interval dt.

In ordinary elastic theory, the relationship between strain e_{ij} and stress σ_{ij} for uniform isotropic material is

$$\sigma_{ij} = K\theta\delta_{ij} + 2\mu(e_{ij} - \frac{\theta}{3}\delta_{ij}) \tag{2}$$

where $\theta = e_{11} + e_{22} + e_{33}$ and K and μ are elastic contants related to Young's modulus E and Poison's ratio σ by

$$K = \frac{E}{3(1-2\sigma)} \text{ and } \mu = \frac{E}{2(1+\sigma)} \tag{3}$$

When there is some temperature change T, there will be thermal expansion $T\beta$ in every direction, where β is the linear thermal expansion coefficient. So a strain $T\beta\delta_{ij}$ is added to the original strain if the stress is kept unchanged. Replacing e_{ij} by $e_{ij} - T\beta\delta_{ij}$ in Eq. (2) we obtain a modified strain-stress relation

$$\sigma_{ij} = K(\theta-3\beta T)\delta_{ij} + 2\mu(e_{ij} - \frac{\theta}{3}\delta_{ij}). \tag{4}$$

With this modification, the equation for a longitudinal wave becomes

$$\rho \, \frac{\partial^2 \theta(x,t)}{\partial t^2} = (K + \frac{4}{3}\mu) \, \nabla^2 \theta(x,t) - 3\beta K \nabla^2 T(x,t) \tag{5}$$

where ρ is the density of the medium.

Making Fourier transforms of Eqs. (1) and (5) and eliminating $T(x,\omega)$ from them, one obtains

$$(\nabla^2 + b^2)(\nabla^2 + a^2) \, \theta(x,\omega) = -\nu\nabla^2 Q(x,\omega) \tag{6}$$

where

$$a^2 = \frac{i\omega C_P}{K_H} \qquad b^2 = \frac{\omega^2 \rho}{K+4/3\mu} = \frac{\omega^2}{V_s^2} \qquad \nu = \frac{3\beta K}{(K+4/3\mu)K_H}$$

and V_s is the velocity of sound.

By the method of Green's functions, the solution of Eq. (6) is found to be

$$\theta(x,\omega) = \frac{\nu}{4\pi(a^2-b^2)} \int Q(x',\omega) \, \frac{a^2 e^{ia|x-x'|} - b^2 e^{ib|x-x'|}}{|x-x'|} \, dx'$$

a has an imaginary part so $e^{ia|x-x'|}$ dies away at large values of $|x|$ thus

$$\theta(x,\omega) = \frac{-\nu b^2}{4\pi(a^2-b^2)} \int Q(x',\omega) \, \frac{e^{ib|x-x'|}}{|x-x'|} \, dx' \tag{7}$$

is good enough.

Equation (7) is quite general and can be used for any form of $Q(x,\omega)$, but the integration is not easy to carry out. However, for a range of ω, an approximation may be made.

To get some idea of the size of the "hot spot" or the size of $Q(x,t)$ within which the contribution to the total energy deposition is important, we   estimate as   follows: the   depth   that electromagnetic wave can penetrate into metal is

$$\delta \approx \sqrt{\frac{C^2}{\omega \sigma_c}}$$

where C is the speed of light and σ_c is the conductivity of the metal. Then we note that near the monopole track the major frequency is

$$\omega \approx \frac{V_m}{r}$$

where V_m is the velocity of the monopole and r is the distance from the monopole. By setting $\delta = r$, we get the size of the hot spot.

$$\delta \approx \frac{C^2}{V_m \sigma_c} \approx 10^{-5} \text{ cm}$$

which is 1/1000 of the wavelength of 10 MHz sound wave. It can be shown that when the size of the hot spot is much smaller than the wavelength, it is a good approximation to treat $Q(x,t)$ as a delta function.

$$Q(x,t) = \delta(x)\delta(y)\delta(z - V_m t)V_m \left(\frac{dE}{dz}\right) \tag{8}$$

where (dE/dz) is the energy loss per unit length of monopole track, V_m is the velocity of monopole, and we have assumed that the monopole moves along z-axis. This gives

$$Q(x,\omega) = \frac{1}{2\pi} \left(\frac{dE}{dz}\right) \delta(x)\delta(y)e^{iwz/V_m} \tag{9}$$

Substitution of Eq. (9) into (7) gives a Hankel function and by using an asymptotic expansion of the Hankel function for large argument one gets

$$\theta(x,\omega) = \frac{3i}{8\pi} \frac{dE}{dz} \frac{\beta K \omega \sqrt{(\pi/2r|\omega|}\sqrt{V_s^{-2}-V_m^{-2}}}{V_s^2 C_P(K+(4/3)\mu)(1+i\omega K_H/V_s^2 C_P)} e^{i\omega(z/V_m+r\sqrt{V_s^{-2}-V_m^{-2}})\pm i(\pi/4)} \tag{10}$$

where $r = \sqrt{x^2 + y^2}$; "$\pm$" is "+" when ω is positive and "−" when ω is negative.

The heat conduction gives a term $(i\omega K_H/V_s^2 C_p)$ in the denominator, otherwise we would have some trouble of divergence when we transform back to $\theta(x,t)$. However, for aluminum at $\omega = 10^7 \sec^{-1}$ this term is only about 10^{-5} so we see that in the frequency region we are able to detect, the heat conduction has no effect and can be neglected.

By neglecting heat conduction and also by noting $V_m \gg V_s$, Eq. (10) is further simplified. Noting that the pressure is simply $P = E\theta$, we finally get a simple formula

$$P(x,\omega) = \frac{\pm i}{B\pi} \frac{\beta}{C_m} \left(\frac{1+\sigma}{1-\sigma}\right) \left(\frac{dE}{dz}\right) \sqrt{\frac{\pi V_s |\omega|}{2r}}\, e^{i\omega(z/V_m + r/V_s) \pm i\pi/4} \qquad (11)$$

where C_m is the specific heat of unit mass while C_p is that of unit volume.

Assuming the bandwidth of the transducer and amplifier is $\Delta\omega = \omega_{HIGH} - \omega_{LOW}$ then the effective pressure the transducer responds to is

$$P(x,t) = \int_{\omega_{LOW}}^{\omega_{HIGH}} e^{-i\omega t} P(x,\omega)\,d\omega + \int_{-\Omega_{HIGH}}^{-\omega_{LOW}} e^{-i\omega t}\, P(x,\omega)\,d\omega \qquad (12)$$

At the time $t = z/V_m + r/V_s$ all the frequency components have the same phase. This corresponds to the signal peak, so the peak pressure is

$$P_{peak} = \sqrt{2} \int_{\omega_{LOW}}^{\omega_{HIGH}} |P(x,\omega)|\,d\omega \quad \frac{1}{8\pi} \left(\frac{1+\sigma}{1-\sigma}\right) \sqrt{\frac{\pi V_s \omega}{r}} \left(\frac{dE}{dz}\right) \Delta\omega \qquad (13)$$

Using the physical constants for 2124 aluminum, and assuming $(dE/dx) = 1$ GeV/cm with $\omega = 2\pi \times 10^7 \sec^{-1}$, $\Delta\omega = 3 \times 10^7$, $r = 100$ cm, we get

$$P_{peak} = 1.2 \times 10^{-2} \text{ dynes/cm}^2 \qquad (14)$$

Akerlof has calculated the thermoacoustic signal of a slow magnetic monopole as[14]

$$P_{peak} \approx \frac{\beta}{C_m} \left(\frac{dE}{dz}\right) \frac{1}{\Delta t^2} \sqrt{\frac{\Delta t}{t}} \qquad (15)$$

where Δt is the width of the pulse and t is the time for signal to reach the transducer. This equation agrees with Eq. (13) if $t \approx r/V_s$ and $\Delta t \approx 1/\omega \approx 1/\Delta \omega$.

Sources of acoustic noise. Considering the very small signal, to evaluate this technique, it is important to understand where the technique will be fundamentally limited due to acoustic noise. In order to minimize this problem we have chosen to work at high frequency (~10 MHz). This is about the highest frequency where the acoustic pulse will pass through the aluminum with small absorption.

The primary frequency of the acoustic pulse generated when a monopole passes through the medium is much higher (>100 MHz); however, that frequency is absorbed with some of the energy becoming thermal and some re-emitted at lower frequency. Below ~15 MHz, the acoustic pulse will pass through the medium. This inefficiency of turning the energy loss into an acoustic pulse is the fundamental limitation in this method. We have estimated that the final signal (calculated in the previous section) is comparable or greater than the acoustic noise from the aluminum medium itself. This has encouraged us to pursue the technique to determine where it will ultimately be limited.

At present, the real limit should come from the noise in the transducer itself. In our present detector, transducers, and electronics, we expect this noise to be about 50 times greater than the monopole signal. We are empirically determining this level and studying various methods such as focusing the acoustic pulse for improving this S/N ratio.

A study of the intrinsic limit due to acoustic noise has been made by Akerlof[14] with the conclusion that this technique is practically unfeasible. We have not become convinced of this limit, as witnessed by the experimental work proceeding at Caltech. We agree that our present detector will not be able to reach the necessary sensitivity; however, we hope to determine whether it will be possible with various improvements, such as acoustic focusing, different choice of materials (possibly magnetostrictive or ferromagnetic), working at low temperature, and using improved acoustic detectors.

For the purpose of this review and objectivity, we repeat below Akerlof's main arguments.

Intrinsic limits for acoustic detection: (Argument due to C. Akerlof[15]) "The basic formalism for computing an acoustic displacement function is given by Eqs. (28) and (29) of Akerlof.[14] As a source function we will use the following piecewise quadratic form:

$$
s(t) = \begin{cases}
0, & t<0 \\[2mm]
t^2, & 0\leq t<\frac{1}{4}\Delta t \\[2mm]
-(t-\frac{1}{2}\Delta t)^2+\frac{1}{8}\Delta t^2, & \frac{1}{4}\Delta t\leq t<\frac{3}{4}\Delta t \\[2mm]
(t-\Delta t)^2, & \frac{3}{4}\Delta t\leq t<\Delta t \\[2mm]
0, & \Delta t\leq t
\end{cases} \tag{16}
$$

With this choice, s(t) is a reasonable model for an impulse function and meets the requirement that the first derivative be everywhere continuous.

"The displacement function $u_r(r,t)$ is derived from s(t) by application of the Green's-function method:

$$
u_r(r,t) = \frac{a_0}{r}[f(r,t)-2f(r,t-\frac{1}{4}\Delta t)+2f(r,t-\frac{3}{4}\Delta t)-f(r,t-\Delta t)] ,
$$
$$\tag{17}$$
$$
f(r,t) \equiv \left[ct\sqrt{c^2t^2-r^2} + r^2\ln(ct-\sqrt{c^2t^2-r^2}) -r^2\ln(r)\right] \theta(ct-r).
$$

The amplitude of the displacement function, a_0, is chosen to correspond to the maximum thermal expansion of the medium at the time all of the energy has just been absorbed. Thus, a_0 is fixed by the condition

$$
\frac{\Delta V}{\ell} = \frac{\beta}{c_p\rho}\frac{dE}{dx} = 2\pi r u_r (r,\Delta t)\big|_{max} \tag{18}
$$

The maximum for the right-hand side of Eq. (18) occurs for r = 0.6673 cΔt, which yields the following expression for a_0:

$$
a_0 = 0.657 \frac{(\Delta V/\ell)}{c^2\Delta t^2} . \tag{19}
$$

"Assuming no attenuation, the acoustic signal will propagate radially outwards indefinitely,...with amplitude decreasing as $1/\sqrt{r}$. The total kinetic energy in this pulse can be evaluated from

$$E = \frac{\rho}{2} \int \left(\frac{\partial u}{\partial t}\right)^2 2\pi r \ell dr \ , \tag{20}$$

where ℓ is the length of the track in the medium. This integral can be computed analytically:

$$E = \frac{3\pi}{2} \ln\left(\frac{27}{16}\right) \rho \ell a_0^2 c^4 \ \Delta t^2 = 1.064 \ \frac{\ell}{\rho \Delta t^2} \left(\frac{\beta}{c_p} \frac{dE}{dx}\right)^2 \ . \tag{21}$$

"In the initial epoch of the acoustic-wave development, the energy is deposited in a very short time within a radius of the order of 1000 Å of the particle track. The frequency of the acoustic pulse is correspondingly high, much higher than can be easily transported by the medium. Thus the initial acoustic waves are continually reabsorbed, heating an ever larger cylindrical core. As a result, the general behavior of the acoustic pulse will be similar to the pulse shape described by Eq. (17), but shifted up in wavelength. This rescaling will continue until the attenuation length of the characteristic frequency is comparable to the physical dimensions of the detector.

"The spectral structure of the initial acoustic pulse can be computed analytically in the limit that the radial distance r is much greater than the wavelength:

$$\tilde{u}(\omega) = \lim_{r \to \infty} \frac{1}{\sqrt{2\pi}} \int_{-\infty}^{\infty} \sqrt{r} \ u_r(r,t) \ e^{i\omega t} dt \tag{22}$$

$$= 8a_0 \ c^{3/2} \omega^{-5/2} \sin^2 \left(\frac{1}{8}\omega \Delta t\right) \sin\left(\frac{1}{4}\omega \Delta t\right) e^{i(1/2)\omega \Delta t + i(3/4)\pi} \ .$$

The total kinetic energy is related to the spectral distribution by

$$E = \pi \rho c \ell \int_{-\infty}^{\infty} \omega^2 \tilde{u}(\omega) \tilde{u}(-\omega) d\omega \tag{23}$$

"For our purposes, the important issue is the fraction of the energy available at frequencies much less than $1/\Delta t$.... As can be seen from Eqs. (22) and (23), the fraction of the available energy at low frequencies scales as ω^4. This behavior is independent of the detailed shape of the pulse leading edge. Thus, acoustic detectors with high-frequency cutoffs sacrifice the major share of the total energy. We can also infer from Eqs. (22) and (23) that only later epochs of the rescaled acoustic pulse are important for the generation of the low-frequency components.... As an example, for a pulse that has evolved to a 10-times-greater wavelength, the energy available at the lower frequencies is correspondingly 100 times greater although the total energy is 100 times less.

"From the above discussion we can see that the characteristic frequency, and thus the available signal energy, is closely linked to the acoustic attenuation properties of the medium. For a metal several mechanisms are responsible for acoustic loss. These include thermoelastic effects, dislocation damping, Rayleigh scattering from randomly oriented metal grains, and electron-phonon scattering. For the frequencies of interest, most of these effects lead to attenuation coefficients which increase as f^2. The one exception is Rayleigh scattering which exhibits an f^4 dependence for acoustic wavelengths greater than the grain size. Electron-phonon scattering is only important for very pure metals at low temperatures. Since the density of crystalline defects depend critically on metallurgical treatment, the acoustic attenuation in normal metals can vary considerably. Available data indicate that aluminum is one of the most acoustically transparent metals, but above 10 MHz the attenuation length diminishes to less than 1 m with the onset of Rayleigh scattering.

"With this background we can now make some strong statements about the acoustic detectability of grand-unified-theory monopoles. Two basic strategies for signal detection are conceivable. The simplest is one-shot sampling of the acoustic pulse as it travels past an array of transducers. In this case, the total detectable energy is just the signal energy itself. This scheme takes advantage of the high-energy spectral components but requires elaborate filtering to isolate the signal from the thermally excited sea of phonons. No matter how carefully the matched filter is designed, the maximum signal-to-noise ratio that can be achieved is given by $E/(1/2kT)$. Unfortunately, even this optimum value is of the order of unity for devices of practical interest. In principle, one might gain by successively sampling the same wave front with many layers of transducers. Unfortunately, as will be shown below, the size of the detector and the number of tranducers required would be extravagant.

"The second sampling technique relies on measuring the acoustic wave many successive times in a high-Q device such as a gravity-wave detector. In principle, the signal-to-noise ratio can be increased by a factor of Q since the amplitude of the signal can be coherently added over many cycles. This improvement in signal processing comes with two penalties. The obvious disadvantage is that the detecting frequency must be low to coincide with a low-harmonic excitation of the detector structure; otherwise high-Q performance cannot be attained. Secondly, the rapid frequency variation of acoustic attenuation leaves most of the acoustic energy at frequencies far above the fundamental. These intermediate frequencies will not absorb within the interior volume but will be damped by interaction with the detector surface and support structure. This reduces the fraction of the energy available for coherently exciting the fundamental vibration modes.

"To put some numerical flesh on these arguments, we will optimistically assume the monopole energy loss to be 1 GeV/cm for $\beta \simeq 10^{-3}$. For higher-velocity monopoles, there is no virtue in acoustic techniques since the energy loss in organic scintillators should be easily detectable. The slowest monopole of interest has a β of 3.8×10^{-5}, the escape velocity from the surface of the earth. This is several times the speed of sound in typical solids so the acoustic wave front must propagate with the bow-wave geometry of Cerenkov radiation. In an aluminum block with characteristic length of 1 m the maximum useful frequency is limited to about 10 MHz by the decreasing acoustic attenuation length. Thus, from Eq. (21), the total acoustic energy within the entire detector is, at most, 0.45 eV and so at 300° K the maximum achievable signal-to-noise ratio is 35 if the entire spectrum of signal energy is measured and the matched filter rejects all competing normal modes. These last two assumptions are exceedingly difficult to meet in practice. The magnitude of the indices of the normal modes of interest is given by n $\ell/c\Delta t$ where c is the velocity of sound in the medium. For ℓ and Δt given above, n 2000. Extraction of all the signal energy requires more than n^2 transducers. Since the number of modes is proportional to n^3, about 10^{10} modes each contribute 1/2kT to the thermal background noise. The matched filter must provide a rejection factor of 10^{10} to yield a marginally detectable signal. This must be performed 10^7 times per second for all possible orientations of the incident monopole track. Given the attendant technical difficulties, room-temperature acoustic detection of magnetic monopoles is categorically infeasible.

"There is a slightly greater hope if one could lower the device temperature. At 1 millidegree the signal-to-noise ratio would be of the order of unity within a single wide-band transducer. However, for a detector that is sensitive to only the low frequency modes, the decrease in signal strength must be matched by a corresponding increase in the resonant Q. For a device like the Stanford gravity-wave detector, effective noise temperatures of 1 microdegree must be reached while retaining efficient coupling of the higher-frequency waves into the fundamental modes. This goal, although not impossible, will require considerable effort.

"One might wonder if some other medium than aluminum might

yield more energetic acoustic pulses. Examination of Eq. (18) shows that this is not likely. The ratio β/c_p is insensitive both to material and temperature since the two parameters depend on the same properties of the crystal lattice.[14] Similarly, the ratio $(dE/dx)/\rho$ cannot vary radically either. Because of the high electron Fermi energy and acoustic transparency, aluminum is probably the best material for which the energy loss can be calculated with any degree of confidence.

"In all of this discussion, consideration of the problems of actually performing the acoustic measurements has been ignored. In real life these difficulties will significantly degrade the performance from the ideal sketched above. In view of the intrinsically unpromising nature of the thermoacoustic technique further work in this area seems pointless."

Transmission in medium

In a previous section we obtained a formula for P_{peak} and know that is is proportional to

$$\frac{\beta}{C_m} \frac{(1+\sigma)}{(1-\sigma)} \sqrt{V_s} \left(\frac{dE}{dz}\right)$$

Since (dE/dz) is proportional to the conductivity the pressure is proportional to

$$\frac{\beta\sqrt{V_s}}{C_m\rho_r} \frac{1+\sigma}{1-\sigma}$$

where ρ_r is the resistivity of the metal. The intensity of a sound wave $I = (1/2)/(P^2/Z)$ is proportional to the following quantity

$$\frac{1}{Z} \left(\frac{\beta\sqrt{V_s}}{C_m\rho_r} \frac{1+\sigma}{1-\sigma}\right)^2$$

where Z is the acoustic impedance of the material. By comparing this quantity for different materials we can get some guide for choosing the best for our experiment. The following table shows the related quantities of several common materials and the calculated relative intensities, I, in arbitrary units. We also include some other quantities for later use.

As shown by Table 2, silver is much better than other materials, but it is too expensive. Copper is better than aluminum, but its acoustic impedance is too high to match the

Table 2. A Comparison of Materials

	Al	Cu	steel	Ag	Hg	
Thermal Exp. Coeff. β	25	16.6	9.6	19	60.6*	$10^{-6}/°K$
Resistivity ρ_r	2.65	1.67	9.71	1.59	98.4	$m\Omega/cm$
Velocity of Sound V_s	6420	5010	5790	3650	1451	m/s
Specific Heat C_m	0.215	0.092	0.106	0.0566	0.0331	$cal/°K \cdot g$
Poisson's Ratio σ	0.355	0.37	0.29	0.38	0.5	
Acoustic Imped. Z	1.73	4.46	4.5	3.8	1.9	$10^6 g/s \cdot cm^2$
Relative Intensity I	73	145	0.86	496	5.4	
Transmission Coeff. α_t	0.99	0.75	0.75	0.81	0.98	
Relative Signal $I \cdot \alpha_t$	72	109	0.65	401	5.3	

transducers. The transducers we have are made of lead meta-niobate which has acoustic impedance $Z_t \approx 1.5 \times 10^6 g/s\ cm^2$, so it matches aluminum but not copper. Taking the transmission coefficient

$$\alpha_t = \frac{4Z_t Z}{(Z+Z_t)^2}$$

into account we have the last row of Table 2. Now, copper is only a little bit better than aluminum but it's much heavier and costs much more, so we decided on aluminum for our prototype. The choice of material is not restricted to solids; in fact, liquids have the advantage of having much larger thermal expansion coefficients. The only highly conductive liquid, mercury, turns out to be much worse than aluminum as shown in Table 2.

Of the aluminum alloys, we have chosen 2124-T851 which is very stiff and has small grain size so that the attenuation of high frequency sound is small.

We have measured the attenuation length in the following way. We put a transducer on each end of an aluminum bar of a specific length, one transducer is then driven with short 10 MHz tone bursts and the output voltage of the other transducer is measured. The transducers are then put on another bar of the same alloy but of different length and the measurement is repeated. The results of this test are summarized in Table 3.

From this it's easy to find out the attenuation length is about

Table 3. Attenuation of 10 MHz in Aluminum

Length (inches)	16	28	40
Output Voltage (mV)	38	36	31

L = 118 inches = 3 m

This means that when the signal travels across the disk it only attenuates 25%.

Detector

Transducers. Ultrasonic transducers must be used to convert the energy of a thermoacoustic pulse into electronically measurable form. The transducers currently in use are K.B. Aerotech "alpha series" of 10 MHz transducers. These consist of a polished aluminum face cemented to a lead meta-niobate ($PbNb_2O_6$) element with bonded electrodes, with a soft rubber backing material. These transducers are .5 inch in diameter, and designed to have an impedance of 50 Ω at 10 MHz, where their response peaks (see Fig. 2).

Acoustic coupling. For optimum detection sensitivity, it is desirable to match the acoustic impedance of the transducer to that of the aluminum disk, and to couple them in a way which will not affect their performance. The acoustic impedance of the transducers is basically just the acoustic impedance of the aluminum face, and provides a good match to the disk. Therefore the dominant acoustic loss at the interface comes from the presence of a couplant.

For a couplant of acoustic impedance Z' separating two regions of acoustic impedance Z, the expression for the transmission coefficient of a longitudinal wave at normal incidence is[16]

$$\alpha_t = \left[\cos^2(k'a) + \left(\frac{Z'^2 + Z^2}{2Z'Z} \right)^2 \sin^2(k'a) \right]^{-1}$$

where a is the thickness of the couplant, and k' is the wave number of the wave in the couplant. Using the value of $Z = 1.7 \times 10^7$ kg$\cdot$m^{-2}s^{-1} for aluminum, and $Z' = 2.4 \times 10^6$ kg m^{-2}s^{-1} for glycerine (a common couplant) it can be seen that for a_t = .9 or better, it is required that the couplant thickness be less than about 3 microns.

In practice, phenyl salicylate is used as the couplant. It

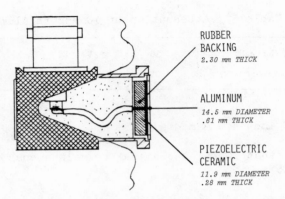

Fig. 2.   The transducers used in this detector are sensitive from 8
         to   12 MHz,   with   a   peak   response   at   10 MHz.   The
         piezoelectric ceramic is lead meta-niobate and is   11.9 mm
         in   diameter and .28 mm thick.   The wear plate is made of
         aluminum which has been polished on both sides and is   one
         10 MHz   wavelength   (.6 mm)   thick.   The ceramic has a soft
         rubber backing to dampen ringing.   These   transducers   are
         coupled   to   the   disk   using   phenyl   salicylate.   The
         interface thickness is kept thin to minimize   transmission
         losses.

may be melted at a moderate temperature on the surface to be
coupled to, and then allowed to recrystallize after the transducer
is applied. It is expected that the acoustic impedance of the
phenyl salicylate is close enough to that of glycerine that the
couplant thickness criterion above will apply to it also.

     It is possible to obtain high transmittance by making the
thickness of the couplant an integral number of half wavelengths,
however the attenuation of sound in glycerine (and other such
couplants) at 10 MHz and the desire for wide bandwidth indicate the
use of as thin a film of couplant as possible.

     Angular response. Because the radius of the transducers is
about   10 times the wavelength of the sound waves in aluminum, the
transducers are directional in response. If a is the radius of the
transducer, and k is the wavenumber of the sound wave, then the
response of the transducer may be approximated by

$$R(\theta) = \frac{\sin[ka \cdot \sin(\theta)]}{ka \cdot \sin(\theta)}$$

where θ is the angle of the incident from normal, and R(θ) is the
response of the transducer, which has been normalized: R(θ) = 1.

Using the values of .635 cm (.25 inch) for the radius of the transducer and 98 cm^{-1} for the wave number of a 10 MHz longitudinal wave in aluminum, one obtains a value of about 6 degrees for the width of the primary response lobe.

Sensitivity. At present there is no direct means to measure the absolute sensitivity of the transducers, due to the lack of calibrated 10 MHz ultrasonic sources. By using the piezoelectric efficiency of the lead meta-niobate material, however, an estimate of the sensitivity may be made.

From power transfer considerations, an equation for the receiving sensitivity S may be written as

$$S = \varepsilon_r \sqrt{\frac{Z_e A}{Z_a}}$$

where ε_r is the receiving efficiency, Z_e is the electrical load resistance, A is the area of the transducer face, and Z_a is the acoustic impedance of the medium. Putting in the values for the transducers now in use yields a sensitivity of 1.3μV/dynes·cm^{-2}, into 50 Ω. The actual input impedance of the preamps used with the transducers is about 200 Ω, so this figure for the sensitivity is improved by a factor of two.

Electrical properties. The most important electrical characteristic of the transducer is its output impedance in the frequency band of interest. The output impedance of the "alpha" series transducers may be characterized as a 50 Ω resistance in parallel with a capacitor of about 300 pF. These are "lump sum" values of distributed electrical and acoustical impedances, but are reasonably accurate across the bandwidth of the transducer.

Proper amplification of the low-level signals from the transducers has required a custom preamplifier design. The low output impedance and the frequency (10 MHz) dictate the use of a bipolar transistor input stage, with four transistors in parallel to lower the input impedance and raise the gain of the input stage. This is followed by an LC tuned JFET amplifier and then an output driver. At present, the preamplifiers are battery powered to help avoid the RF pickup from an external power supply.

The equation for thermal noise in a signal source with output impedance R is[17]

$$e_n = \sqrt{4kTR\Delta f}$$

where e_n is the RMS noise voltage of the source, T is the temperature in Kelvin, k is Boltzman's constant, and Δf is the bandwidth in Hertz. For the ultrasonic transducers being used, the amount of thermal noise generated is about $1nV \cdot Hz^{-1/2}$. Thus for a 4 MHz system bandwidth (which is typical), the input noise from the transducer is about $2\mu V$ RMS.

The measured equivalent input noise levels of the preamps are about $5-7\mu V$ RMS, as measured by comparison to a diode shot noise source. It is apparent that the noise from the preamplifier input dominates that of the transducer.

The overall gain of the preamplifier is about $100\pm20\%$, which is enough to boost the signal to a level where standard amplifiers are adequate for further amplification.

The waveform produced by the transducer in response to an acoustic signal is found to be a damped ringing at the fundamental frequency of the transducer. The internal damping constant of the lead meta-niobate material is 1.3,[18] but the rubber backing increases it further. Generally, a short duration "spike" input will result in a tripolar pulse out of the transducer, if the amplitude of the input is not too large.

Performance of Prototype

Setup. To date we have operated in the transducer trigger mode. One disk was instrumented with one transducer pair acting as the trigger. This trigger was used to initiate a $250\mu s$ event gate. A 10 MHz clock is gated with the event gate, and fed into the counting inputs of an octal latching scaler. The other transducer signals acted as stops for this scaler. At the end of an event gate, a LAM signal was sent to the PDP 11/34 that was used for data acquisition. The data acquisition program was used to reject spurious events and perform track fits. The discriminator thresholds for the trigger transducers were set to about $10\mu V$, as referred to the transducer outputs. At this level the singles rates were 1 KHz. The stop transducer thresholds were set slightly higher to reduce the singles rate to about 100 Hz. This rate was caused by the preamplifier input noise. The coincidence gate for the two trigger transducers was set 200 ns wide. The trigger rate was about 4 Hz, of which 5% were accidentals.

Results. For diagnostic purposes, we required that the computer print out any events where both scalars were stopped during an event gate. We found that we could operate in this mode for periods of several hours without an event, but these periods would be ended by bursts of events. The track fit program placed these events on a radial line from the center of the disk to the trigger transducer pair, with the majority of events at the center of the disk. These spurious events were thought to be electronic pickup which the preamplifiers receive simultaneously.

To test this idea, a veto transducer was installed with the same amplification and ground connections, but with no acoustic contact to the disk. This transducer was used as a veto for the other channels. This addition eliminated the spurious events at the center. To eliminate the remaining events, additional trigger logic is being installed to veto events were multiple pulsing occurs. Isolation transformers, more power supply filtering and optoisolators are also being added.

Future improvements. As previously stated, the output of a transducer for a monopole pulse should be about .05μV. Our thresholds were set at about 10μV which implies we are presently a factor of 200 off in our sensitivity. To gain this factor the following improvements are being considered. The PNP transisitors in the input stage of the preamp were not optimized, and it is likely we can reduce the threshold by a factor of two by carrying out this optimization. Another factor of two could be gained by using larger area transducers. The transducer ceramic is thin (.28mm) and brittle, so our expectations of gaining by this method are modest. The operating temperature of the preamps and transducers could be lowered in order to minimize the thermal noise.

Acoustic lenses could be installed which would enhance our sensitivity by increasing the effective area of the transducers. If this proves feasible, we could conceivably gain an order of magnitude. Other options include using a material that couples to the monopole magnetostrictively as well as by eddy currents, optimizing the geometry, and developing trigger schemes that will allow higher singles rates.

DUMAND AS AN ACOUSTIC MONOPOLE DETECTOR

Porter, in a submission to this workshop, has looked at the possibility of using a Dumand-type array for monopole detection.[19] He evaluates an acoustic array, which was once considered for Dumand but has been given up in favor of a phototube array detecting Cerenkov radiation.

Porter notes that Dumand was designed to detect the hadronic shower from high-energy neutrino cascades. These cascades are approximately ~10m long in H_2O and the shock wave is spherical. A monopole will deposit its energy through the entire depth. This line source will give a shock wave that is cylindrical. This means the signal will fall as $r^{-1/2}$. For a cylinder of water (L = 1 cm, R = 10^{-4}cm) the pressure $\Delta p = 2 \times 10^4$ dynes/cm^2. Since this falls off like $r^{-1/2}$, at 1 km, the signal would still be ~0.6 dynes/cm^2 which would be easily detectable.

Unfortunately, just as in the eddy current detector using aluminum (described in the previous sections), the primary pulse is very high frequency ($\sim 10^8$ Hz) and mostly absorbed. He calculates that the pressure due to the $0-10^6$ Hz components that are not badly attenuated in water, $\Delta p_{100m} \sim 2.4 \times 10^{-5}$ dynes/cm^2. This signal is impossible to detect in a single transducer. The only possibility would be coherent detection using many detectors in a very large array.

CONCLUSIONS

The development of methods for building very large area may be important to approach the astrophysics bounds in future monopole detectors. This paper reviews the present understanding of the prospects of using acoustic techniques for monopole detection. Present techniques appear to be about a factor of 10^2 away from the required sensitivity. Although some improvements can be envisioned, whether the necessary sensitivity can be obtained remains unclear.

ACKNOWLEDGEMENTS

The author would like to acknowledge help and discussions with his collaborators R. G. Cooper, C. E. Lane, and G. Lin at Caltech on much of this work and C. Akerlof and N. A. Porter for their submissions to this workshop.

REFERENCES

1. P.A.M. Dirac, "Quantised Singularities in the Electromagnetic Field," Proc. Royal Soc. London A 133, 60 (1931).
2. P.A.M. Dirac, "The Theory of Magnetic Poles," Phys. Rev. 74, 617 (1948).
3. G. 't Hooft, "Magnetic Monopoles in Gauge Theories," Nucl. Phys. B 79, 276 (1974).
4. A. Polyakov, "Particle Spectrum in the Quantum Field Theory," Pis'ma Zh. Eksp. Teor. Fiz. [JETP Lett.] 20 [20] p. 430 [194] (1974) [1974].
5. M. J. Longo, "Massive Magnetic Monopoles: Indirect and Direct Limits on Their Number Density and Flux," Phys. Rev. D 25, 2399 (1982).
6. M. S. Turner, E. N. Parker, and T. J. Bogdan, "Magnetic Monopoles and the Survival of Galactic Magnetic Fields," Phys. Rev. D 26, 1296 (1982).
7. B. Cabrera, "First Results from a Superconductive Detector for Moving Magnetic Monopoles," Phys. Rev. Let. 481378 (1982).

8. E. Loh, Electronic Cosmic Ray Searches, Talk presented at the
    Racine Monopole Workshop, Racine, Wisconsin, September
    1982.

9. V. P. Martem'yanov and S. Kh. Khakimov, "Slowing-Down of a
    Dirac Monopole in Metals and Ferrmomagnetic Substances,"
    Zh. Eksp. Teor. Fiz. [Sov. Phys. JETP] 62 [35] p. 35 [20]
    (1972) [1972].

10. S. P. Ahlen and K. Kinoshita, "Calculation of the Stopping
    Power of Very Low Velocity Magnetic Monopoles," Phys. Rev.
    D, (submitted).

11. K. Hayashi, "Stopping Power Formula for a Cosmological
    Monopole," 81-0222, Kinki University, Osaka, Japan (1981).

12. S. P. Ahlen, "Theoretical and Experimental Aspects of the
    Energy Loss of Relativistic Heavily Ionizing Particles,"
    Rev. Mod. Phys. 52 (1), 121 (1980).

13. J. G. Learned, "Acoustic Radiation of Charged Atomic Particles
    in Liquids: An Analysis," Phys. Rev. D 19 (11), 3293
    (1979).

14. C. W. Akerlof, "Limits on the Thermoacoustic Detectability of
    Electric and Magnetic Charges," Phys. Rev. D 26, 1116
    (1982).

15. Partially reproduced from C. W. Akerlof, Intrinsic Limits for
    Acoustic Detection of Magnetic Monopoles, Phys. Rev. D 27,
    1675 (1983).

16. J. Blitz, Fundamentals of Ultrasonics, Butterworths, London
    (1963).

17. C. D. Motchenbacher and F. C. Fitchen, Low-Noise Electronic
    Design, John Wiley (1973).

18. H. Krautkrämer, J. Krautkrämer, W. Grabendörfer, and L.
    Niklas, Ultrasonic Testing of Materials, Springer-Verlag,
    New York (1969), trans. B. W. Zenzinger.

19. N. A. Porter, Possible Detection Methods for Fast GUT Magnetic
    Monopoles; submitted to the Racine Monopole Workshop,
    Racine, Wisconsin, September 1982.

BINDING OF MONOPOLES IN MATTER AND SEARCH IN

LARGE QUANTITIES OF OLD IRON ORE

David B. Cline

Physics Department
University of Wisconsin-Madison
Madison, Wisconsin 53706

INTRODUCTION

The symmetry between electric fields and magnetic fields in Maxwell's equations, but the lack of abundant free magnetic charge compared to electric charge, has captured the attention of several generations of physicists. In 1931 Dirac went one step further; the existence of free magnetic charge (Dirac monopole) can provide a reason for the quantization of electric charge (eg = $n\hbar/2c$).[1] A preview of the Dirac formula was provided earlier by J. J. Thompson who discussed the angular momentum of the electromagnetic field, quantization of which gives the Dirac condition.

Many scientists have worked on the monopole problem in the following years. It is notable that there was no argument for a particular mass of the Dirac monopole. Experimental seaches were shooting in the dark. In 1974 't Hooft[2] and Polyakov[2] showed that magnetic monopoles exist as solutions in many non-abelian gauge theories (including Grand Unified Theories);[2] the massive vector bosons in gauge theories provide a mass scale for the monopoles. The interest in this subject has grown in the past few years. One of the striking features of unified theories of weak and electromagnetic interactions and of the Grand Unified interaction is that relations between masses, mixing angles and coupling constants are derived.[2] For example, the mass of the intermediate bosons follow from the Weinberg angle and the Fermi coupling constant G. The first concrete suggestion of monopole mass was

$$M_m \simeq (\alpha)^{-1} M_W \simeq 13 \text{ TeV} \tag{1}$$

in models of weak electromagnetic unification. The success of QCD
and remarkable similarity of the weak, electromagnetic and strong
interaction has led to the concept of Grand Unification; in this
case the unifying mass is the mass of the X, Y leptoquark bosons.
The corresponding monopole mass is now expected to be

$$M_m \simeq (\alpha)^{-1} M_X \simeq 10^{16}-10^{17} \text{GeV} \tag{2}$$

since it is thought that $M_X \simeq 10^{14}-10^{15}$ GeV. The large masses
"predicted" for the magnetic monopole immediately changes one's
evaluation of the previous searches for monopoles for two reasons:

A. The experimental signature or production yield in
    cosmic-ray interactions for 13 TeV or $10^{16}-10^{17}$ GeV
    monopoles is likely to be different from that expected for
    light and hence very relativistic monopoles--this had been
    one of the key signatures for monopoles in the cosmic
    rays.

B. The production rate of 13 TeV monopoles in cosmic-ray
    interactions or $(10^{15}-0^{16})$ GeV monopoles in the early
    universe is unpredictable.

There are constraints on the number of free magnetic charges in the
Galaxy due to the existence of galactic magnetic fields and from
limits to the number of particles in the universe with very high
mass (due to the total amount of mass in the universe). These
constraints limit the number of monopoles to a small fraction of
the number of nucleons and immediately lead to, at best, very small
fluxes of monopoles in the cosmic radiation.[3-6]

BINDING OF MONOPOLES TO ELECTRONS, ATOMS
AND NUCLEI; PICKUP OF NUCLEI ON SLOW MONOPOLES[7]

An early work on this was Malkus.[8] The effective potential for
the radial motion of a monopole-particle system in one partial wave
is of centrifugal form, with the coefficient $\ell(\ell + 1)$ replaced by
an eigenvalue of the operator $(m \, \vec{r} \times \vec{v})^2 - g\nu\vec{S}\cdot\hat{r}$, where $\vec{S}$ is the spin
of the particle, g its gyromagnetic ratio and ν is the product of
its charge and the magnetic charge of the monopole. [Dirac's
quantization condition says that $\nu = $ (integer)/2, in units of $\hbar$.]
In an angular state in which the eigenvalue "$\ell(\ell + 1)$" is negative,
the "centrifugal" potential is attractive; if the eigenvalue is
$<-1/4$, there are an infinite number of bound states. There is no
simple general expression for the least value of "$\ell(\ell + 1)$", but in
one case there is: If $S \leq \nu$, "$\ell(\ell + 1)$" equals $\nu(1 - gS)$ in the

(unique) state of least total angular momentum [$j_{min} = \nu - S$], and if g>0, this is the least eigenvalue. Hence, if $gS - 1 > 1/4$, at least one partial wave has an infinite number of bound states.

This last condition is not satisfied by the electron which has $gS - 1 \simeq \alpha/2\pi = .001$ unless the monopole's magnetic charge, in units of 1/2e, is enormous. Kazama, Yang and Goldhaber[9] found, however, that there is always one bound state, at E = 0, if the electron is given a positive pointlike anomalous dipole moment. This is of course not realistic, since the anomalous moment of the electron comes from radiative corrections (Schwinger) so that it is "soft", with a spatial scale of m_e^{-1}; it is also sensitive to magnetic field strength on the scale m_e^2/e. There is a problem in principle in the monopole-electron system, because the effective radial potential vanishes in the state of lowest j, allowing the electron and monopole to meet; this is a no-no for a point charge and point monopole. Jackiw and Rebbi treated this problem using a 't Hooft monopole, which has a finite extent and again found a bound state at E = 0. Such deeply bound states, for quarks as well as leptons, concern the structure of the monopole and the Callan-Rubakov effect. We shall restrict ourselves here to more weakly bound states.

Consider now the interaction of a monopole with a hydrogen atom. There are many partial waves in which the potential for radial motion is dominated by the $\vec{S}\cdot\hat{r}$ term, so $V \simeq \pm 1/4m_e r^2$ (taking the monopole to have "unit" strength, $\hbar/2e$). As in all cases where the particle interacting with the monopole is an extended body, the short-range interaction depends on the response of the body to the monopole's magnetic field. But a bound can be put on the energy of the lowest bound state if one judges that beyond a distance r_o the body is little disturbed by the magnetic field, so that the "centrifugal" potential holds for $r > r_o$; at worst, the potential for $r < r_o$ can be taken to be an infinite "hard core". The binding energy E_o of the ground state in this potential is then a lower bound on the true ground state binding energy. So, for the hydrogen atom, if we take r_o = Bohr radius, the ground state binding energy is of the order of a Rydberg. It is more difficult to estimate the binding, if any, to a hydrogen molecule. Here there is no long-range attraction; at short-range is it energetically favorable to excite the molecule to a triplet state, which is attracted to the monopole?

Nuclei with sufficiently large magnetic moments will bind to a monopole.[10] For a proton, taking the "hard core" radius r_o to be 1 fermi yields $E_o \sim 0.015$ MeV; for Al_{27}, taking $r_o = 1.4 \times (27)^{1/3}$ fermi yields $E_o \simeq .5$ MeV. A monopole passing through matter containing such nuclei will eventually capture one, the most likely inelastic process for a slow monopole (e.g., a "cosmic monopole" with $v = 10^{-3}c$) being electric dipole radiation. The radiative capture cross section for a nucleus of charge Q and mass M into a state of binding energy B is roughly $\sigma_\gamma \simeq Q^2 (B/M) \cdot (\pi/k^2)$. Observe the $k^{-2}(\propto v^{-2})$ velocity dependence and the rather weak dependence on B. A "cosmic" monopole ($v = 10^{-3}c$) passing through the earth's crust has a mean free path for capturing an Al_{27} nucleus (the most abundant nuclide with a magnetic moment in the crust) of <10Km. This nucleus is permanently bound to the monopole (unless of course the monopole eats it via the Callan-Rubakov effect!), because $v = 10^{-3}c$ means a c.m. energy of $\leq .02$ MeV for any collision with further nuclei. There is zero cross section for capturing a second nucleus, since the long-range interaction is then Coulomb repulsion. The above estimate for σ_γ does not apply to a proton, because the only attractive partial wave is J = 0 (assuming a "unit" strength monopole) and so the k^{-2} term in σ_γ (one factor k^{-1} comes from the attractive r^{-2} potential) would have to come from a J = 0 → J = 0 radiative transition, which doesn't exist. This is relevant to the question of whether monopoles in the early universe would capture protons (at the time that neutron-proton capture was occurring).

Monopoles with velocity $\beta > 10^{-5}$ are expected to pass through the earth if they remain as free monopoles. Once an Al_{27} atom or nucleus has been picked up, if the atom is stripped, the stopping probability in the traversal through the earth will increase. Thus it is possible that the flux of monpoles passing up through the earth is decreased relative to the flux at the surface of the earth. However, for example, the Baksan monopole search was sensitive to monopoles coming up from below or down through the ~300 m overburden (the expected charge pickup probability in the overburden is .3 km/10 km ~ .03 and rapidly increases with increasing zenith angle).[15]

The binding of monopoles to magnetic structures or to individual nuclei or atoms will have important consequences for the search for small concentrations of monopoles in matter. Other estimates of the binding of monopoles to matter are to be found in Ref. 16.

The binding of monopoles to nuclei may give alternate techniques to search for monopoles. However, for very massive monopoles, the expected signal passing through superconducting devices should not differ appreciably from that of a free monopole.

BINDING IN FERROMAGNETIC MATTER

There have been several calculations of the binding energy of a magnetic monopole to magnetic domains, the earliest being that of Goto.[11] Of course, closer to a monopole than 300 Å, its field exceed the interior magnetic field of a domain, and closer than

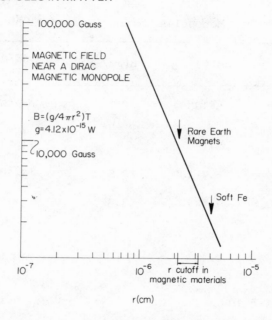

Fig. 1. Magnetic field near a Dirac magnetic pole.

100 Å, its field exceeds the saturation field of ferromagnetic matter (Fig. 1). However, the force on the monopole can be found by consideration of the stress tensor farther away, where its field is weak. The crucial result is that the force on the monopole is due to $\vec{H}$, not $\vec{B}$; hence a monopole can be bound to a place where $\nabla \cdot H \neq 0$. (A simple argument for $\vec{H}$ to be the force field is that if a monopole goes around a closed circuit, no work can be done, because the ferromagnet has no energy to give. On the microscopic level the reason is that a monopole cannot pass through electrons.) Kittel and Manoliu[12] refined Goto's calculation to include the ferromagnetic exchange interaction. Their estimate for the binding energy to a single domain is ~50 eV, which is not inconsistent with the classical caluculation of Goto,[11] which includes the full domain structure.

There is also an image force on a monopole outside an unmagnetized ferromagnetic medium caused by the induced magnetization, with a potential minimum at the boundary. Figure 2 shows the forces due to the image charge on the monopole.

POSSIBLE SOURCES OF SLOW MONOPOLES NEAR THE EARTH

In order to carry out a broad search for massive monopoles, it is useful to have some picture of where slow, massive monopoles

Table 1. Monopoles in the Sun and Planets*

	R(cm)	M(g)	p(gm/cm^2)	$\beta\dfrac{dE/dx}{stop}$	β_{escape}
Earth	6.4×10^8	$6*10^{27}$	~2.7	$\sim4\times10^{-5}$	$\sim2\times10^{-5}$
				$\rightarrow10^{-4}$	
Saturn	6×10^9	5.7×10^{29}	~2.5	$\sim3\times10^{-4}$	
Jupiter	7.1×10^9	1.9×10^{30}	3.3	$\sim3\times10^{-4}$	
				$\rightarrow10^{-3}$?	
Sun	7×10^{10}	2×10^{33}	160	$\sim10^{-2}$	$\sim10^{-4}$

Galaxy $\beta_{escape} \simeq 10^{-3}$

Virgo Cluster $\beta_{escape} \simeq 3\times10^{-3}$

*From D. Ayres et al., Proc. of the 1982 DPF Summer Study of Elementary Particle Physics and Future Facilities, Snowmass, p. 603 (1982).

probable trapped monopole. K. Schatten[17] has used the measurements of the magnetic field near the moon to put very restrictive limits on the difference between the number of north and south monopoles in the moon.

OLD IRON ORE BODIES AS POTENTIAL COLLECTORS
USING LARGE PROCESSING PLANTS FOR MONOPOLE SEARCHES

To date the total amount of matter processed in a search for magnetic monopoles is about 100 kg. In order to carry out a sensitive experiment to extend the limits appreciably, a method has been proposed in which $\simeq10^6$ tons of material, 10^9 times more material than in previous experiments, is processed.[18] The search for monopoles trapped in matter requires that very slow monopoles be detected ($\beta < 10^{-5}$c). We emphasize that the technique outlined here will apply to a large range of monopole masses (m $\sim10^7$-10^{17} GeV) and is not constrained to the presently popular GUT monopoles.

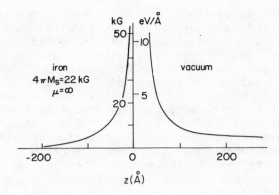

Fig. 2.   Forces due to the image charge on   a   monopole   near   iron
          medium.

that may stop in the surface of the earth may originate.   There are
three possible sources.

1.   Slow cosmic monopoles that pass   through   the   earth   and   lose
     energy   by   eddy   current   heating in the core--these monopoles
     could be on orbits that oscilate back   and   forth   through   the
     earth   and   have turning points of the orbits near the surface.
     In favorable cases the monopoles may be trapped in matter   near
     the surface.

2.   Monopoles trapped in meteorites that   collide   with   the   earth
     will   be   lost   in   the entry into the earth's atmosphere; this
     could provide a very small but low   velocity   source   of   heavy
     monopoles.   Since   the   meteors   come   from   very   old bodies,
     presumably formed during the time the solar system was   formed,
     this possibility extends the potential collector lifetime to ~4
     billion years.

3.   During the past (0-4) billion years, monopoles trapped in stars
     could   have   been   released through supernova explosions.   Thus
     the   local   monopoles   flux   might   have   a   time   variation.
     Potential   collectors would integrate over this vast period and
     be sensitive in a different way than present-time detectors.

     For these reasons and because it may be   possible   to   process
vast   quantities   of   iron   ore   by industrial techniques, it seems
possible to search for massive monopoles in old iron ore   (see   the
next section).

     There may be monopoles trapped in other planets and   the   sun.
Table 1   gives   the   initial   monopole velocity that will lead to a

The direct detection of monopoles with low velocity ($v/c \leq 10^{-5}$) must be carried out by interaction with bulk electromagnetic systems--i.e., superconducting coils in which magnetic flux is trapped. We can attempt to estimate the possible numbers of trapped monopoles in 10^6 tons of iron ore by the following argument: 10^6 tons of ore corresponds to an area of 1 km × 1 km and a depth of $\simeq 1/2$ meter. The material was exposed for 2×10^9 years--thus the integrated flux of monopoles through this area is related to the limits of the flux in cosmic rays. One half of the monopoles would pass throgh this area coming up from the earth. We estimate the capture probability from the ratio of the thickness of the material to that of the earth. These estimates indicate that less than one monopole should be trapped in a ton of iron ore if the Parker bounds are correct. Of course, if there are local sources of monopoles with large fluxes, this estimate could be far too low.

Because the density of trapped monopoles in iron ore should be proportional to the amount of time that the ore has been below the Curie point, it is important to choose an ore from a body of the greatest possible geological age. Of the ore being mined now, there are 3 major types that have been around for a significant period. Figure 3 shows a "history" of iron ores on earth.

Volcanic iron ore is simply iron ore brought to the surface from a depth of approximately 15-20 kilometers by a volcano which spewed it over the surrounding area. This type of ore is the youngest and is approximately 40-60 million years old. Not many deposits of this ore are currently being mined in the continental U.S. with any significant throughput, however.

Contact metamorphic iron ore is formed when hot magma intrudes from below into a pocket of softer rock, particularly limestone, passing through it and leaving iron ore behind after it cools. This type is generally mined in Utah and is approximately 200-600 million years old.

Sedimentary iron ore is the oldest and most stable type and is formed in the following way: First a volcanic iron ore deposit is formed. This eventually is weathered with the iron carried to shallow seas or lagoons where it ends up as sediment through a series of chemical reactions. Eventually the sea floor is raised up and the iron is now in the sedimentary rock. This type is found extensively in Wisconsin and Minnesota where it has spread in great sheets approximately 1.8-2.2 billion years old. It is also the type most likely to have trapped monopoles in it due to the long period it has been around. It is also mined in large quantities which makes it ideal for a monopole search. Typically large industrial plants process $>10^6$ tons of ore per year and heat them above the Curie temperature.

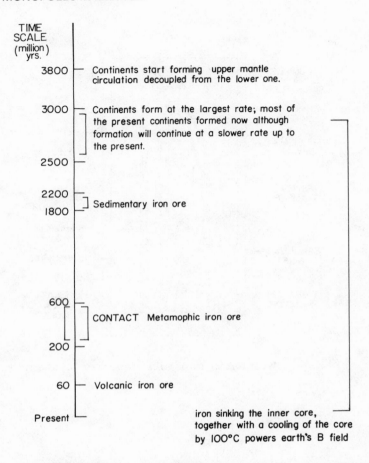

Fig. 3. Iron ore history of the earth. Note that approxmately one
         fourth of the energy used to heat up the earth comes from
         gravitational collapse, the rest from radioactivity.

        A prototype experiment to search for monopoles in old iron ore
heated   above   the   Curie   temperature   is being carried out by the
Wisconsin group at Black River Falls, Wisconsin.[18,19] The   initial
detector is shown in Fig. 4; it consists of 4 superconducting coils
and 3 SQUIDS surrounded by a superconducting shield.   Future
searches using >10^7 tons of iron ore might be carried out at large
plants using such a detector. Figure 5 shows the coil arrangement
which has two magnetometers and one gradiometer.

THE WISCONSIN MONOPOLE DETECTOR

        In the pioneering experiments of the Alvarez group,[13]

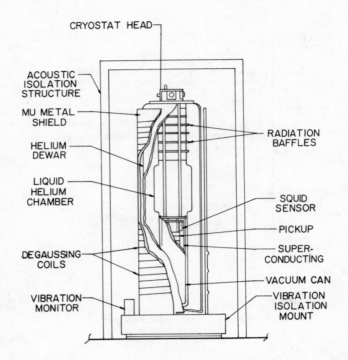

Fig. 4. Wisconsin monopole detector system.

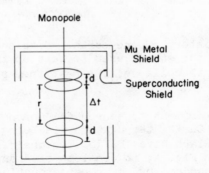

Fig. 5. Coil arrangement for Wisconsin detector using two
magnetometers and one gradiometer. The gradiometer
consists of two counterwound coils coupled to one SQUID.
Magnetic decoupler shields are placed between the
magnetometer coils and the gradiometer. The coil
diameters are 10 cm while d = 10 cm and r = 20 cm.

monopoles were searched for using small superconducting coils. The use of a SQUID magnetometer to search for magnetic monopoles was one of the earliest applications of the SQUID, and the technique has since been further refined. It is inherently appealing because the passage of a single monopole through a superconducting circuit would produce a flux change of two quanta ("fluxons"), while the output of a SQUID is periodic in flux with a period of one fluxon ($2_\times10^{-15}$ Weber). Hence as long as the noise level can be kept sufficiently low, the passage of a monopole through a superconducting coil magnetically coupled to a SQUID should give a large, unique, and unmistakable signal, a DC level shift of two periods. SQUID magnetometers now commercially available are sensitive at the millifluxon level. Thus SQUID-based monopole detectors operate in a regime in which sensitivity can actually be sacrificed in the interest of noise reduction.

Nonetheless, in a magnetometer designed to admit a material sample, the noise problem can be quite severe. As a benchmark, it should be noted that the earth's magnetic field amounts to about a million fluxons per square centimeter. Thus previous monopole searches by this technique have utilized coils with apertures of less than a square centimeter, and have accordingly been restricted to the study of very small samples, with an aggregate mass of a few tens of grams.

The detectors can be divided up into those that are used to search for cosmic monopoles ($\beta \geq 10^{-4}$) and those that are used to search for monopoles bound to matter or released from matter in the earth's gravitational field ($\beta << 10^{-5}$). An example of the former detector is the one used by Cabrera in which extremely low magnetic fields were maintained by the use of inflated super conducting lead "balloons".[14] This detector needs to be rather isotropic in order to increase the solid angle for cosmic monopoles. But a search for monopoles with masses comparable to the Grand Unification scale, falling from a moderate height, invites the use of a very different sort of magnetometer. Such monopoles, with or without a "retinue" of bound atoms, can freely penetrate ordinary solids. Even superconductors are penetrable, for as a monopole approaches a superconducting surface, its field surpasses the superconductor's critical field long before forces capable of having a significant effect on its motion have been generated. Furthermore, the origin (at least the point of origin) of these monopoles is known and thus the angular acceptance can be increased and the requirements of extremely low magnetic fields can be relaxed. Furthermore, several coils can be put in coincidence. This is the technique used for the Wisconsin monopole detector[18],[19] (Fig. 4).

We now discuss the size limitation of superconducting coil detectors. The largest SQUID magnetometers built to date were experimental antennas for VLF radio reception by submarines. Their

apertures approach a square meter. They differ from the device
needed for monopole detection in two important respects: First,
they are coupled to the SQUID through a resonant system of modest
bandwidth, tuned to a 3 KHz carrier; second, magnetic shielding was
restricted to a conductive layer to damp out high-frequency eddy
currents, plus the natural shielding provided by seawater.

A new DC SQUID is being developed by SHE Corp. Using this
SQUID, it should be possible to construct superconducting monopole
detectors with a diameter of 1m or more. This seems to be the
limitation size for the near future.

Figure 6 shows the signals expected in the Wisconsin monopole
detector for a falling monopole. The signature is extremely
characteristic and hopefully will be distinguished from the large
sources of magnetic dipole fluctuations in the vicinity of the
detector.

It is clear that future searches for monopoles in industrial
plants will provide a severe constraint on the detector. In this
sense the Wisconsin detector can serve as a prototype for future
experiments.

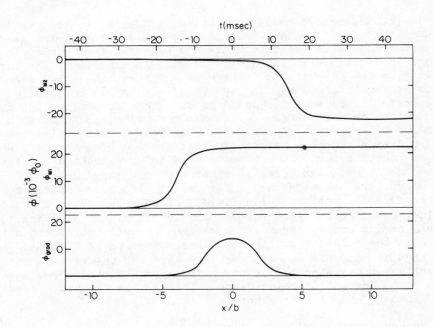

Fig. 6. System response for Wisconsin detector for a monopole
         intersecting all four loops.

ACKNOWLEDGEMENTS

We wish to thank G. Goebel and the members of the Wisconsin monopole detector group for many helpful discussions.

REFERENCES

1. P.A.M. Dirac, Proc. R. Soc. London, A 133, (1931); Phys. Rev.
   74, 817 (1948).
2. G. 't Hooft, Nucl. Phys. 879, 276 (1974); A. Polyakov, ZhETF
   Pis. Red. 20, 430 (1974); (JETP Lett. 20, 194 (1974));
   J. C. Pati and A. Salam, Phys. Rev. Lett. 31, 661 (1973);
   H. Georgi and S. L. Glashow, Phys. Rev. Lett. 32, 438
   (1974); H. Georgi, H. R. Quinn, S. Weinberg, Phys. Rev.
   Lett. 33, 451 (1974). P. Langacker, "Grand Unified
   Theories," 1981 Int. Symp. on Lepton and Photon
   Interactions; J. P. Preskill, Phys. Rev. Lett. 43, 1365
   (1979); M. B. Einhorn, D. L. Stein, and D. Toussaint, Phys.
   Rev. D 21, 3298 (1980); F. A. Bais and S. Rudaz, Nucl.
   Phys. B 170, 507 (1980); A. H. Guth, Phys. Rev. D 23, 347
   (1981); A. H. Guth and E. Weinberg, Phys. Rev. D 23, 876
   (1981); P. Langacker and S.-Yi. Pi, Phys. Rev. Lett. 45, 1
   (1980); M. Daniel, G. Lazarides, and Q. Shafi, Phys. Lett.
   91B, 72 (1980); F. A. Bais, Phys. Lett. 98B, 437 (1981) and
   F. A. Bais and P. Langacker, TH, 3142-CERN; S. Bludman,
   1981 Renc. de Moriond, UPR-0175T.
3. G. Lazarides, Q. Shafi, and T. F. Walsh, Phys. Lett. 100B, 21
   (1981); E. N. Parker, "Cosmic Magnetic Fields," Oxford
   (1979); M. J. Longo, "Massive Magnetic Monopoles: Indirect
   and Direct Limits on their Number Density and Flux,"
   UM-HE-81-33 (1981); E. N. Parker, Ap. J. 160, 383 (1970);
   S. A. Bludman and M. A. Ruderman, Phys. Rev. Lett. 36,
   (1976).
4. For magnetic monopole reviews, see P. Goddard and D. Olive,
   Rep. Prog. Phys. 41, 91 (1978); C. N. Yang in Proceedings
   of the 19th International Conference on High Energy
   Physics, S. Homma, M. Kawaguchi and H. Miazawa, eds., Tokyo
   (1978); W. Marciano and H. Pagels, Phys. Rep. 36C, 137
   (1978); A. Jaffe and C. Taubes, "Vortices and Monopoles,
   Structure of Static Gauge Theories," Birkhauser, Boston
   (1980); Monopoles in Quantum Field Theory: Proceedings of
   the Monopole Meeting, Trieste, N. Craigie, P. Goddard and
   W. Nahm, eds., World Scientific Publishing Co., Singapore
   (1982).
5. Y. Zeldovich and M. Khlopov, Phys. Lett. B 79, 239 (1978); J.
   Preskill, Phys. Rev. Lett. 43, 1365 (1979).
6. E. N. Parker, Ap. J. 163, 225 (1971), ibid., 186, 295 (1971);
   M. S. Turner, E. N. Parker and T. J. Bogdan, Phys. Rev. D
   26, 1296 (1982).

7. This description was provided by C. Goebel, private
    communication.
8. W.V.R. Malkus, Phys. Rev. 83, 899 (1951).
9. Y. Kazama and C. N. Yang, Phys. Rev. D 15, 2300 (1977), C. N.
    Yang, Bound States for the e-g System, Proceedings of the
    Conference on Monopoles in Quantum Field Theory, Trieste,
    Dec. 1981; C. Goebel, private communication.
10. Papers on monopole-nucleus bound states are W.V.R. Malkus,
    loc. cit.; C. Calucci, Capture of Nucleons by a Monopole,
    Lett. Nuovo Cimento 32, 201 (1981); C. Goebel, Interaction
    of Monopoles with Atoms, Electrons and Nuclei, in
    Proceedings of the Monopole Seminars, D. Cline, ed.,
    Madison (1981); C. Goebel and T. Kirkman, private
    communication.
11. E. Goto, J. Phys. Soc. Japan 19, 1412 (1958).
12. C. Kittel and A. Manoliu, Phys. Rev. B 15, 333 (1977).
13. R. Ross et al., Phys. Rev. D 8, 698 (1973).
14. B. Cabrera, Phys. Rev. Lett. 48, 1146 (1982).
15. A. Chudakov, private communications to D. Cline and P.
    McIntyre.
16. E. M. Purcell et al., Phys. Rev. 129, 2326 (1961); E. Amaldi
    et al., CERN report 63-13.
17. K. H. Schatten, Search for Magnetic Monopoles in the Moon,
    Phys. Rev. D 1, 2245 (1970); K. H. Schatten, NASA preprint
    (1982).
18. D. Cline et al., "A New Experimental Technique to Search for
    Relic Magnetic Monopoles," Univ. of Wisconsin proposal to
    the U.S. Department of Energy, 1982; D. Cline, "The Search
    for Magnetic Monopoles," presented at the Meeting on
    Experimental Tests of Unified Theories, Venice, March 1982.
19. M. Simmons, R. Sager, B. Lindquist, private communication.

MONOPOLE ENERGY LOSS AND DETECTOR EXCITATION MECHANISMS

S. P. Ahlen

Department of Physics
University of California
Berkeley, California 94720

INTRODUCTION

The prediction of the existence of supermassive magnetic
monopoles[1,2] in grand unification theories (GUTs),[3,4] and of the
copious production of these monopoles in the early universe,[5] have
led to a revived interest in the theory of the electromagnetic
interactions of moving magnetic monopoles with matter.   In
particular, there is great interest in the rate at which monopoles
lose energy in various types of astrophysical objects such as the
earth and sun so that a determination of the likelihood of
primordial monopoles being trapped by these objects can be made.
There is even greater interest in the question of whether or not
the quantity and quality of energy lost by magnetic monopoles in
conventional types of particle detectors is adequate for their
detection. In the pre-GUT era (roughly prior to 1979), there was
little doubt about the answers to such questions. The monopole
mass was expected to be sufficiently small so that acceleration of
the monopoles to relativistic velocities by the Galactic magnetic
fields was thought to be inevitable. For such velocities it is
easy to show that the ratio of monopole stopping power to that for
a relativistic singly charged electric particle would be $\sim(g/e)^2$
where e is the charge of an electron and g is the magnetic charge
of the monopole. Since various quantization arguments require that
$g = n\hbar c/2e$ where $n = \pm 1, \pm 2, \ldots$, the stopping power of a monopole
would be truly enormous. It would be more than large enough to
enable monopoles to be easily detected with almost any kind of
instrument which could detect electric particles. Furthermore, the
rate of energy loss in the earth would be large enough to stop a
considerable fraction of monopoles so that searches for monopoles
trapped in matter in the earth might be expected to be meaningful.

259

However, the large mass which GUT monopoles are expected to have (10^{16} GeV/c^2) would reduce their velocity attained by Galactic field acceleration to ~10^{-3} c. Until recently, the electromagnetic interactions of monopoles having such a small velocity with the material of either astrophysical objects or particle detectors were only poorly understood. It will be the purpose of this review to clarify this issue, so that a realistic assessment of the experimental evidence pertaining to GUT monopoles can be made.

REVIEW OF THE STOPPING POWER OF MONOPOLES WITH LARGE VELOCITY

The energy loss problem for fast monopoles has been discussed by several authors.[6-10] For projectile velocities v = βc, which are much larger than typical orbital electron velocities, ~αc (α being the fine structure constant) it is legitimate to divide the collisions between the projectile and the absorber material's electrons into two classes, the so-called close and distant collisions. In the close collisions the energy transfers are so large that the electrons can be regarded as free and the straightforward application of two-body kinematics and the use of the appropriate scattering cross section yields the rate of energy loss, i.e., the stopping power, for these collisions. The energy lost in each distant collision is generally so small that it is not legitimate to consider the electrons in distant collisions as free. However, since the majority of these collisions are of a grazing nature, it is legitimate to consider the excitation of an atom as a perturbation by the electric field of the projectile. For most distant collisions the impact parameter is large enough that one can assume the dipole approximation. The approximations for the close and distant collisions are valid whether the projectile has an electric or a magnetic charge. The key distinction between these two types of particles arises from the velocity dependence of their lab frame (the frame in which the material is at rest) electric fields. For monopoles, the magnitude of the electric field is reduced from that of an electric particle with the same velocity and charge (in cgs units) by a factor β. For β >> α it is essentially the lab frame electric field which determines the interaction between the projectile and the electrons. Since stopping power scales as the square of the electric field, it is apparent that the ratio of fast monopole stopping power to that of an electric projectile is ~$(g\beta/Ze)^2$ where g is the monopole charge, Ze is the charge of the electric particle, and -e is the electronic charge. More precisely, for β >> α, it has been shown that the stopping power for electric particles is:[11,12]

$$\frac{dE}{dx}\bigg|_e = \frac{4\pi NZ^2 e^4}{mc^2\beta^2}\left[\ln\frac{2mc^2\beta^2\gamma^2}{I_e} - \beta^2 - \delta_e/2\right], \quad \beta \gg \alpha \qquad (1)$$

and for magnetic particles is:[10]

$$\frac{dE}{dx}\Big|_m = \frac{4\pi N g^2 e^2}{mc^2} \left[\ell n \frac{2mc^2\beta^2\gamma^2}{I_m} - \frac{1}{2} + K(|g|)/2 - \delta_m/2 \right], \quad \beta \gg \alpha \quad (2)$$

where $K(|g|) = 0.406$ (0.346) for monopoles with $|g| = \hbar c/2e(\hbar c/e)$, $\gamma^2 = 1/(1-\beta^2)$, N is the number density of electrons, m is the electron mass and $I_{e,m}$ and $\delta_{e,m}$ are the mean ionization potential and density effect corrections for the electric and magnetic projectiles. We have assumed the atoms of the stopping material to be light enough and β to be large enough so that shell corrections[11],[12] can be neglected.

It has been shown[10] that although I_m and δ_m are fundamentally different from I_e and δ_e, in practice the distinction is small and actually vanishes in the limit of very small density such as for gases. Thus, we can set $I_m \simeq I_e = I$ and $\delta_m \simeq \delta_e = \delta$. In Eq. (2) we have neglected the Bloch correction which has been previously proposed for monopoles.[10] This has been done since it has been found[13],[14] that this correction is quite dependent on high order QED effects which have not been worked out for monopoles. In any case, the correction is not likely to be more than several percent in stopping power which is the level observed for relativistic uranium.

LIMITATIONS OF THE LARGE VELOCITY APPROXIMATIONS

The principle requirement for the validity of Eqs. (1) and (2) is that there exists a sufficiently small fraction of collisions that do not satisfy either the close or distant collision approximations that negligible error is incurred by failure to treat this intermediate class of collisions properly. The validity of the close collision approximation depends on the ratio of the kinematically limited energy transfer $\epsilon_m = 2mc^2\beta^2\gamma^2$ to the characteristic atomic energy I. For $\epsilon_m \gg I$ there will clearly be a large number of collisions for which it is legitimate to neglect atomic binding. The validity of the distant collision approximation depends on the ratio of the adiabatically limited maximum impact parameter $b_{ad} \simeq \hbar\gamma v/I$ to the Bohr radius a₀ which is a measure of atomic radius. For $b_{ad} \gg$ a₀ there is a large number of atomic collisions in which excitation is not forbidden by adiabaticity and in which the dipole approximation is valid. Taking I ~ 100 eV, it is found that the validity of the close collision approximation requires $\beta \gg 0.010$ and that of the distant collision approximation requires $\beta \gg 0.027$. It has been found,[15] for example, that significant deviations from Eq. (1) occur for electric particle velocities smaller than 0.04 c. Thus, one must not take seriously the rapid decline to zero for Eqs. (1) and (2) as $\epsilon_m \to I$, since this merely reflects the complete breakdown of the

assumptions that led to their derivation. In order to determine
the correct value of stopping power for $\beta < 0.01$, one must adopt a
different approach.

THE FERMI GAS APPROXIMATION

A great deal of work has been done[16-19] on the stopping power
of slow electrically charged particles. The most successful model
for this problem has been that in which the properties of the
stopping material have been approximated by those of a free gas of
electrons. This is clearly appropriate for interactions with
conduction electrons in metallic absorbers. It is also not
difficult to understand the success of the model in correctly
predicting the rate of energy loss of particles in heavy atoms.
For low enough velocities, adiabatic considerations constrain
projectiles to interact only with the particular atom through which
it is passing at any given time. For atoms with large enough
atomic numbers ($Z \gtrsim 10$), one can think of the atom as consisting
of a number of cells in which electrons are moving in regions
characterized by locally uniform potentials. This is in fact the
basis of the Thomas-Fermi atomic model.[20] In considering the
dynamics of the interaction between a projectile and the electrons
one is, therefore, allowed to neglect the electron-atom
interactions. As long as one is careful to confer the electrons
with a velocity distribution appropriate for a degenerate Fermi gas
and to take into consideration the fact that electrons can only
undergo transitions to states which are not already occupied (i.e,
to regions outside of their local Fermi sphere), the
projectile-electron interactions can be described by two-body
kinematics such as was done for the close collisions in the high
velocity limit. This is equivalent to what I shall refer to as the
binary encounter approximation (BEA) which can, and has been,
applied to physical systems other than Fermi gases of electrons.
For example, by conferring an electron gas with the velocity
distribution appropriate to that of hydrogenic atoms, one can
calculate the cross section for ionizing K-shell electrons by
electric projectiles[21,22] by integrating the interaction cross
section over energy transfer using the K edge as a lower limit.
Although the BEA seems to be purely classical, it has been shown to
be essentially equivalent[23,24] to the plane wave Born approximation
(PWBA), which in turn is equivalent[25] to the impact parameter
method (IPM) in which the first Born approximation is used to treat
the excitation of an atom by the classical electric field of a
projectile. The relationship of stopping power calculations to the
BEA, and thus to the implicit assumption of the validity of a
perturbation approach, is most apparent in the calculation of Fermi
and Teller[16] which is in fact a prototypical BEA calculation.
Those calculations of stopping power based on the dielectric
properties of electron gases[17-19] are not so clearly related.

However, as the derivation of the dielectric and magnetic permeabilities obtained there explicitly require a perturbative approach, it is not surprising that the results obtained agree fairly well with the more elementary approach of Fermi and Teller.

From the above we conclude that there are two requirements for the validity of the Fermi gas approximation for the treatment of the stopping power of charged particles:

1) The free electron gas approximation must be valid for describing the properties of the system under consideration. This will clearly be the case for conduction electrons in metals. It will also be true for heavy atoms ($Z_2 \gtrsim 10$) for which the Thomas-Fermi description is valid, and for projectile velocities large enough so that the kinematic limit to energy transfer from a heavy projectile (mass >> m) to an electron with a characteristic atomic velocity v_F (which can be taken to be the Fermi velocity) is large compared to the energy level spacing of the atom. The kinematic limit to energy transfer (for v_F << c) is $2mv(v+v_F)$. Setting this equal to 2 eV and assuming $v_F = \alpha c$, we find that v must be larger than $\sim 3 \times 10^{-4}$ c in order to neglect the finite atomic energy spacings. Brandt and Reinheimer[18] have made progress toward understanding in detail the nature of the energy loss of heavy electric particles near such velocity thresholds by considering the dielectric properties of an idealized system characterized as a degenerate, free Fermi gas separated from the continuum by an energy gap E_G. As $E_G \rightarrow 0$, the results of Ref. 18 approach those of Lindhard[17] and others in which a pure Fermi gas was assumed. For $E_G \neq 0$, the results of Ref. 18 indicate that there is a threshold velocity for electronic energy loss given by the two-body kinematics expression above. For a velocity twice the threshold, the stopping power is $\sim 1/2$ that obtained if $E_G = 0$ is assumed. Of course, one would almost certainly obtain the same results from a Fermi-Teller approach with suitable limits of integrations on scattering angles imposed. An attractive feature of the calculation of Ref. 18 is that the assumed model for the stopping material should apply to a wide range of condensed media. Although it is easy to convince oneself from studies of K-shell excitation cross sections[26] that inner shell electrons do not participate in energy loss processes for projectiles with velocities <<αc, it can be argued[27] that the valence gas of electrons which participates in energy loss processes at low velocity should be the same as that which takes part in low energy plasma excitations of solids. The number density of such electrons can be determined by studies of characteristic energy loss spectra of energetic electrons transmitted through thin films.[28] These spectra are insensitive to the conductivity of the solid, and seem to reflect rather the valence properties of the atom involved. In particular, the number density of those electrons which take part in plasma excitations of metals typically exceeds by a large factor the number density of conduction electrons.[28]

2) In order to apply the BEA to evaluate stoping power in
Fermi gases one must also require the validity of a perturbation
approach. For those situations in which the electrons are
delocalized (such as for conductors, aromatic rings in organic
scintillators, or excitons or electron-hole pairs in crystals) this
seems justified for the case of massive singly charged electric
projectiles such as protons. In such a case, the proton would be
one of many positive charges with which the electron interacts.
Furthermore, the effective charge of the proton would be at most
the same as that of any other postive charge which interacts with
the delocalized electrons. One can get an idea of the situation
for monopoles by comparing the interaction cross section of
monopoles and electrons with that for protons and electrons. To a
good approximation the former is given by (see Ref. 29 for the
exact result)

$$\frac{d\sigma}{d\Omega}\bigg|_m = \frac{g^2 e^2}{4m^2 v_F^2 c^2 \sin^4 \theta/2} \tag{3}$$

while the latter is given by the Rutherford cross section

$$\frac{d\sigma}{d\Omega}\bigg|_e = \frac{e^4}{4m^2 v_F^4 \sin^4 \theta/2} . \tag{4}$$

In the above equations we have assumed $v \ll v_F$ so that v_F should be
used as the relative velocity in the cross sections. The angle θ
denotes the center of momentum frame electron scattering angle.
The ratio of the monopole to proton cross section,

$$\frac{d\sigma}{d\Omega}\bigg|_m \bigg/ \frac{d\sigma}{d\Omega}\bigg|_e = \left(\frac{g}{e}\right)^2 \left(\frac{v_F}{c}\right)^2 \tag{5}$$

is a measure of the ratio of the interaction strength of monopoles
with electrons to that of protons with electrons. For Dirac
monopoles, $g = e/2\alpha$. Since $v_F \sim \alpha c$ we see that the above ratio is
~1/4, so that Dirac monopoles have comparable interaction strengths
as protons. Thus, it is reasonable to assume that if a
perturbation approach is valid for protons it will also be valid
for monopoles. Therefore, similar techniques used to evaluate
stopping powers of protons in delocalized electron gases should
yield valid results for monopoles in such systems. Furthermore,
since the effect of a proton or monopole on an electron in a heavy
atom ($Z_2 \gtrsim 10$) should be small compared to that of the nucleus, the
BEA should be valid for both types of projectiles. Therefore, the
calculations of stopping powers of both in Thomas-Fermi atoms
should be valid.

RESULTS FROM FERMI GAS CALCULATIONS

Electric Projectiles

The Fermi-Teller calculation[16] of stopping power of heavy electric particles in matter yielded the result:

$$\frac{dE}{dx}\bigg|_e = \frac{4}{3\pi} \frac{m^2 Z^2 e^4 v}{\hbar^3} \ln(1/\theta_{min}), \quad v \ll v_F \tag{6}$$

where θ_{min} was the minimum allowed electron scattering angle. This was assumed to be determined by screening and was given by $\theta_{min} \sim (\alpha c/v_F)^{1/2}$. The above result is similar to one obtained from an approximation of an equation derived by Lindhard,[17] in which the dielectric properties of the Fermi gas were used to determine the braking force on the projectile, ZeE, where E is the electric field acting on the projectile due to polarization of the medium. This force is simply the stopping power and it is found to be

$$\frac{dE}{dx}\bigg| = \frac{4}{3\pi} \frac{m^2 Z^2 e^4 v}{\hbar^3} [\ln\pi\left(\frac{v_F}{\alpha c}\right)^{1/2} + \frac{\alpha c}{\pi v_F} + (\ln\pi - 1)/2], \quad v \ll v_F \; . \tag{7}$$

Equation (7) has been calculated for protons in silicon and is plotted in Fig. 1. In the calculation, $v_F = \hbar(3\pi^2 N)^{1/3}/m$ has been determined by assuming the electrons to be uniformly distributed throughout the material. Comparison of this result with experiment is made in Fig. 1 where the measured stopping powers[15] are shown as open circles. The agreement is quite good. Also shown are experimental data for proton stopping powers at the Bragg peak and at large velocities. The Bethe-Bloch stopping power curve (Eq. (1)) is also shown. Note the premature decline of Eq. (1). As emphasized earlier, this is due to the failure of the high velocity approximations. It is worth pointing out the Z^2 dependence on projectile charge in Eqs. (6) and (7), which follows from the perturbation approach. The observed dependence on projectile atomic number is closer to a linear one. However, this is due to screening by electrons bound to the projectile.[36] For ions as light as hydrogen the orbital radius of such a bound electron would actually be large compared to the spacing of other electrons and the shielding effect should not be significant for protons. Experimental verification of this point has been pointed out elsewhere.[27] Finally, we emphasize that experimental data on the stopping power of protons in silicon do not extend to velocities low enough to reveal any threshold effects which might be expected from two-body kinematics ($\beta \sim 10^{-4}$ for a band gap of 1.1 eV).

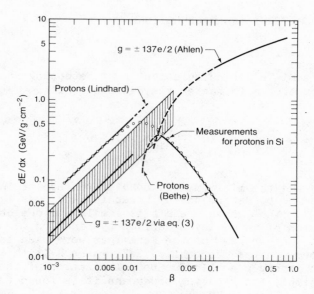

Fig. 1. Stopping powers in silicon for protons and for monopoles
with g = ±137e/2. Solid lines are calculations which are
taken from Ref. 11 (high velocity proton curve labeled
Bethe), Ref. 10 (high velocity monopole curve labeled
Ahlen), Ref. 17 (low velocity proton curve labeled
Lindhard) and Ref. 31 (labeled as Eq. (3) of that work).
Dashed lines are extrapolation of the various theories
into regions of questionable validity. Note that the
Bethe theory as shown does not include shell corrections.
The shaded region indicates estimated range of errors for
the slow monopole stopping power. The open circles are
the averaged values of high quality measurements of proton
stopping power in silicon and are taken from Ref. 15.

Magnetic Particles

Ahlen and Kinoshita[31] have extended the technique of Lindhard
to calculate the stopping power of magnetic monopoles in Fermi
gases. In particular, they have used the fact that $dE/dx|_m = gH$
where H is the magnitude of the magnetic field induced at the
monopole by eddy currents. Note that is is H and not the flux
density B which is to be used. This follows from the energy
conservation equation of the generalized Maxwell equations:

$$\frac{c}{4\pi} \vec{\nabla} \cdot (\vec{E} \times \vec{H}) + \frac{1}{4\pi}\left(\vec{H} \cdot \frac{\partial \vec{B}}{\partial t} + \vec{E} \cdot \frac{\partial \vec{D}}{\partial t}\right) = -\vec{J}_m \cdot \vec{H} - \vec{J} \cdot \vec{E} \qquad (8)$$

where $\vec{J}_m$ and $\vec{J}_e$ are magnetic and electric particle current densities respectively. It is worthwhile to note that if one erroneously assumed the force to be gB, it would be found that $dE/dx|_m = 0$ which is clearly impossible. By using the relation

$$\vec{B}(\vec{k},\omega) = \mu(k,\omega)\vec{H}(\vec{k},\omega) \qquad (9)$$

and by taking $\mu(k,\omega)$ from the work of Lindhard,[17] it was found[31] that

$$\frac{dE}{dx}\bigg|_m = \frac{2\pi Ng^2 e^2 v}{mc^2 v_F} \left[\ln \frac{1}{Z_{min}} - 1/2\right] \qquad v \ll v_F \qquad (10)$$

where Z_{min} is determined by eddy current losses for the case of conductors and is given by

$$Z_{min} = \hbar/(2mv_F\Lambda), \quad \text{conductors} \qquad (11)$$

where Λ is the conduction mean free path given approximately by[32]

$$\Lambda \approx 50aT_m/T , \qquad (12)$$

and where a is the lattice constant, T_m is the melting temperature and T is the temperature of the metal. Note that stopping power increases as the temperature is reduced, due to the increase in the conduction electron mean free path. Physically, this is due to the fact that Λ is a measure of the lateral extent of the electron wave packets. Thus, $1/\Lambda$ is a measure of the minimum allowed electron scattering angle which enters into Eq. (10) much as θ_{min} occurred in Eq. (6). In applying Eq. (10) for conductors, one should assume N to be given by the density of conduction electrons. However, there will be an additional contribution to the stopping power due to bound electrons. This, in fact, is the only component for non-conducting media. To obtain the value for this, one should use Eq. (10), assume N to be given by the total density of non-conduction electrons, and let $\Lambda \to a_0$, which is roughly the "mean free path" of an electron bound to an atom. We have followed this prescription to calculate the stopping power of a Dirac monopole in silicon. This is displayed in Fig. 1 along with the result for high velocity obtained from Ref. 10. The cross hatched region corresponds to a possible range of errors for the calculation. Part of this is due to the lack of consideration of the dynamical effects of the electron's magnetic moment in the derivation of Eq. (10). It is known that this cannot amount to much more than a 50% increase in stopping power as this is the difference between a Fermi-Teller type of calculation,[31] which includes the monopole-magnetic moment interaction, and Eq. (10).

We will subsequently discuss some prior work on monopole stopping power in which results at odds with Eq. (10) were obtained. In particular, some results were quoted in which other than linear dependences of stopping power on velocity in Fermi gases were obtained. At this time, it should be pointed out that the linear velocity dependence can be obtained from very general arguments. For example, it follows directly from Ohm's law and the scaling of monopole to proton cross sections. Equation (10) has also been derived[33] as an example of Brownian motion in which the force exerted on the monopole is examined as a form of friction. Finally, Eq. (10) has been derived independently[34] by G. W. Ford. Thus, I think it is fair to say that there is no controversy associated with the application of Eq. (10) to conducting media.

FERMI GASES WITH ENERGY GAPS

Ahlen and Tarlé[35] have estimated the response of plastic scintillators to monopoles by considering the valence electrons of the material as a Fermi gas with an energy gap of ~5 eV, the first excited electronic energy level of a benzene ring. This has been justified on the basis of extrapolation of results obtained with slow protons in scintillators. A summary of data and theory for the energy loss and scintillation response of a typical plastic scintillator, NE110, are given in Fig. 2. S_e is the electronic stopping power and has been extrapolated to very low velocities on the basis of data from Ref. 15 (open circles) and Ref. 36 (solid circles). S_n is the nuclear stopping power calculated from the LSS theory[37] and $S_{TOT} = S_p + S_e$. Finally, S_G is the stopping power which corresponds to exciting the 5 eV energy gap and which was calculated according to the model of Brandt and Reinheimer.[18]

On the right hand side of the figure are data on the scintillation efficiencies (dL/dE) of NE110 which have been obtained from L vs. E data[38] for the response of the scintillator to recoil protons from neutron exposure. The data have been fit to the standard Birks[39] expression for saturation,

$$\frac{dL}{dE} = \frac{A}{1+B \ dE/dx} \tag{13}$$

which has been observed to describe the core component of scintillation over a wide range of conditions.[40] For proton energies less than 100 keV, a value of B = 0.014 cm/MeV is required (this is larger than the value required for greater than 100 keV due to the narrower column of excited material left by low velocity projectiles which result in an enhanced probability for non-radiative quenching reactions).

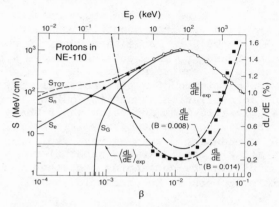

Fig. 2.   Contributions to the stopping power of slow protons (left
          scale) and comparison of calculated and experimental
          scintillation conversion efficiencies (right scale). Data
          for S_e, (open circles from Ref. 15, solid circles from
          Ref. 36) have been linearly extrapolated to low velocities
          and have been added to S_n from Ref. 17 to obtain S_{TOT}. For
          scintillation efficiency, the solid squares have been
          obtained from experimental data[38] on the response of
          scintillators to recoil protons produced by neutron
          exposure. The horizontal line marked $<dL/dE>_{exp}$ is the
          ratio L/E for 10 keV protons as measured in Ref. 38.

        Equation (13) was used to calculate[35] the yield of
scintillation photons, L, by a 10 keV proton in NE110. The
expression

$$L = \int \frac{dL}{dE} (S/S_{TOT}) dE_{TOT} \qquad (14)$$

was used for this calculation. By assuming $S = S_G$, a value of $L_G =$
40.3 eV was obtained which compares favorably with the experimental
value[38] L_{exp} = 37.0 ± 3.7 eV. Assuming $S = S_e$, the less favorable
value L_e = 45.5 eV was obtained. This suggests that as far as
scintillation is concerned, the electronic energy lost by protons
is accurately accounted for by the BEA in a Fermi gas with an
energy gap. This, in fact, is what would be expected since
scintillation involves principally the excitation of delocalized π
electrons in aromatic rings for which the perturbation
approximations should be valid. It is interesting to note that the
experimental values for S_e are not consistent with the existence of
a threshold velocity associated with an energy gap. This does not
necessarily conflict with our above statement regarding
scintillation, however. More will be said about this later.

Ahlen and Tarlé[35] have used the above results to calculate the scintillation yield of a Dirac monopole in NE110. The results are shown in Fig. 3. Curve B is for a bare monopole while curve A is for a monopole bound to an electric particle with charge e. If monopoles are bound to heavier nuclei,[41-3] the signal would be even larger. Note the existence of a threshold velocity at $\beta \sim 6 \times 10^{-4}$ above which the scintillation signal is quite large compared to that of a relativistic muon and below which the signal is absent. The threshold is due to the two-body kinematic constraint for $E_G = 5$ eV. By reducing the energy gap (for example, with commercially available acrylic based naphthalene scintillators or with exotic scintillators containing pentacene fluor molecules) it would be possible to reduce the threshold.

CLASSICAL ELECTRON GASES

In order to understand the dynamics of monopoles in the universe, it will be necessary to calculate their rate of energy loss in stars. For example, Dimopoulos et al.[44] have speculated that the possible observation by Cabrera[45] of a Dirac monopole could be reconciled with the Parker limit[46] if the sun could serve as a trap for Galactic monopoles. This would require sufficient energy loss while traveling through the solar material to bind the monopoles to the solar system. Since the fully ionized material of the sun is too hot to be considered degenerate, and, in fact, is sufficiently hot so that it can be considered a classical gas (see Fig. 4), the previous calculations of energy losses in Fermi gases are invalid. This can be seen most clearly by realizing that in a fully degenerate electron gas, it is impossible for a projectile to gain energy from the gas. This is not so for a classical gas. In

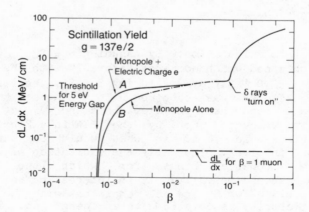

Fig. 3. Estimates of scintillation yield for magnetic monopoles. See text for a description of the different curves.

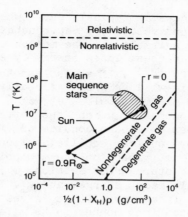

Fig. 4. Region of temperature-density space suitable for various
         ideal gas approximations. Note that for all points within
         our   Sun,   the   nonrelativistic,   nondegenerate   gas
         approximation is valid. X_H is the hydrogen mass fraction
         of the gas.

this case the mechanism for energy transfer from a projectile to
the gas is similar to the Fermi acceleration mechanism of cosmic
rays in which high energy particles gain more energy, on average,
than they lose in collisions with magnetic irregularities in the
Galaxy since head-on collisions are more likely than overtaking
collisions. In the problem of monopole energy loss, the monopole
takes   the   role   of the magnetic irregularity and the electrons in
the sun take the role of the cosmic rays. Tarlé and Ahlen[47] have
carried out the calculation of energy loss by assuming the BEA
whereby a Maxwell-Boltzmann velocity distribution was assumed.   As
might be expected, dE/dx is linear in velocity for $\beta < \alpha$ and it
approaches a constant value as $\beta \to 1$. The value of dE/dx in GeV
cm^2/g depends strongly on the region of the sun under
consideration. For example, near the surface of the sun ($0.9 R_\odot$)

$$\frac{dE}{dx} = 100 \text{ GeV cm}^2 \text{ g}^{-1} \beta, \quad r = 0.9 \text{ R}, \quad \beta \lesssim \alpha, \tag{15}$$

while at the center

$$\frac{dE}{dx} = 10 \text{ GeV cm}^2 \text{ g}^{-1} \beta, \quad r = 0, \quad \beta \lesssim \alpha, \tag{16}$$

The calculations of Ref. 47 can be used to calculate the
probability for the sun to capture monopoles via the mechanism of

Ref. 44. Figure 5 shows a typical case where a monopole with impact parameter b, and velocity v_∞ approaches the sun. In Fig. 5, the conditions are suitable for the monopole to be drawn into the sun through which it passes, loses energy by electronic collisions, and departs with velocity v_e. If v_e is less than or equal to escape velocity, the monopole will be gravitationally bound, as required by the mechanism of Dimopoulos et al.[44] By performing the appropriate integration in which the effects of gravity and electronic energy loss have been considered (magnetic field effects and energy loss by nuclear collisions contribute very small components to the net force on the monopole), Tarlé and Ahlen[47] have determined the capture cross section, G_c, of the sun as a function of monopole mass, charge, and v_∞. If G_c is multiplied by the isotropic flux of monopoles, it gives the rate at which monopoles are trapped into bound orbits. Values for G_c are shown in Fig. 6. It can be seen that only under a very limited range of conditions is the result close to that obtained if one assumes a completely absorbing nongravitating sun. For small velocities ($\beta \lesssim 10^{-4}$) it is usually legitimate to equate G_c with that expected of a completely absorbing gravitating sun, as was done by Dimopoulos et al. However, for large velocities the energy of monopoles with masses in the GUT range (10^{15}–10^{17} GeV/c^2) are so large that electronic collisions are inadequate to trap them in the solar system. The threshold value of v_∞ for trapping monopoles with mass 10^{17} GeV/c^2 is 9×10^{-4} c, for 10^{16} GeV/c^2 is 4×10^{-3} c, and for 10^{15} GeV/c^2 is 2.5×10^{-2} c. Since velocities of Galactic monopoles relative to the solar system must be of the order of 10^{-3} c or larger, it is quite possible that the Dimopoulos et al. argument may become untenable if the expected monopole mass rises above 10^{17} GeV/c^2.

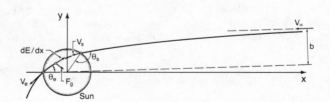

Fig. 5.   Schematic of mechanism by which a monopole may get trapped by our solar system. It approaches the sun with impact parameter b, velocity v_∞. If it passes through the sun, it will lose energy by collisions with solar electrons to emerge with a velocity v_e less than that it entered with. If v_e is smaller than escape velocity, the monopole is bound to the sun.

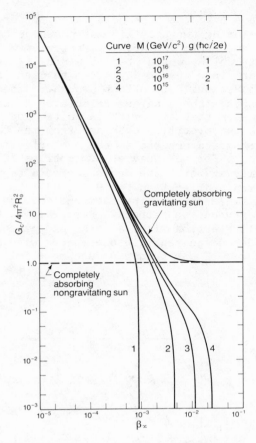

Fig. 6.   Cross sections G_c for capturing monopoles of varying mass, charge by our sun. When G_c is multiplied by an isotropic flux of monopoles, one obtains the rate at which monopoles are captured into the solar system.

STOPPING POWER IN LIGHT ATOMS

There are two difficulties with the calculation of the stopping power of monopoles (or electric particles) passing through isolated atoms with $Z_2 < 10$: 1) it is not legitimate to make the simplifying approximation that the atomic electrons can be treated as a degenerate Fermi gas; 2) the assumption that the monopole-electron (or proton-electron) interaction is a perturbation will probably be invalid, particularly as $Z_2 \to 1$. The data shown in Fig. 7 emphasize the second point. These are measured cross sections[26] for exciting K-shell electrons in carbon as a function of velocity for a variety of projectiles. The point labeled β_0 corresponds to the threshold velocity for excitation one would calculate assuming two-body kinematics where the electron is free and at rest. This is about a factor of 2 larger than the projectile velocity which would be obtained if the electron's motion were taken into account. It can be seen from Fig. 7 that the data for H and He ions fall off at roughly these velocities. The solid lines through the data are calculated cross sections from the BEA and it is seen that agreement is quite good. The finite cross sections at low velocities obtained from the calculations and substantiated by the data are due to the velocity spread of the K-shell electrons. For the heavier projectiles (C, N, O, Ne, Ar, Kr, Xe) the velocity at which the cross sections begin their rapid decline is ten times smaller than for H and He. This is due to the fact that the BEA fails for the interactions involved due to the inadequacy of perturbation methods. For example, the interaction of a carbon K-shell electron with a passing carbon nucleus can hardly be considered as negligible compared with the interaction with its own nucleus. These phenomena are more fully described in

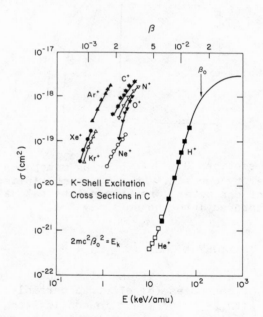

Fig. 7. Data from Ref. 26 for the excitation of carbon K-shell electrons by various projectiles as a function of projectile velocity. β_0 is the threshold velocity for this, assuming the kinematics for a collision of the projectile with a free electron at rest. The curve through the H[+] and He[+] data is from the binary encounter approximation.

two review articles on the subject.[21],[22] For the present purpose, it is sufficient to be aware of the possibility for reactions to proceed at velocities below BEA thresholds if there is reason to believe that perturbation techniques are inappropriate, and if overall energy-momentum conservation is allowed.

On the basis of the above discussion it would not be surprising if protons or monopoles could lead to the excitation of hydrogen atoms for velocities smaller than v_1 where $2mv_1(v_1+\alpha c) = 10.2$ eV, or $v_1 = 1.2 \times 10^{-3}$ c. Of course, the requirement of overall energy-momentum conservation implies that any excitation would be impossible for monopole velocities less than v_2 where $1/2 \, m_p v^2 = 10.2$ eV, or $v = 1.5 \times 10^{-4}$ c. However, since the region between 10^{-4} c and 10^{-3} c is a very important one with regard to searches for GUT monopoles, it is of extreme interest to ascertain the response of atoms to monopoles in this regime. A very important step in this direction has been made by Drell et al.[48] who calculated the cross section for exciting hydrogen atoms by slow monopoles. Strictly speaking their calculation applies to no actual physical system since the dipole approximation which was employed should be valid only for $\beta \leq 10^{-4}$ at which velocities there can be no excitation of hydrogen. Thus, they had to neglect the recoil of the proton in the calculation. Nevertheless, the results obtained have potentially great impact since they reveal that excitation probabilities are considerably enhanced over that expected from the BEA. In fact, they found the stopping power to be proportional to velocity, just as it was in conductors, Thomas-Fermi atoms and in classical plasmas such as the sun. The mechanism for the stopping power is basically simple. The presence of the monopole causes the splitting of energy levels by the Zeeman effect of the strong radial magnetic field of the monopole. If the monopole was moving infinitely slowly, there could be no net excitation unless the levels cross, which they do only for zero impact parameter. However, for monopoles with finite velocity, the uncertainty principle allows energy fluctuations of the order $v\hbar/a_0 = 3.7$ eV$(\beta/10^{-3})$ during the passage of a monopole with velocity βc. Thus, if the first excited energy level is separated from the ground state by only several tenths of an eV by the Zeeman splitting mechanism, excitation is allowed at $\beta \sim 10^{-4}$ by means of the smearing effect of the Heisenberg principle. Note that this is essentially a statement of the adiabatic theorem.

It is not clear how this result will affect searches for monopoles. As stated above, it actually relates to a non-existent physical system. It is suggestive, however, that a similar effect may apply to realistic systems, such as rare gas atoms which could be used to search for monopoles through their fluorescence, or to the σ bonding electrons of scintillators which could be excited by a Zeeman mechanism with the subsequent transfer of their excitation to the delocalized π electrons which could result in

scintillation.[49] Further work in these areas would be of great
value. In addition, it would be useful if an experimental
technique could be used as a test of the computational techniques
employed in Ref. 48. Although it is not possible to do so with a
monopole beam, there actually exist data on the interaction of
protons with carbon which have not yet been satisfactorily
explained. It is not inconceivable that a variant of the level
splitting method could resolve this problem, which was mentioned
previously in connection with the data for S_e in Fig. 2. These
data have been calculated from data on the electronic energy losses
of very low velocity (down to $\beta = 6 \times 10^{-4}$) protons in vapor
deposited carbon foils.[36] Since carbon prepared in this way should
have an energy gap,[50] one would expect, on the basis of the BEA,
that threshold effects should become important at such a small
velocity. That they do not suggests either that the yield of
excited molecules ceases to be proportional to the electronic
energy loss of a projectile at small velocities (which could be
dissipated as some type of mixed mode excitation analogous to
Cerenkov radiation) or that the BEA is invalid for carbon. It has
been observed previously that the data on the scintillation yield
for NE110 is suggestive of the validity of the BEA for the π
electrons. The delocalized character of these electrons supports
this conclusion. However, for localized, outer shell electrons in
carbon, or for σ bond electrons in NE110, it would not be too
surprising if the perturbation approach fails for protons or
monopoles. If the experimental results for protons in carbon can
be substantiated by the level splitting type of calculation, it
would be a great triumph and would be very supportive of the
monopole calculations.

SUMMARY OF ELECTRONIC STOPPING POWER CALCULATIONS

     The results of the various stopping power calculations for
Dirac monopoles which were discussed in the previous sections are
shown in Fig. 8. For hydrogen and silicon, the high velocity
components have been calculated according to Ref. 10. At low
velocities, the dashed lines for silicon and hydrogen correspond to
threshold velocities due to monopole-electron kinematic constraints
for the 1.1 eV band gap in silicon and to monopole-atom kinematic
constraints for the 10.2 eV first excited state of a real hydrogen
atom. It is worthwhile to note how well the low velocity limits
for hydrogen and silicon extrapolate to connect with their high
velocity components. Note also that the enhancement by a factor of
~10 of stopping power (in GeV cm^2 g^{-1}) in hydrogen compared to
silicon is not surprising since at large velocities stopping power
in the former is much larger than in the latter due to differences
in the neutron to proton ratio and in the mean ionization
potential.

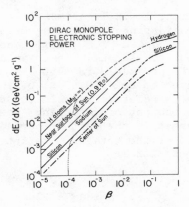

Fig. 8. Calculations of stopping power of monopoles with the Dirac
charge in a variety of materials. The calculation at low
velocities for hydrogen is from Ref. 48, those for the sun
are from Ref. 47, those at low velocity for hydrogen and
silicon are from Ref. 10. The dashed lines at low
velocity for hydrogen and silicon are due to threshold
effects   from   monopole-atom   and   monopole-electron
kinematics respectively. Thus, the hydrogen curve is
based on the assumption of an infinitely massive proton.

The stopping power in sodium (at room temperature) is seen to
be a factor ~3 larger than in silicon. This is due primarily to
the long mean free path for the sodium conduction electrons
(310 Å). This results in the contribution to stopping power by the
single conduction electron per sodium atom being nearly 6 times
larger than that due to the remaining ten electrons combined.

In the calculations of the stopping power in the sun, no
assumptions regarding the ratio of the monopole velocity to the
electron velocity were made, other than requiring the validity of
non-relativistic dynamics. The turn-over of the curves are,
therefore, calculated rigorously. Since such was not the case for
the Fermi gas calculations, it is reassuring to observe that the
change from the regime of linear dependence on velocity to the
plateau region occurs roughly where required to match the low
velocity Fermi gas and hydrogen calculations with the large
velocity calculations.

In order to assess the likelihood of monopoles being stopped
in the earth should they strike it, it is of interest to know the
stopping power of monopoles in the earth's interior. Since the
earth's crust, hydrosphere, and atmosphere constitute less than 1%

of the mass of the earth, it is legitimate to consider only the core and mantle in considering the slowing of GUT monopoles. The iron-nickel core[51] is solid for radii less than 1240 km and is liquid from 1240 km out to 3483 km. The mean density of the core is 11 g/cm^3, its temperature is 4000° K and its electrical conductivity is believed to be[51] $(0.6 \pm 0.2) \times 10^{16}$ s^{-1}. This conductivity implies a relatively small conduction mean free path (2.2 Å) so that the component of stopping power from conduction electrons is not so great in the earth's core (23% of the total) as it was for sodium (85% of the total). The total stopping power in the core for Dirac monopoles is

$$\frac{dE}{dx} = (24.5 \text{ GeV cm}^2 \text{ g}^{-1})\beta, \quad g = 137e/2, \text{ core}. \tag{17}$$

The mantle consists of most of the earth for radii greater than 3483 km. It has a density of ~5 g/cm^3 and is a very poor conductor (σ ranges from ~10^7 s^{-1} near the earth's surface to ~10^{11} s^{-1} near the core-mantle interface[52]). As such, stopping power is restricted essentially to that component in Thomas-Fermi atoms which is calculated to be

$$\frac{dE}{dx} = (21.5 \text{ GeV cm}^2 \text{ g}^{-1})\beta, \quad g = 137e/2, \text{ mantle}. \tag{18}$$

At this time it is not clear whether Eqs. (17) and (18) apply at velocities smaller than several times 10^{-4} c. At such velocities kinematic constraints from the BEA would prevent atomic excitation. However, in view of the results from Ref. 36 and Ref. 48, it seems likely that the linear dependence on β of electronic energy loss continues to low velocities, even though this may not result in a production of excited atoms. In any case, it is definitely true that in the core the stopping power will be at least as large as $(5.7 \text{ GeV cm}^2 \text{ g}^{-1})\beta$ which is the conduction electron component. In the following it will be assumed that the stopping power in the earth is given by $(23 \text{ GeV cm}^2 \text{ g}^{-1})\beta = S$. Magnetic fields in the earth will be neglected. Even if a field as large as 100 G is assumed (an estimate[51] of the azimuthal field in the earth's core), the magnetic force is smaller than that due to S for all velocities greater than 10^{-5} c. By including the effects of gravity and nuclear stopping[31] as well as of electronic stopping, it can be shown that monopoles with mass of 10^{16} GeV/c^2 cannot be trapped by the earth unless their velocity at infinity is less than 6×10^{-5} c. On the other hand, if monopoles have sufficient binding energy and cross section for the capture of a nucleus in the earth's crust, as has been suggested[43] for aluminum, then for velocities as large as 10^{-3} c the monopole-nucleus composite system would be stopped in the earth. This is clearly very important for searches

for monopoles which have been trapped in the earth.   If   monopoles
do not bind to nuclei it is unlikely that they would ever be
stopped by the earth.

DETECTION OF SLOW MONOPOLES THROUGH
ELECTRONIC EXCITATION AND IONIZATION

The focus of the above discussion has been to a large extent
concerned with the amount of energy lost by the monopole, and, with
the exception of the discussion on plastic   scintillators,   not so
much with the nature of the energy deposited in the absorbing
medium. However, this is the feature which is crucial to the
question of whether or not GUT monopoles can be detected and
identified with conventional types of charged particle detectors.
It is worthwhile to dispel at the outset of this section any notion
that it is not possible to detect electronic excitation for
velocities $\lesssim \alpha c$ as has been suggested in the past. This can be done
by presenting counter examples based on experiments with electric
projectiles. For example, the ionization of photocathode materials
with work functions ~2 eV has been observed for heavy ions with
β ~ 10^{-4} through the phenomenon of phototube afterpulsing.[53] In
addition, channeltrons having work functions ~4 eV (Ref. 54) have
been   used[55] to   detect He$^+$ ions with good efficiency for
β ~ 5×10^{-4}. As discussed earlier, there is indirect evidence that
polyvinyltoluene   based   organic scintillators, for which the
strongest low lying electronic excitation level is ~5 eV, will emit
significant quantities of scintillation light down to β ~ 6×10^{-4}.
It is not an accident that in the above examples the lowest
velocity in units of 10^{-4} c at which an observation of an effect
was made was so nearly equal to the energy gap in units of eV.
This is just what one calculates for BEA kinematics: $2mvv_F = E_G$
with v_F ~ αc. As the above systems utilized delocalized electrons
as the detection medium, the perturbation approximation and hence
the BEA should be valid and so the BEA threshold should apply.   In
Fig. 7 data on K-shell excitation have been shown which prove that
sub-BEA threshold velocities for electronic excitation are possible
if the projectile-electron interaction is sufficiently disruptive.
The data on proton stopping power in Fig. 2 suggest the existence
of a similar phenomenon for the excitation of carbon bonding
electrons.

One would expect that in any system for which the
electron-projectile interaction could be regarded as a perturbation
for an electric particle, such would be the case for a magnetic
monopole.   In this event, the BEA will apply and the threshold
velocity for detectability will be given roughly by E_G(eV) 10^{-4} c.
On the other hand, if the perturbation approach is invalid, one
would expect that projectiles could be detected at lower
velocities.   This is exemplified by the calculation of Drell et

al.[48] in which the excitation cross section of a hydrogen atom with
an infinitely massive proton would remain finite to arbitrarily
small velocities. This calculation could be used to justify
searches for slow monopoles with gaseous scintillation or
proportional counters in which the atomic excitation would be
detected either directly through fluorescence or indirectly through
ionization via the Penning effect. In principle, the threshold
velocity of detectability of monopoles with such systems is given
by $1.5 \times 10^{-4}\ c\sqrt{E(10eV)/A_2}$ where E is the atomic excitation energy
in units of 10 eV and A_2 is the atomic number of the gas atom or
molecule which collides with the monopole. For light atoms, such
as hydrogen or helium, for which fairly reliable calculations of
excitation cross sections should be possible, kinematics
constraints are not much less restrictive than apply to those of
low band gap condensed matter detectors such as silicon diodes, or
pentacene based scintillators. One advantage of the latter type
detectors is the ability to easily calculate the near threshold
response. On the other hand, for heavy gas atoms for which the
threshold effects are less severe, it may be more difficult to

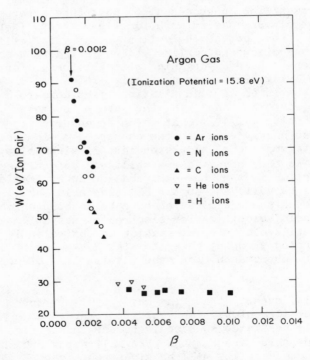

Fig. 9. Values of ionization yield per total energy deposited for
         various projectiles in argon gas as a function of
         projectile velocity (from Ref. 57).

calculate the level splitting to justify the ability to observe excitations at low velocity.

Ullman[56] has done pioneering experiments to search for exotic, slowly moving, supermassive particles. In fact, his experiments were initiated prior to the GUT monopole era. However, the proportional counters used in the experiments required the ionization of argon for the detection of a monopole. At this time it is not clear how the ionization yield of argon depends on velocities (even for electric projectiles) for velocities less than 10^{-3} c. Data on the W value for Ar are shown in Fig. 9 for a variety of projectiles.[57] Experimental conditions require heavier projectiles to get to lower velocity so that protons were used only for $\beta > 0.004$, while Ar projectiles extend down to $\beta = 0.0012$. Note that the W value is the ratio of the total energy of an ion which is lost in the gas to the number of ionizations produced. The increase in W for $\beta < 0.004$ reflects the increased difficulty of ionizing a gas at low velocities. However, Ullman[56] has argued that this is due primarily to the increased contribution of nuclear stopping power to the total stopping power at lower velocities. By assuming LSS cross sections, Dennis[58] has found that if this effect is considered, the resulting W_e values (electronic energy loss per ion pair) are relatively constant. However, it has been observed[59,60] that the LSS cross sections are typically too large by a factor of ~2 so that a substantial part of the increase of W in Fig. 9 must be due to an increase in W_e. Thus, the data do not rule out the existence of a threshold velocity at the BEA threshold of ~10^{-3} c for electric projectiles in argon, and, in fact, suggest that such a threshold may exist. This could be regarded as somewhat surprising in view of the K-shell excitation cross sections of Fig. 7 where it was observed that for projectiles of comparable atomic number as the target, sub-BEA threshold excitation was possible. One might think that for Ar projectiles passing through Ar gas, a similar effect might occur for the ionization yield. However, it should be noted that larger impact parameters will generally be responsible for net ionization, which involves outer shell electrons, than for inner shell excitation. Since the effective charge of Ar at $\beta \sim 10^{-3}$ is less than 1, it is likely that insofar as excitation of outer shell electrons is concerned, the interactions of Ar ions will not be significantly different from those of H ions (this would not be the case for K-shell excitation since the small impact parameters required for such processes would greatly reduce the screening of the charge of the heavy projectile by bound electrons). Since the passage of a proton through a heavy atom can be regarded as a perturbation (this should even apply to outer shell electrons which view the potentials of a large number of electrons and the potential of the nucleus in addition to that of the proton and an electron which may be attached to it), it might be expected that the BEA should apply for net ionization, and that the corresponding velocity thresholds would exist. One would expect the same thresholds to apply for monopoles. Since Ullman's experiment[56] was restricted to searching for particles with $\beta < 1.2 \times 10^{-3}$, it is quite possible that he would have failed to observe them even if they existed and passed through his detector. It should be noted that the mechanism of Drell et al.[48] would be of no help in permitting subthreshold ionization, as the continuum is too far from the ground state of an atom to be reached by level splitting by the monopole's magnetic field. In fact, in view of the comments made above it may be that subthreshold <u>excitation</u> of heavy atoms is not even allowed, at least at levels which could be detected easily.

To summarize, there are two means by which one can search for monopoles with velocities smaller than 10^{-3} c. One can use low band gap condensed materials which can be excited by monopoles through monopole-electron collisions with velocity thresholds $^{\sim}E_G(eV)10^{-4}c$. Such materials include semiconductor devices, inorganic scintillators, organic scintillators and infrared phosphors.[61] Or one can use gas scintillators which can conceivably be excited by the Zeeman mechanism with velocity thresholds $^{\sim}1.5 \times 10^{-4}c\sqrt{E(10eV)/A_2}$. It is possible that such mechanisms could reduce the threshold velocities of plastic scintillators and track etch detectors to $^{\sim}3 \times 10^{-5}$ c by means of interactions with the binding electrons of carbon atoms. Searches based on direct ionization of atoms by monopoles are not likely to succeed for $\beta < 10^{-3}$ although ionization could possibly be used as an indirect signature of the passage of a slow particle by means of the energy transfer from an excited atom or molecule (via the Zeeman mechanism) to a molecule with a lower lying ionization potential.

There are several ways in which one could, in principle, distinguish a slowly moving magnetic monopole from a slowly moving electric particle. For example, most electronic detectors would be capable of providing a velocity signal in addition to a total energy loss signal which would be about four times smaller for a Dirac monopole compared to a singly charged electric projectile in energy gap detectors. On the other hand, it may be possible to construct gaseous detectors which would detect only magnetic particles, by the Zeeman mechanism, below a threshold velocity. However, this would have to be demonstrable through detailed calculations of the energy loss of electric particles at low velocity in which it would be shown that excitation could not be induced by these particles.

It has been suggested[62] that the yield of triplet states in scintillators might be considerably enhanced for a low velocity monopole compared to that for an electric particle and that this could be used to discriminate against electric particles on the basis of the scintillation pulse shape.[39] However, by assuming the BEA and by taking the appropriate linear combinations of helicity flip and non-flip scattering amplitudes for electron-monopole scattering,[29] it has been calculated[35] that the total number of excited singlet states will always exceed the number of triplet states for the case of a Dirac monopole. Furthermore, any triplet states that are produced would not be observed in plastic scintillators or in poorly prepared liquid scintillators due to quenching by trace quantities of oxygen.[63]

Finally, it should be noted that all that has been said above would have to be modified if the monopole is carrying along with it a nucleus such as aluminum.[43] In such a case one should be able in principle to calibrate all monopole detectors with low energy heavy ion accelerators. This has never been done at the velocities of interest due to technical difficulties and to lack of interest. However, it now seems worthwhile to pursue such studies. From evidence quoted above, it seems likely that a monopole-Al composite would display the usual BEA thresholds in energy gap detectors and would have signals much larger than those of a bare monopole above threshold. As for the atomic-molecular systems, it is not clear what the signal would be for a low velocity monopole-Al composite. Since the force due to the Al will be much larger than that of the monopole in close collisions with atomic electrons, it seems probable that the interaction with Al will dominate. From studies of K-shell excitation at the velocities (Fig. 7), it appears that excitation may occur at velocities smaller than determined by BEA kinematics.

DETECTION OF MONOPOLES THROUGH
ENERGY LOSS IN CONDUCTORS AND SUPERCONDUCTORS

The linear dependence of monopole energy loss on velocity in conductors as given by Eq. (10) is well established and there is no reason to suspect the existence of velocity thresholds. This fact has been used to motivate the consideration of thermo-acoustic detection of monopoles[64],[65] in metals. At this time, however, it seems that there exist[65],[66] rather severe noise problems which limit the technique to monopole velocities significantly larger than those for which excitation sensitive devices are eminently suitable.

One might imagine the stopping power of monopoles in superconductors to be truly enormous (by letting $\Lambda \to \infty$ in Eq. (10)). However, the stopping power in superconductors will not be significantly different from that in ordinary conductors at large velocities due to the fact that the intense magnetic field of the monopole will exceed the critical field out to a relatively large impact parameter.[67] Assuming a critical field of 1 kG, one finds that for distances less than r = 570 Å from a Dirac monopole the material will be in the non-superconducting state. Since this distance is of the order of conduction mean free paths for good conductors at room temperature, which serve as effective maximum impact parameters in Eq. (10), it is apparent that the energy loss of a monopole in the cylinder of quenched superconductor will be of similar magnitude to that in metals at room temperature. In particular, it will depend linearly on velocity and will be of the order of 100 MeV cm^2 g^{-1} at $\beta = 10^{-3}$. There is an additional component of stopping power for superconductors however.[68] This is due to the field energy generated in the quenched cylinder by the supercurrents which have been set up to prevent an electric field from being induced in the unquenched part of the superconductor by the monopole. If a Dirac monopole passes through a superconductor of thickness x, 2 flux quanta ($\phi = 2\pi\hbar c/e$) thread the quenched cylinder after the monopole is gone. The magnetic field in the cylinder is given by $\phi/(\pi r^2)$ and the field energy by $x\phi^2/(8\pi^2 r^2) = x\hbar^2 c^2/(2e^2 r^2)$. This yields a stopping power of 42 MeV/cm. This is a small fraction of stopping power at $\beta \sim 10^{-3}$ but dominates for velocities smaller than 10^{-4} c.

One might wonder if a constant component of stopping power also exists for ordinary conductors. The existence of such a term would not have shown up in the analysis of the stopping power of monopoles presented earlier. This is due to the fact that in the BEA all of the energy transfer is assumed to manifest itself in the kinetic energy of the scattered electrons. This assumption is quite explicit in the Fermi-Teller approach. It is implicit in the Lindhard approach insofar as a perturbation approximation has been assumed to derive the dielectric and magnetic permeabilities so that only two-body collisions are considered. This implies that it would be impossible for the magnetic field energy to be nearly as large as that of the kinetic energy transferred to electrons since the induction of large magnetic fields is essentially a coherent phenomenon. To see this, note that the ratio of the total magnetic field energy of an electron traveling in a circle of radius R to the kinetic energy of the electron is $T_F/T_K \sim r_e/R$ where r_e is the classical radius of the electron. This ratio is always much less than 1. For a group of n electrons moving incoherently, the ratio T_F/T_K is still r_e/R, while for a group of electrons moving in unison in the same circle it is n times as large. This is due to the field energy increasing by a factor n^2 while the kinetic energy changes only by the factor n. To estimate the ratio of field energy to kinetic energy of electrons left in the wake of a monopole moving through a conductor, we need to estimate n and R. The former is given by the product of the conduction electron density N with the volume V which contains the maximum number of electrons which can contribute coherently to an induced magnetic field by their azimuthally induced motion. By symmetry V will be cylindrical in shape. Its length will be $v\Lambda/v_F$, which is the distance behind the monopole at which an electron can possess memory of the monopole's passage before it is forgotten through a dissipative collision with a lattice impurity. The cylinder radius can be no larger than $\Lambda v/v_F$ beyond which individual electrons collide with lattice impurities prior to the time it takes to induce a collective motion. For typical conductors, $N \sim 1/(5a_0)^3$ so that currents of electrons moving coherently will be confined to radii between ~ 1 Å and $\Lambda v/v_F$. Since $\Lambda \sim 10^2$-10^3 Å, it is reasonable to assume that $R \sim \Lambda v/2v_F$ for $v \ll v_F$. With this prescription it is found that $T_F/T_{K_3} \sim 0.05(\beta/10^{-3})^2(10^{-3}\Lambda/a_0)^2$ which is much less than 1 for $\beta < 10^{-3}$. Thus, it is safe to conclude that there will not be a constant component to stopping power as there is for superconductors.

OTHER CALCULATIONS OF THE STOPPING POWER OF GUT MONOPOLES

    The issue of the detectability of GUT monopoles by conventional types of charged particle detectors has been shrouded in confusion by a number of calculations which have been in error or have been misinterpreted. In this section a number of these calculations will be discussed. Although most of these remain unpublished, they have received widespread distribution in preprint form. It is important that the flaws in these calculations be indicated so that decisions on the best means of pursuing the search for GUT monopoles will not be made on the basis of erroneous information.

    There have been three calculations on the stopping of monopoles in conductors which differ from the results presented here.[8,65,69] The equations of Refs. 8 and 69 were presented without details so that it is impossible to evaluate them except on the basis of the results. Although each of these results for stopping power displayed a linear dependence on monopole velocity, neither had the logarithmic dependence on conduction mean free path and hence on the temperature of the metal. In fact, the result from Ref. 8 had no temperature dependence at all, which can be taken as manifest evidence for its being incorrect. The equation for the stopping power in conductors from Ref. 69 depended on the conduction mean free path as $\Lambda^{1/2}$ rather than as $\ln(2k_F\Lambda) - 1/2$ as in Eq. (10). It is difficult to see how the former dependence can be correct although, as said previously, there is no way to isolate the error made in the absence of details. Although the stopping power derived by Akerlof[65] was in greater error than those values obtained in Refs. 8 and 69, his attempt was extremely valuable. This is due to the fact that he presented a detailed account of his assumptions and calculations which allowed a determination of his error to be made. This error was in assuming a purely frequency dependent conductivity in evaluating the energy deposited in the conductor by eddy currents. It was pointed out in Ref. 31 that it is necessary to take into account the wavelength dependence as well for the case of the energy loss of particles moving more slowly than the Fermi velocity. It is probable that the error made in Ref. 8 was also the assumption of a frequency dependent conductivity. Another error must also have been made which accounts for the difference between the results of Refs. 8 and 65.

There has been, to the best of my knowledge, just one attempt besides that of Drell et al.[48] to calculate the stopping power of monopoles in hydrogen atoms.[70] Aside from the rather crude manner in which the problem was set up in Ref. 70, there were two flaws in the calculation, the results of which indicated a much more rapid decline at low velocities than obtained in Ref. 48. The first flaw was the assumption of a perturbation approximation which has been shown here to be inappropriate for monopoles in light atoms. The second was the assumption that monopole stopping power scales as $(g\beta/Ze)^2$ compared to that of electric projectiles. This has been shown here to be valid for $\beta \gg \alpha$. However, at small velocities the correct scaling is given by the ratio of monopole to electric projectile cross sections which yields the factor $(gv_F/Zec)^2$. This error has been made by others and these will be discussed shortly.

Figure 10 contains results of a number of calculations which have been made of the stopping power of monopoles in non-conducting solids. These are for silicon (solid curves) and plastic (dashed curves). The incredible disparity between the calculations for the same material as shown in Fig. 10 should be contrasted with the remarkable similarity of results for stopping power in a wide variety of absorbers in Fig. 8. Most of the calculations shown in Fig. 10 are in error. For example, the curve labeled T has been

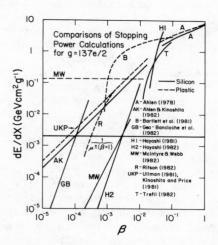

Fig. 10. Comparisons of stopping power calculations for Dirac monopoles in nonconducting solids. See text for discussion.

obtained[71] by assuming that the impulse approximation, in which one neglects the motion of the atomic electrons while the projectile passes by, is valid for the stopping power of slow monopoles. This is clearly invalid for $\beta \ll 10^{-2}$. The curve labeled H2 (Ref. 72) and the solid curve labeled MW (Ref. 62) are incorrect due to an erroneous application of the $(g\beta/Ze)^2$ scaling law. The curve labeled B (Ref. 73) is based on an invalid extrapolation of an expression for primary ionization based on the response of plastic track detectors to fast heavy ions. The dashed curve labeled MW (Ref. 62) is an estimate of the stopping power of monopoles in plastic scintillator assuming a velocity independent cross section for exciting triplet levels in aromatic rings. This is in principle analogous to the Zeeman splitting calculation of Ref. 48. However, lack of a detailed calculation prevents it from being taken too seriously. The curve labeled R (Ref. 74) is an attempt to moderate the stopping power of monopoles by the BEA threshold effect by performing an appropriately limited integration over a Thomas-Fermi carbon atom. One problem is the paradoxical nature of this term in that carbon atoms should not be very suitable for the Thomas-Fermi approximation. Another problem is that no attention was paid in Ref. 74 to the threshold limitations of proton stopping power from which the monopole stopping power was obtained by scaling the cross sections. The curve labeled UKP (Refs. 56 and 75) was obtained by the somewhat lucky guess that monopole stopping power should be the same order of magnitude as proton stopping power. Limitations of this curve, and the one labeled AK (Ref. 31) would only arise if there are low velocity thresholds at $~10^{-4}$ c for silicon and $~5 \times 10^{-4}$ c for plastic. The curves labeled H1 (Ref. 76) and GB (Ref. 69) are based on classical calculations of the stopping power of monopoles in atoms. Although I have not tried to reproduce their results, I question the classical approximation and am, therefore, not surprised, for example, that the curve labeled H1 fails to smoothly connect the solid curves labeled A (Ref. 10) and AK, which should be valid in silicon for $\beta > 0.05$ and $10^{-4} < \beta < 10^{-2}$ respectively. To conclude, the only curves that I trust in Fig. 10 are those labeled UKP (and then, perhaps only for $\beta > 5 \times 10^{-4}$), AK ($\beta > 10^{-4}$), and A.

ACKNOWLEDGEMENTS

In arriving at my views of the energy loss of slow monopoles which I have summarized here, I have benefited from conversations with a number of people. In this regard I would especially like to thank Carl Akerlof, Blas Cabrera, Sid Drell, Bill Ford, Ray Hagstrom, Kay Kinoshita, Peter McIntyre, Don Morris, Mark Mueller, Steve Parke, Greg Tarlé and Jack Ullman. I would also like to extend my gratitude to the number of people who made the initial, although often erroneous, calculations of GUT monopole stopping power. I learned much from their mistakes, and the physics community should be grateful to them for boldly entering the field of the energy loss of charged particles in matter, which is so full of complexities and subtleties. They should not be faulted for errors they have made as these have actually hastened what I believe to be the development of a fairly well established state of knowledge for the stopping power of slow monopoles.

REFERENCES

1. G. t'Hooft, Nucl. Phys. B79, (1974).
2. A. Polyakov, JETP Lett. 20, 194 (1974).
3. H. Georgi and S. Glashow, Phys. Rev. Lett. 32, 438 (1974).
4. H. Georgi, H. R. Quinn and S. Weinberg, Phys. Rev. Lett. 33, 451 (1974).
5. J. P. Preskill, Phys. Rev. Lett. 43, 1365 (1979).
6. H. J. D. Cole, Proc. Cambridge Philos. Soc. 47, 196 (1951).
7. E. Bauer, Proc. Cambridge Philos. Soc. 47, 777 (1951).
8. V. P. Martim'yanov and S. Kh. Khakimov, Sov. Phys. JETP 35, 20 (1972).
9. S. P. Ahlen, Phys. Rev. D 14, 2935 (1976).
10. S. P. Ahlen, Phys. Rev. D 17, 229 (1978).
11. U. Fano, Ann. Rev. Nucl. Sci. 13, 1 (1963).
12. S. P. Ahlen, Rev. Mod. Phys. 52, 121 (1980).
13. S. P. Ahlen, G. Tarlé and P. B. Price, Science 217, 1139 (1982).
14. S. P. Ahlen and G. Tarlé submitted to Phys. Rev. Lett. (1983).
15. H. H. Andersen and J. F. Ziegler, Hydrogen Stopping Powers and Ranges in All Elements (Pergamon, New York) (1977).
16. E. Fermi and E. Teller, Phys. Rev. 72, 399 (1947).
17. J. Lindhard, Mat. Fys. Medd. Dan. Vid. Selsk. 28, No. 8 (1954).
18. W. Brandt and J. Reinheimer, Can. J. Phys. 46, 607 (1968).
19. P. M. Echenique, R. M. Nieminen and R. H. Ritchie, Solid State Commun. 37, 799 (1981).
20. L. D. Landau and E. M. Lifshitz, Quantum Mechanics 2nd edition (Pergamon Press, New York), p. 241 (1965).
21. W. E. Meyerhoff and K. Talubjerg, Ann. Rev. Nucl. Sci. 27, 279 (1977).
22. J. D. Garcia, R. J. Fortner and T. M. Kavanaugh, Rev. Mod. Phys. 45, 111 (1973).
23. D. R. Bates and W. R. McDonough, J. Phys. B 3, L83 (1970).
24. D. R. Bates and W. R. McDonough, J. Phys. B 5, L107 (1972).
25. H. A. Bethe and R. W. Jackiw, Intermediate Quantum Mechanics (Benjamin, New York), p. 326 (1968).
26. R. C. Der, T. M. Kavanagh, J. M. Khan, B. P. Curry and R. J. Fortner, Phys. Rev. Lett. 21, 1731 (1968).
27. A. Mann and W. Brandt, Phys. Rev. B24, 4999 (1981).
28. L. Marton, L. B. Leder and H. Mendlowitz, Advances in Electronics and Electron Physics 7, 183 (1955).

29. Y. Kazama, C. N. Yang and A. S. Goldhaber, Phys. Rev. D 15,
     2287 (1977).
30. J. Lindhard and M. Scharff, Phys. Rev. 124, 128 (1961).
31. S. P. Ahlen and K. Kinoshita, Phys. Rev. D 26, 2347 (1982).
32. J. M. Ziman, Principles of the Theory of Solids (Cambridge
     University Press), p. 190 (1964).
33. W. L. Schaich, private communication (1982).
34. G. W. Ford, Phys. Rev. D 26, 2519 (1982).
35. S. P. Ahlen and G. Tarlé, Phys. Rev. D 27, 688 (1983).
36. S. H. Overbury, P. F. Dittner, S. Datz and R. S. Thoe, Rad.
     Eff. 41, 219 (1979).
37. J. Lindhard, M. Scharff and H. E. Schiøtt, Kgl. Danske
     Videnskab. Selskab., Mat. Fys. Medd. 33, No. 14 (1963).
38. J. A. Harvey and N. W. Hill, Nucl. Instr. and Meth. 162, 507
     (1979).
39. J. B. Birks, The Theory and Practice of Scintillation Counting
     (Pergamon Press, Oxford) (1964).
40. M. H. Salamon and S. P. Ahlen, Nucl. Instru. and Meth. 195,
     557 (1982).
41. W. V. R. Malkus, Rev. 83, 899 (1951).
42. D. Sivers, Phys. Rev. D 2, 2048 (1970).
43. C. Goebel, Proc. Monopole Seminars, Madison (1981).
44. S. Dimopoulos, S. L. Glashow, E. M. Purcell and F. Wilczek,
     Nature 298, 824 (1982).
45. B. Cabrera, Phys. Rev. Lett. 48, 1378 (1982).
46. M. S. Turner, E. N. Parker and T. J. Bogdan, Phys. Rev. D 26,
     1296 (1982).
47. G. Tarlé and S. P. Ahlen, to be published (1983).
48. S. D. Drell, N. M. Kroll, M. T. Mueller, S. J. Parke and M. A.
     Ruderman, Phys. Rev. Lett. 50, 644 (1983).
49. J. B. Birks, Photophysics of Aromatic Molecules (Wiley, New
     York) (1970).
50   J. Kakinoki, K. Katoda, J. Hanawa and T. Ino, Acta.
     Crystallogr. 13, 171 (1960).
51. E. N. Parker, Cosmical Magnetic Fields (Clarendon Press,
     Oxford), Chapter 20 (1979).
52. W. M. Kaula, An Introduction to Planetary Physics (Wiley, New
     York), p. 147 (1968).
53. G. A. Morton, H. M. Smith and R. Wasserman, IEEE Trans. Nucl.
     Sci. NS-14, No. 1, 443 (1967).
54. D. S. Evans, Rev. Sci. Instr. 36, 375 (1965).
55. B. Tatry, J. M. Bosqued and H. Reme, Nucl. Instr. and Meth.
     69, 254 (1969).
56. J. D. Ullman, Phys. Rev. Lett. 47, 289 (1981).
57. J. A. Phipps, J. W. Boring and R. A. Lowry, Phys. Rev. A 135,
     36 (1964).
58. J. A. Dennis, Rad. Eff. 8, 87 (1971).
59. H. Grahmann and S. Kalbitzer, Nucl. Instru. and Meth. 132, 119
     (1976).

60. W. D. Wilson, L. G. Haggmark and J. P. Biersack, Phys. Rev. B
     15, 2458 (1977).
61. R. Hagstrom private communication (1982).
62. P. McIntyre and R. Webb, Texas A & M University Preprint
     (1982).
63. F. D. Brooks, Nucl. Instr. and Meth. 162, 477 (1979).
64. B. Barish, Proceedings of the Magnetic Monopole Workshop,
     Racine, Wisconsin, edited by P. Trower and R. Carrigan, to
     be published (1982).
65. C. W. Akerlof, Phys. Rev. D 26, 1116 (1982).
66. C. W. Akerlof, submitted to Phys. Rev. D (rapid
     communications) (1982).
67. B. Cabrera, private communication (1982).
68. D. Morris, private communication (1982).
69. J. Gea-Baneloche, K. Cahill, D. Rossbach and A. Comtet,
     University of New Mexico Preprint, (June, 1982).
70. S. Geer and W. G. Scott, CERN Preprint (April, 1981).
71. J. S. Trefil, University of Virginia Preprint (1982).
72. K. Hayashi, Kinki University Preprint (February, 1982).
73. D. F. Bartlett, D. Soo, R. L. Fleischer, H. R. Hart, A.
     Mogro-Campero, Phys. Rev. D 24, 612 (1981).
74. D. Ritson, Stanford University Preprint (1982).
75. K. Kinoshita and P. B. Price, Phys. Rev. D 24, 1707 (1981).
76. K. Hayashi, Lett. Nuovo Cim. 33, 324 (1982).

ELECTRONIC COSMIC RAY MONOPOLE SEARCHES

Eugene C. Loh

Department of Physics
University of Utah
Salt Lake City, Utah 84112

INTRODUCTION

Searches for magnetic monopoles of the order of a few GeV in mass are well documented by Carrigan,[1] Carrigan et al.,[2] Craven et al.,[3] and are reviewed by Longo.[4] I will limit this paper to reviewing current and future searches for massive magnetic monopoles of the 't Hooft[5]-Polyakov[6] type with e/2α as the magnetic charge, an idea originated by Dirac.[7]

In the simple GUT theory[8] the magnetic monopoles are produced at the grand unification energy scale set by the lepto-quark boson mass. The magnetic monopole mass is expected to be $m_m \approx m_x/\alpha$. The best guess of m_x is about 10^{15} GeV[9], a value also related to the proton lifetime, and the grand unification fine structure constant α is about 1/50.

Since no observed object in the universe is hot enough to produce these monopoles, it is commonly accepted[10] that monopoles can only have been produced at the very early epochs ($\leq 10^{-36}$ s) of the Big Bang. The predictions of monopole density made with the standard theory exceed the present matter density by ~11 orders of magnitude. This prediction clearly violates a number of astrophysical observations such as the existence of galactic magnetic field,[11] galactic stability,[12] and expansion of the universe,[13] and can also be ruled out by all the existing monopole search experiments. Much effort is now being expended on phase transition studies[14] in the very early universe which hopefully will lead to a more realistic prediction for the monopole density. In this picture, the galactic magnetic field provides the acceleration mechanism and a guide with which the flux limit and velocity limit of the magnetic monopoles can be estimated.

FLUX LIMITS

If the mass of the monopole is 10^{17} GeV or less, the monopoles are accelerated by the galactic magnetic field to a velocity above 10^{-3} c and ejected from the galaxy. This follows from the fact that the galactic magnetic field is of the order of a few microgauss and coherent over a distance of at least 10^{21} cm. The flux of the monopoles must be such that the magnetic field must not be drained off in a period short compared to the regeneration time.[15] Details of this limit are worked out by Turner, Parker and Bogdan.[16]

If the mass of the monopole is greater than 10^{17} GeV, the monopole can be bound to the galaxy for a period of time. In this case, these monopoles will have the virial velocity of the galaxy ~10^{-3} c, and the flux is again limited by the argument of the draining of the galactic field. For monopole masses around 10^{19} GeV, the monopole can be bound for the lifetime of the galaxy. For a monopole of this mass, the stability of the galaxy imposes a lower limit than the limit set by the galactic magnetic field. The galactic stability limit is based on a set of observations on the mass distribution of elliptical and spiral galaxies similar to our galaxy.[13]

Recently, Dimopoulos, Glashow, Purcell, and Wilczek[17] raised the interesting possibility that monopoles can be captured by stellar systems. Although the capture mechanism is not well understood, the flux of Cabrera's[18] magnitude can exist without violating the magnetic fields of the solar system and of the sun. The earth would encounter monopoles gravitationally bound to the solar system at not much more than its own orbital velocity, 10^{-4} c.

Figure 1 shows the various limits stated above. Based on these arguments, the recent searches cover the velocity region from 10^{-2} c to 10^{-4} c.

MONOPOLE ENERGY LOSS

Monopole energy loss in matter has been calculated for a wide range of velocities.[19] There is general agreement that for $v \geq 10^{-3}$ c the calculations by Ahlen and Kinoshita[20] represent a lower bound on the energy loss. For still smaller velocities, calculations are being done on simple atoms and preliminary results on the hydrogen atom are reported by S. Parke[21] at this conference. Parke et al. predict an energy loss of ~40 times the ionization of a relativistic muon (I_{min}) in hydrogen for monopole velocity of 10^{-4} c. The energy loss is enhanced by the crossing of hydrogen atomic levels induced by the monopole. It is likely that such

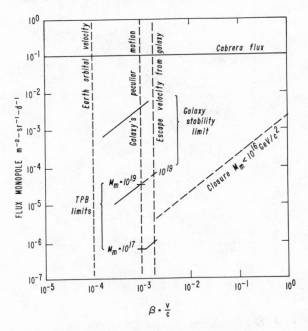

Fig 1. Monopole flux limits.

enhancements also occur in scintillators and noble gases such as argon. However, in the absence of rigorous calculations in the materials used in the detectors, a calculation by Ritson[22] (Fig. 2) is used to normalize the velocity sensitivity limit of the experiments reported. Whenever possible, the detector threshold setting of each experiment[23] is given so as to enable the reader to reinterpret the result when new and more rigorous energy loss calculations become available.

PRESENT MAGNETIC MONOPOLE SEARCHES

1. The University of Michigan[24] search uses a horizontal stack of five scintillation counters (Fig. 3). The detector is set to trigger on particles depositing 1/100 of the minimum ionization energy (I_{min}) in 4 out of 5 counters with particle velocity in the region $3 \times 10^{-4} < \beta < 1 \times 10^{-2}$. No event is recorded which penetrates all five counters. An upper limit of 0.1 $m^{-2}sr^{-1}d^{-1}$ is obtained (90% c.l.).

2. The Utah-Stanford search[25] is located in the Mayflower Mine near Heber, Utah. The overburden of the detector is 5×10^4 g cm^{-2}. The 2.7 m^2sr scintillation counter detector (Fig. 4) is arranged in 3 double layers of 4 counters each with an extra

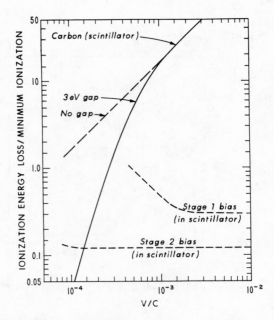

Fig. 2. Monopole energy loss in carbon.

layer of 4 counters. The electronics is set to detect
particles in the velocity range of 1×10^{-4} c to 0.01 c and to
detect particles depositing more than 0.12 I_{min} in each of the
6 layers of counters. A monopole candidate must also satisfy
spatial and temporal reconstruction criteria. No events have
been observed. The upper limit of the monopole flux is 1×10^{-2} m^{-2} sr^{-1} d^{-1} (90% c.l.).

3. The Bologna group[26] has enlarged the detector used in an
earlier monopole search[27] (Fig. 5). The new scintillation
counter detector has an area-solid-angle product of 38 m^2 sr
with the threshold of each counter set to detect particles
depositing 20 times I_{min}. No magnetic monopoles have been
detected. Several velocity regions were explored in different
runs. The results are summarized in Fig. 10.

4. The Tokyo group has two different monopole searches[28] in
progress with two different arrays. The first search[29]
utilized a 1.1 m^2 sr scintillation counter detector (Fig. 6).
This experiment is sensitive to monopoles in the velocity
region of $0.01^{-2} < \beta < 0.1$. Two layers of the counters
generated triggers, and 6 layers of shower counters with bias
set at 1.2 I_{min} were used to remove accidental coincidences.
No candidate event has been detected. An upper flux limit of

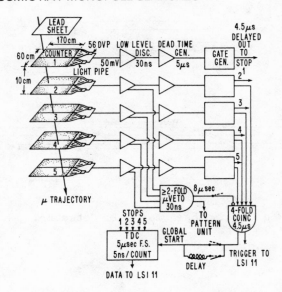

Fig. 3. University of Michigan monopole detector.

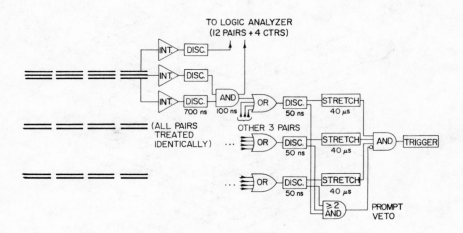

Fig. 4. Utah-Stanford monopole detector.

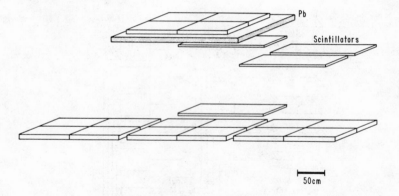

Fig. 5.   Bologna monopole detector.

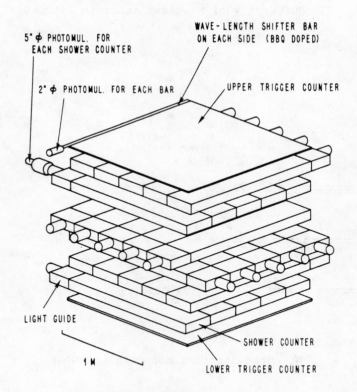

Fig. 6.   Tokyo monopole detector.

1.2×10^{-2} m^{-2} sr^{-1} d^{-1} was obtained. Since then the threshold of the system has been lowered to 1/40 of I_{min} and the detector size enlarged to 1.42 m^2 sr. A flux limit of 2×10^{-2} m^{-2} sr^{-1} d^{-1} is now obtained for $2 \times 10^{-4} < \beta < 10^{-3}$ (90% c.l.).

The Tokyo group also installed a new scintillation hodoscope counter array[28] in the Kamioka Mine (Fig. 7). The new detector has an area-solid-angle product of 22 m^2 sr. An event is defined as a slow particle ($3 \times 4^{-4} < \beta < 0.1$) depositing 1/16 I_{min} in all 6 layers. The monopole flux upper limit is 3.54×10^{-3} m^{-2} sr^{-1} d^{-1} (90% c.l.).

5. The experiment in the Baksan Mountain[30] uses the existing cosmic-ray detector (Fig. 8). A full description of the detector is reported in the 1979 International Cosmic Ray Conference.[31] For the monopole search, the energy threshold is set to detect 1/4 I_{min} particles. An upper limit of 1.3×10^{-5} m^{-2} sr^{-1} d^{-1} (90% c.l.) for monopoles in the velocity region $4 \times 10^{-3} < \beta < 5 \times 10^{-2}$ is reported.

6. The nucleon-decay experiment located in the Soudan Mine[32] uses a large proportional counters array to detect the presence of slowly moving particles (Fig. 9). The 71.6 m^2 sr detector is capable of identifying particles which traverse the proportional counters with velocity between 3×10^{-4} and 3×10^{-2} c, and ionize at least half as much as a relativistic muon. A trigger is generated by signals in 3 consecutive horizontal proportional counter layers. No monopole has been detected. An upper monopole flux limit of 3.5×10^{-4} m^{-2} sr^{-1} d^{-1} (90% c.l.) is obtained.

7. The Japan-India collaboration[33] used the nucleon-decay apparatus in the Kolar gold mine to search for a monopole flux. The 208 m^2 sr proportional counter array is set to detect particles which deposit more than 2.5 times I_{min} in five out of eleven consecutive proportional counter layers. An upper limit of 3×10^{-5} m^{-2} sr^{-1} d^{-1} (90% c.l.) is obtained for monopole velocity $10^{-3} < \beta < 1$.

8. A large detector (180 m^2 sr) made of a single layer of CR39 plastic was used by the Berkeley group[35] in a search for fast monopoles. From the calculations of Ahlen,[36] it is estimated that for monopole with velocity greater than 0.02 c a track can be detected in the plastic. No candidate is observed in the plastic. The limit of monopole flux is set as 8.7×10^{-5} (95% c.l.).

The results of all the searches are listed in Table 1 and plotted in Fig. 10.

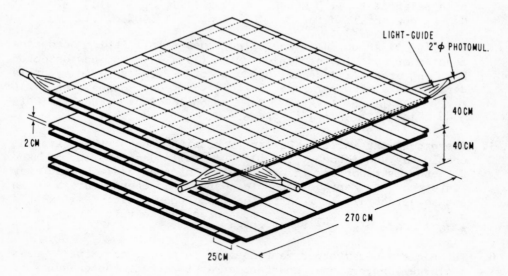

Fig. 7.   Tokyo-Kamioka Mine monopole detector.

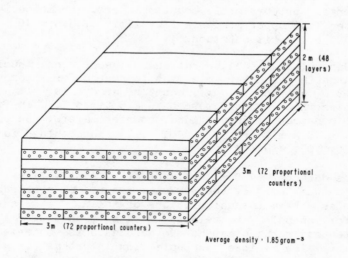

Fig. 8.   Soudan Mine monopole detector.

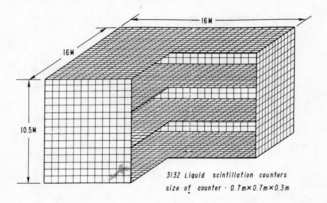

3132 Liquid scintillation counters
size of counter · 0.7m×0.7m×0.3m

Fig. 9.   Baksan Mountain (U.S.S.R.)

Table 1.   Summary of Experimental Results.

Exp	Detector	Sensitivity	Estimated Electronic β Acceptance	m^2-sr^{-1}-day	m^2 sr
BNL	Prop.	2 I_{min}	3×10^{-4}, 1.2×10^{-3}	< .03	1.3 π
Michigan	Scin.	~1/100 I_{min}	3×10^{-4}, 10^{-2}	< .1	3.16
Utah Stanford	Scin.	1/20 I_{min}*	9×10^{-5}, 10^{-2}	< 1×10^{-2}	2.7
Tokyo I	Scin.	1.2 I_{min}	.01, 0.1	< 1.3×10^{-2}	1.1
II	Scin.	1/40 I_{min}*	2×10^{-4}, 5×10^{-3}	< 2.4×10^{-2}	1.4
III	Scin.	1/16 I_{min}*	3×10^{-4}, ------	< 3.5×10^{-3}	2.2
Bologna	Scin.	20 I_{min}	2×10^{-3}, 0.5	see plot	38
Soudan I	Prop.	1/2 I_{min}	3×10^{-4}, 3×10^{-2}	< 1.8×10^{-4}	71
Baksan	Scin.	1/4 I_{min}	~4×10^{-3}, $5^6 10^{-2}$	< 1.3×10^{-5}	
Kolar Gold Mine	Prop.	2.5 I_{min}	1×10^{-3}, 1	2.5×10^{-5}	
Berkeley	Cr 39	z/β ≥ 30	.02, 1	1.3×10^{-4}	90

*Integrating amp used.

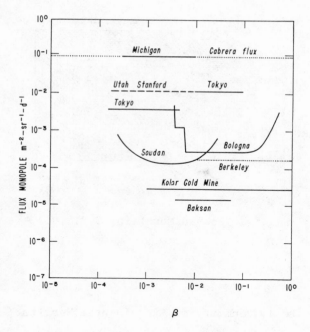

Fig. 10. Summary of experimental results.

NEW MONOPOLE SEARCHES

     Several groups are modifying existing detectors to search for
magnetic monopoles. Those detectors being modified are: the CHARM
detector at CERN, the EAGLES 3 detector at FNAL, and the Fly's Eye
detector at Utah. The CHARM detector consists of 78 layers of 3m ×
3m × 0.03m scintillation counters. The counters are separated by a
distance of 0.2 m.   The electronics have been modified to detect
slowly moving particles ($10^{-3} \lesssim \beta < 0.5$) depositing more than 1/3
I_{min} in 3 consecutive counter layers.

     The EAGLES detector consists of 24 layers   of   4m × 4m
proportional counters. The trigger threshold will be sensitive to
particle energy deposition of about 20 times I_{min}.

     The Fly's Eye detector at Utah[39] consists of sixty-seven 62"
diameter f/1 mirrors located in an area ~80m × 60m. The light from
each mirror is viewed by fourteen 3.5" diameter photomultiplier

tubes through hexagonal light collecting cones. This detector scans the entire night-sky for high energy (> 10^{17}eV) cosmic-ray induced showers. When a distant shower is detected, the time of arrival information of the light signals is used in conjunction with the geometry of the track to estimate the cosmic-ray shower energy. This detector will be able to look for events resulting from monopole-antimonopole annihilations as suggested by Hill and Schramm in this Conference.

A group at Berkeley[40] is in the process of constructing a large 1.5m × 2m × 0.076m acrylic scintillation counter viewed by sixty 3" diameter photomultiplier tubes. Pulse height and pulse shape information from the tubes will be used to identify slowly moving particles. A Cherenkov veto counter placed on top of the scintillation counter will be used to reject fast particles.

Research groups at the University of Pennsylvania[41] and Texas A & M University[42] are designing large scintillation counter arrays to be placed in underground mines for monopole detection. The University of Pennsylvania group is designing a detector resembling an 8m × 8m × 16m hollow box with a total of 200 scintillation counters covering all 6 sides of the box. Each counter is a 0.4m × 0.3 × 8m teflon lined PVC box filled with an extremely transparent mineral oil-based liquid scintillator developed by the group. Two 5" hemi-spherical photomultiplier tubes are used to view the ends of the box. This detector will cover the existing tetrachloroethylene solar neutrino tank in the Homestake Mine. Preliminary calculations show that the sonic pulse in the solar neutrino tank can be used as an additional signature for monopole identification.

The group at Texas A & M University is designing a large 3-layer scintillation counter to be placed in a mine with a 5×10^4 g cm^{-2} overburden. The detector with an effective area of 150 m^2, is made of 108 4' × 8' × 3/8" thick acrylic based scintillation counters placed on the surface of a 24' cube. The ends may be used as a mid-layer to provide redundancy for the purpose of identification. This device is designed to detect particles with large ionization loss (> $4 \times I_{min}$) in the detector while traveling at velocities between $0.04 < \beta < 1.0$.

All of these new magnetic monopole searches are summarized in Table 2.

It is clear that results from the present set of experiments using ionization techniques indicate that monopole flux in the velocity region $10^{-3} < \beta < 0.1$ is less than 10^{-5} m^{-2} sr^{-1} d^{-1}. There exists a great body of ionization detector technology which can be used to make large area detectors for future monopole searches. However, many theoretical calculations are needed in

Table 2. New Magnetic Monopole Search Experiments.

New Detectors	Size	Sensitivity	Trigger Logic	Solid Angle-Area
CHARM Scintillators	3m × 3m × 0.3 × 78 layers	$I_{min}/3$	Any 6 layers	~ 400 m² sr
Eagle 3 Prop. Chambers	4m × 4 m × 20 layers	20 I_{min}	> 4 plane	~ 328 m² sr
Utah Fly's Eye 67-62" D Mirrors	Look for "events" of the type suggested by Hill-Schramm			2×10^9 m² sr (10% duty cycle) $\approx 2 \times 10^8$ m² sr
Berkeley Acrylic Scintillators	1.5m × 2m × 0.076 m viewed by sixty 3" PMT			
Texas A & M Acrylic Scintillators	4' × 8' × 3/8" 108 counters on a 25' cube	4 I_{min}		~ 600 m² sr
U. of Pennsylvania Liquid Scintillators	8m × 8m × 16 m hollow box	$I_{min}/10$		~ 1500 m² sr

order to pick the proper detector which will be sensitive to the
passage of slowly moving (β 10^{-4}) monopoles.

ACKNOWLEDGEMENTS

Many thanks go to Drs. G. Giacomelli, E. Peterson, K. Lande,
R. C. Webb, P. Musset, M. Goodman, P. Price, L. Sulak, C. Hill, and
M. Koshiba for supplying data on recent monopole searches.

The work is supported in part by a grant from the National
Science Foundation.

REFERENCES

1. R. A. Carrigan, Jr., FERMILAB-77/42 2000.00.
2. R. A. Carrigan, Jr., R. E. Craven, and W. P. Trower,
       FERMILAB-81/87 2000.00.
3. R. E. Craven and W. P. Trower, VPI-EPP-81-1.
4. M. J. Longo, Phys. Rev. D 25, 2399-2405 (1982).
5. G. 't Hooft, Nucl. Phys. B 79, 276 (1974).
6. A. Polyakov, JETP Lett. 20, 194 (1974).
7. P. A. M. Dirac, Proc. Roy. Soc. A 133, 60 (1931).
8. For review see P. Langacker, Phys. Rep. 72C, 185 (1981).
9. T. W. B. Kibble, in "Monopoles in Quantum Theory," N. S.
       Craigie, P. Goddard, and W. Nahm, eds., World Scientific
       Publishing co., Pte Ltd., Singapore.
10. J. P. Preskill, Phys. Rev. Lett. 43, 1365 (1979).
11. E. N. Parker, "Cosmical Magnetic Fields," Clarendon Press,
       Oxford (1979).
12. V. C. Rubin, C. J. Peterson, and W. K. Ford, Jr., Ap. J. 239,
       50 (1980); V. C. Rubin, Ap. J. 238, 808 (1980).
13. J. Kristian, A. Sandage, J. Westphal, Ap. J. 221, 383 (1978).
14. A. H. Guth, "The Proceedings of the Nuffield Workshop on the
       Very Early Universe," G. W. Gibbons, S. W. Hawking, and S.
       Siklos, eds., Cambridge University Press, Cambridge (1983);
       J. Ellis, D. V. Nanopoulos, K. A. Olive and K. Tamvakis,
       Ref. TH 3404-CERN.
15. E. N. Parker, Ap. J. 160, 383 (1970).
16. M. S. Turner, E. N. Parker, and T. J. Bogdan, Phys. Rev. D 26,
       1296 (1982).
17. S. Dimopoulos, S. L. Glashow, E. M. Purcell, and F. Wilczek,
       Nature 298, 824 (1982).
18. B. Cabrera, Phys. Rev. Lett. 48, 1378 (1982) and updated value
       reported in this Conference.
19. S. P. Ahlen, Rev. Mod. Phys. 52, 121 (1981); V. P.
       Martem'yanov, and S. Kh. Khakimov, Sov. Phys. JETP 35, 20
       (1972); C. W. Akerlof, Phys. Rev. D 26, 1116 (1982).

20. S. P. Ahlen, and K. Kinoshita, to be published in Phys. Rev. D.
21. S. Drell, N. M. Kroll, M. T. Mueller, S. J. Parke and M. A. Ruderman, SLAC-PUB-3021, 1982 (T/E).
22. D. M. Ritson, SLAC-PUB-2950, 1982 (T/E).
23. It should be noted that the energy deposited in the detector by a monopole is distribued over a time period equal to the time it takes the monopole to traverse the detector. In the case where the response time of the detector is short compared to its transit time, the detector sensitivity is decreased by the ratio of the response time over the transit time. This decrease in sensitivity can be recovered with the aid of an integrating amplifier whose integration time is adjusted to be that of the transit time.
24. J. K. Sokolowski, and L. R. Sulak, unpublished preprint.
25. D. E. Groom, E. C. Loh, H. N. Nelson, and D. M. Ritson, PRL. 50, 573 (1983).
26. R. Bonarelli, P. Capiluppi, I. D'Antone, G. Giacomelli, G. Mandrioli, C. Merli and A. M. Rossi, preprint.
27. R. Bonarelli, P. Capiluppi, I. D'Antone, G. Giacomelli, G. Mandrioli, C. Merli, and A. M. Rossi, Phys. Lett. 112B, 100 (1982).
28. M. Koshiba, S. Orito, Y. Totsuka, M. Nozaki, T. Mashimo, K. Kawagoe, and S. Nakamura, preprint LICEPP, University of Tokyo.
29. T. Mashimo, K. Kawagoe and M. Koshiba, J. Phys. Soc. Japan 51, 3065 (1982).
30. E. N. Alexeyev, M. M. Boliev, A. E. Chudakov, B. A. Makoev, S. P. Mikheyev, Yu V. Sten'kin, Nuovo Cimento Lett. 35, 413 (1982).
31. E. N. Alekseyev, V. V. Alexeyenko, Yu. M. Andreyev, V. N. Bakatanov, A. N. Budkevich, A. E. Chudakov, M. D. Gal'perin, A. A. Gitelson, V. I. Gurentsov, A. E. Danshin, V. A. Dogujaev, V. L. Dadykin, Ya. S. Elensky, V. A. Kozyarivsky, I. M. Kogai, N. F. Klimenki, A. A. Kiryishin, Yu. N. Konovalov, B. A. Makoev, V. Ya. Markov, Yu. Ya. Markov, Yu. V. Malovichkio, N. A. Metlinsky, A. R. Mikhelev, S. P. Mikheyev, Yu. S. Novosel'tsev, V. G. Sborshikov, V. V. Sklyarov, Yu. V. Sten'kin, V. I. Stepanov, Yu. R. Sula-Petrovsky, T. I. Tulupova, A. V. Voevodsky, V. I. Volchenko, and V. N. Zakidyshev, 16th Intl. Cosmic Ray Conference Paper, 10, MN5:4 276 (1979).
32. J. Bartelt, H. Courant, K. Heller, T. Joyce, M. Marshak, E. Peterson, K. Ruddick, M. Shupe, D. Ayers, J. Dawson, T. Fields, E. May, and L. Price, preprint.
33. N. K. Mondal, private communication.
34. M. R. Krishnaswamy, M. G. K. Menon, N. K. Mondal, V. S. Narasimham, B. V. Sreekantan, N. Ito, S. Kawakami, Y. Hayashi, and S. Miyake, "The Second Workshop on Grand Unification," J. P. Leveille, L. R. Sulak, and D. C. Unger, eds., Birkhäuser (1981).

35. K. Kinoshita and P. B. Price, Phys. Rev. 24D, 1707 (1981).
36. S. P. Ahlen, Rev. Mod. Phys. 52, 121 (1980).
37. M. Goodman, private communication.
38. P. Musset, private communication.
39. P. Cady, G. L. Cassiday, J. Elbert, E. Loh, Y. Mizumoto, P.
     Sokolsky, D. Steck, M. Ye, to be published in the DPF
     Summer Study Report, D. Gustafson, ed.
40. P. B. Price, private communication.
41. M. L. Cheery, I. Davidson, K. Lande, C. K. Lee, E. Marshall,
     R. I. Steinburg, G. Sleveland, R. Davis, Jr., and D.
     Lowenstein, preprint, University of Pennsylvania.
42. P. M. McIntyre and R. C. Webb, DOE proposal, DOW-ER40039-4.

SEARCHES FOR MONOPOLES WITH TRACK-ETCH DETECTORS

P. Buford Price

Department of Physics
University of California
Berkeley, California 94720

INTRODUCTION

Unless additional signals in SQUIDs or independent evidence is forthcoming, grand unification monopoles (GUMs) of enormous mass and very low velocity may prove to have been a passing fad. It is appropriate, therefore, to include in this review a discussion of searches for monopoles with conventional masses, capable of being sought at accelerators, as well as searches for GUMs that employ very large area × time factors. To avoid repeating material readily available in other reviews[1,2] or in these proceedings, I shall discuss only recent or planned searches for tracks in dielectric detectors left by monopoles in flight.

THE TRACK-ETCHING TECHNIQUE

Figure 1 shows two of the main features of the track-etch technique. A latent track (Fig. 1a) can be located in one layer of an array of plastic sheets or other track-recording solids by etching for an extended time and detecting the hole (Fig. 1c) with ammonia or some other technique. The ionization rate of the particle that produced the track can be determined by etching another layer for a shorter time and measuring the geometry of the etched cones at the top and bottom of the sheet (Fig. 1b).

As Fig. 2 shows, for the CR-39(DOP) plastic detector, this technique works so well that a charge standard deviation as small as $\sigma_z = 0.06e$ can be achieved for relativistic nuclei with $9 \leq Z \lesssim 30$ by measuring the etch pit diameters at the top and bottom of eight successive sheets.

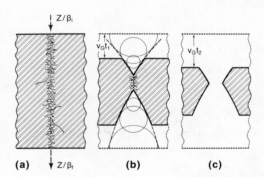

Fig. 1. Track-etching technique for particle identification.
         (a) Sketch showing dense core of radiation-damaged
         material and delta-rays; (b) development of conical
         etchpits at intersection of trajectory with surface; (c)
         development of a hole after prolonged etching.

     Figure 3 shows the response as a function of Z/β, the ratio of
charge to velocity, for a number of track-recording solids. The
response is given in terms of the ratio v of etch rate along the
track, v_T, to general etch rate, v_G, which is related to the etch
pit cone half-angle by $\sin\theta = v_G/v_T = v^{-1}$. In interpreting null
results of searches, it is important to establish the response
curve as a function of Z/β from accelerator calibrations for the
detector used. Because the track etch rate increases gradually
with Z/β, one cannot simply assume that etched tracks of particles
above a well-defined Z/β will be detectable by focusing in the
middle of a sheet and looking for very long etched tracks. It is
absolutely essential to do detailed calibrations if one is trying

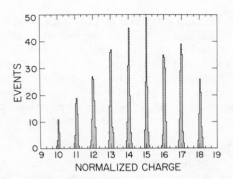

Fig. 2. Example of the extraordinary charge resolution attainable
         with CR-39 plastic track detectors.

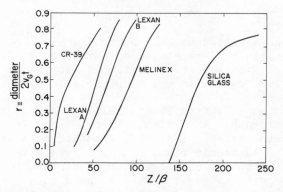

Fig. 3.   Response of several track-etch detectors as a function of
          Z/β. Lexan A and Lexan B refer to the responses of Lexan
          etched in two different reagents.

to detect particles with ionization rate near the minimum value for
which preferential etching occurs. This is the case for very slow
GUMs.

SEARCHES AT ACCELERATORS PRIOR TO 1981

        Figure 4 shows cross section limits for monopole production in
proton-nucleus collisions as a function of monopole mass. The
solid lines refer to direct methods in which poles would be
detected in flight. Dashed lines refer to indirect experiments in
which poles would be stopped and trapped in some material and later
extracted, accelerated and detected.   The dotted lines refer to
cosmic-ray searches, which are even more indirect.

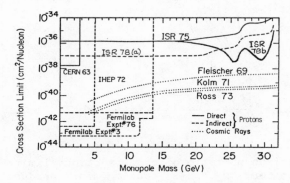

Fig. 4.   Upper limits on cross sections for monopole production in
          proton-nucleus collisions.   Only the direct searches are
          free of assumptions about trapping and extraction
          efficiencies.

RECENT AND PLANNED SEARCHES AT ACCELERATORS

If monopole-antimonopole pairs are produced via a virtual-photon intermediate state, a search using e^+e^- annihilation carries a significant cross sectional advantage over direct searches carried out with pp or $\bar{p}p$ colliders, because the cross section for massive virtual-photon production falls exponentially with mass.[3] Figure 5, from K. Kinoshita's Ph.D. thesis,[4] compares the limits obtained by Kinoshita, Price, and Fryberger,[5] using CR-39 and Lexan detectors at PEP, with limits obtained in proton accelerators and in cosmic rays. The ordinate gives R, the ratio of the cross section limit to the cross section for production of virtual photons with sufficient mass to produce a pair. One sees that the PEP experiment gives a considerably lower cross section limit than did direct searches using proton accelerators.

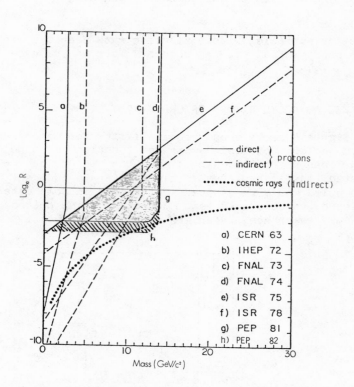

Fig. 5. Comparison of ratio of cross section limit for monopole production to cross section for production of virtual photons' with sufficient mass to produce a monopole-antimonopole pair, for proton experiments (a to f) and e^+e^- annihilations (g and h).

Table 1. Search   for   Highly   Ionizing   Particles   at   PEP
          (E_{CM} = 29 GeV);   K. Kinoshita,   P. B. Price,   and
          D. Fryberger

		1981		1982	Combined
		IR-6	IR-10	IR-10	limit
	$L_i t_i (10^{36}$ cm$^{-2})$	6.1	8.4	30.3	
	pipe thickness (μm)	200	100	200	
magnetic monopole	mass limit (GeV), $g=g_0$	13.7	14.0	13.7	
	$g=2g_0$	9.5	11.5	9.5	
	solid angle (sr)	4.6	2.0	7.0	
	$\sigma(10^{-36}$ cm$^2)$	<1.4	<2.3	<0.18	<0.15
electric, $Z/\beta \geq 20$	solid angle (sr)	1.7	1.7	5.0	
	$\sigma(10^{-36}$ cm$^2)$	<3.7	<2.7	<0.25	<0.22

[QED point cross section = 10^{-34} cm^2 at $\sqrt{s}$ = 29 GeV]

Table 1 summarizes the results of the PEP experiments at the
two interaction regions IR-6 and IR-10, which led to a cross
section limit ~10^{-37} cm^2 (95% CL), about four orders of magnitude
below the QED point cross section at $\sqrt{s}$ = 29 GeV, for monopoles of
mass up to ~14 GeV/c^2.

Table 2 lists other direct searches either in progress or
planned. At this workshop Musset presented the negative results of
a search for monopoles at the $\bar{p}p$ collider, using Kapton plastic
detectors.[6] To my knowledge a response curve for Kapton has not
been published. Musset and coworkers estimated cross section
limits increasing from 10^{-32} cm^2 for M ~40 GeV/c^2 to ~10^{-31} cm^2 for
M ~150 GeV/c^2, assuming that a monopole-antimonopole pair is

Table 2. Recent or Planned Direct Searches for Monopoles

Accelerator	$\sqrt{s}$(GeV)	Detectors	Minimum Z/β	Ref.
PEP (e$^+$e$^-$)	29	CR-39 (and Lexan)	~15 (and 60)	4
SPS ($\bar{p}p$)	up to 540	Kapton	~75?	5
FNAL ($\bar{p}p$)	2000	CR-39 (and Tuffak)	~15 (and 60)	7
SLC (e$^+$e$^-$)	up to 140	CR-39 (and Tuffak)	~15 (and 60)	-
LEP (e$^+$e$^-$)†	up to 140	Kapton?	~75?	-

---

†experiment under negotiation

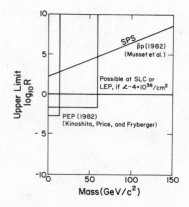

Fig. 6.   Cross section ratio (see caption to Fig. 5) for recent
          e^+e^- and $\overline{p}p$ experiments and ratio attainable at SLC or LEP
          if negative results are obtained with an integrated
          luminosity of $4 \times 10^{36}/cm^2$.

produced via a $\overline{q}q$ collision with the x-distribution for quarks and
antiquarks in the proton and antiproton going as $(1-x)^3$.

     Figure 6 compares the e^+e^- and $\overline{p}p$ experiments in terms of the
same ratio R used in Fig. 5. Clearly it will not be possible to
produce monopoles at the CERN $\overline{p}p$ collider via massive virtual
photon   production   without   orders of magnitude increase in
luminosity. The cross section limit indicated for a SLC or LEP
experiment assumes a null result with an integrated luminosity of 4
$\times 10^{36}/cm^2$ and 4π collection.

TRACK PRODUCTION IN PLASTIC DETECTORS
BY VERY SLOW GRAND-UNIFICATION MONOPOLES

     In other papers in these proceedings various schemes for
detecting   GUMs   at   the   lowest   possible velocities have been
discussed. The most sensitive plastic track detector, CR-39, costs
only about $100 per square meter for an array two layers thick,
whereas a slab of plastic scintillator with phototubes and
electronics would cost nearly 10^2 times as much per unit area. It
is, therefore, worthwhile to examine whether CR-39 could detect
GUMs at as low a velocity as could plastic scintillator.

     As Fig. 7 suggests, plastic scintillator and a plastic
track-etch detector respond very differently to the lateral
distribution of energy deposited by energetic electrons ejected
along the trajectory of an ionizing particle. To see this, first
note that the energy transfer, T, to electrons by a charged

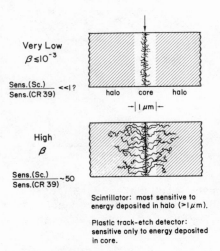

Very Low
$\beta \lesssim 10^{-3}$

$\dfrac{\text{Sens.(Sc.)}}{\text{Sens.(CR 39)}}$ <<1?

halo   core   halo

$\dashv\; 1\,\mu m\; \vdash$

High
β

$\dfrac{\text{Sens.(Sc.)}}{\text{Sens.(CR 39)}}$ ~50

Scintillator: most sensitive to
energy deposited in halo ($>1\mu m$).

Plastic track-etch detector:
sensitive only to energy deposited
in core.

Fig. 7.   Because of radiation quenching in the dense core region
near the particle trajectory, a plastic scintillator is
most sensitive to energy deposited in the halo by
high-energy electrons, whereas a plastic track-etch
detector is sensitive only to energy deposited very close
to the trajectory where the energy density is highest.

particle goes as $dn/dT \propto T^{-2}$, for values of T up to the kinematic
cutoff $T_{max} = 2m_e c^2 \beta^2 \gamma^2$. This leads to a rapid radial decrease in
energy deposited per unit volume by the electrons. At high energy
densities inside a "core" region ~1 μm in diameter nonradiative
processes (radiation quenching) reduce the efficiency dL/dE of
light output by a plastic scintillator, whereas in the "halo"
region the energy density is low enough that radiative processes
dominate.[8] The velocity below which the efficiency of light
production is drastically reduced is ~0.1 c, corresponding to T_{max}
~10 keV, the energy at which ejected electrons deposit a
significant fraction of their energy outside the saturated core.
Ahlen and Tarlé[9] have calculated the light output of a plastic
scintillator as a function of the velocity of a monopole. Their
results are summarized by Ahlen[10] in these proceedings.

A track-etch detector is complementary to a plastic
scintillator in that it is insensitive to energy deposited in the
halo and sensitive only to energy deposited in the core. The
reason is that preferential chemical etching can take place only
where the density of energy deposited by ejected electrons is
extremely large, typically within a hundred Angstroms of the
particle's trajectory. Thus, although CR-39, the most sensitive
track-etch detector, is much less sensitive than is a plastic
scintillator to minimum-ionizing relativistic particles (Z_{min} ~6

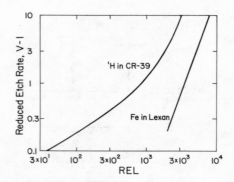

Fig. 8.   Reduced track-etch rate ($v \equiv v_T/v_G$) as a function of restricted energy loss based on data for protons in CR-39 and for ^{56}Fe ions in Lexan, obtained at energies above 0.2 MeV/amu.

for CR-39, Z_{min} <1 for scintillator), its sensitivity to particles of very low velocity, which are incapable of ejecting electrons outside of the core, might actually become competitive with or higher than that of a plastic scintillator.

Studies with a wide variety of ions with different Z and β have shown that at energies above ~0.2 MeV/amu the track etch rate is, to a reasonable approximation, a function of restricted energy loss rate, which refers to the fraction of the total energy loss rate due to electrons with energy less than E_0 (an adjustable parameter found to be ~200 eV for CR-39 and ~10^3 eV for Lexan). This criterion, though simple to apply, is not correct in detail because it gives zero weight to electrons with E >E_0 even though they deposit some of their energy in the core. Figure 8 shows the measured dependence of track etch rate on restricted energy loss (REL) for CR-39 and Lexan at small values of REL relevant to a search for GUMs.

These curves have been used to construct the curves in Fig. 9 showing calculated track etch rate as a function of energy per amu for Fe ions in Lexan, protons in CR-39, and a monopole with the Dirac charge in CR-39. The data points at high energies are consistent with the REL criterion. The three data points for Fe ions at very low energies were obtained by making plastic replicas of etch pits and examining them in a scanning electron microscope[11] as shown in Fig. 10. This procedure was necessary because of the extremely short ranges of the Fe ions. Note that the track etch rates for low-energy Fe are much greater than predicted by the REL criterion. Diamond[12] has found similar enhancements in track etch rate for ^{16}O and ^{12}C ions in Makrofol, a polycarbonate similar to Lexan.

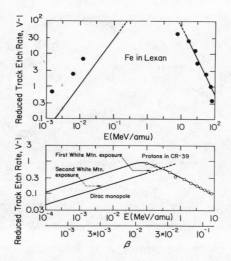

Fig. 9.   Reduced track etch rate as a function of energy per amu
         calculated from the response curves in Fig. 8 for Fe in
         Lexan, protons in CR-39, and a Dirac monopole (g = e/2α)
         in CR-39.   The data points come from various accelerator
         experiments.

Ahlen and Kinoshita[13] have shown that the total energy loss
rate of a monopole at $\beta \lesssim 0.01$ is approximately one-fourth that of
a proton with the same velocity.   At velocities lower than the
kinematic limit for producing electrons of energy E_0, REL is the
same as total dE/dx.   For E_0 = 200 eV, β_0 = 0.014.   The curve in
Fig. 9 showing the expected track etch rate for a monopole assumes
that REL for a monopole is one-fourth of REL for a proton at the
same velocity.

The solid arrow in Fig. 9 indicates the minimum detectable
track etch rate for the method of track scanning used by Kinoshita
and Price[14] in their 1981 search for GUMs using a 15 m^2 array of
CR-39 at White Mountain, CA.   Assuming the correctness of the
response curve of v vs energy, their negative result implies a   95%
CL upper limit of 2 × 10^{-13} cm^{-2} sr^{-1} s^{-1} on the flux of GUMs with
$\beta \gtrsim 0.02$. This result is displayed on the graph of flux upper limit
vs velocity in Fig. 11.   Improvements in the etching and scanning
technique now make it possible to detect tracks at the lower etch
rate indicated by the dashed arrow in Fig. 9.   This rate would
enable monopoles with β at least as low as ~0.005 to be detected.
Two of my students, Steve Barwick and Jim Garnett, are now
analysing a CR-39 array exposed for one year and processed with
this sensitivity.   To establish experimentally the response curve
for protons in CR-39 at very low energies, we are planning to
irradiate CR-39 with deuterons at energies down to a few keV at

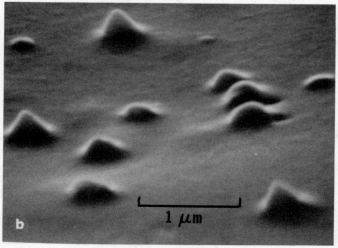

Fig. 10. Scanning electron micrographs of plastic replicas of etch
          pits of very low-energy Fe ions in Lexan (see Ref. 11,
          p. 143). (a) 26 keV/u ^{56}Fe; β = 0.007; Z^* $\approx$ 2; (b)
          3 keV/u ^{56}Fe; β = 0.002; Z^* $\approx$ 1.

Caltech. Deuterons have the advantage over protons that their
range is approximately twice that of protons at the same velocity
and REL. If the REL criterion underestimates the track etch rate
in CR-39 at very low velocities as it does for Lexan and Makrofol,
the minimum monopole velocity detectable with CR-39 will be even
lower than $\beta \approx 0.005$.

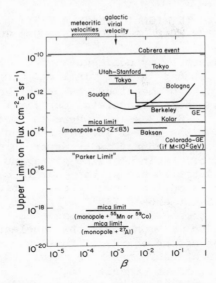

Fig. 11.   Upper limits on flux of Dirac monopoles.   The   Cabrera
limit   is independent of velocity.   The Colorado-GE limit
is from Ref. 30.   The GE limit is from Ref. 31.   The
Berkeley limit at ~2 × 10^{-13} cm^{-2} sr^{-1} s^{-1} is from
Ref. 14.   The limits for monopole-nucleus systems are
discussed in the text and in Ref. 15.   They are based on
the assumption that the mica search, when completed,
shows a negative result.   The other limits are discussed
in the review by E. Loh in these proceedings.

SEARCH FOR COSMIC MONOPOLES AT GROUND LEVELS

        Bartlett et al.[29] are using the large magnetic moment of the
Fermilab 15-foot bubble-chamber magnet for collection and
acceleration of natural monopoles in the atmosphere.   In 100 days
of operation with a polarity such as to gather in geomagnetic field
lines, they obtained an area × time factor of 8.4 × 10^{14} cm^2 sec
for monopoles light enough to be slowed, to be deflected, and to
pass through sheets of Lexan beneath the magnet.   Monopoles with
the Dirac charge and with mass of ~10 GeV/c^2 would be slowed,
gathered, and recorded in Lexan if their initial kinetic energy
were less than ~10^5 GeV; those with mass ~10^2 GeV/c^2 would be
detected in this way if their initial kinetic energy were less than
~10^4 GeV/c^2.   This result is displayed (Colorado-GE) in Fig. 11.
The area × time factor for monopoles with too high a rigidity to be
gathered in by the magnet was 160 times lower.   Monopoles with
velocity ≤0.34 c would not be detected in Lexan with the etching
conditions used in Ref. 29.

USE OF ANCIENT TRACKS IN MICA TO PLACE A VERY
STRINGENT LIMIT ON THE FLUX OF MONOPOLES BOUND TO NUCLEI

With R. L. Fleischer, my colleagues and I[15] are using the
absence of long, etched tracks in large mica crystals to set a
limit about four orders of magnitude below the Parker limit on the
flux of monopoles bound to nuclei. Our limit would apply in the
velocity regime $10^{-4} \lesssim \beta \lesssim 3 \times 10^{-3}$ where the predominant mode of
energy transfer to the medium is via screened Coulomb collisions
with atomic nuclei. Here energy is transferred directly to nuclei,
which recoil out of their lattice sites and leave a chemically
reactive, disordered trail along the particle trajectory. The
so-called "nuclear stopping power" of a bare monopole in this
velocity regime does not exceed $\sim 10^{-3}$ MeV/mg/cm^2, which is several
orders of magnitude too low to produce an etchable track. However,
for a nucleus bound to a monopole the nuclear stopping is an
increasing function of the charge of the nucleus and becomes large
enough to produce an etchable track in mica for $Z \gtrsim 8$. I first
summarize the conclusions Charles Goebel has reached regarding
pickup of a stationary nucleus by a moving monopole and then show
how we arrive at a flux limit based on a search for ancient tracks
in mica. The material in this section is taken from Ref. 15.

An earlier, rigorous calculation by Sivers[16] and Goebel[17] has
shown that a monopole will bind to a nucleus with a gyromagnetic
ratio greater than 2. For a potential energy $V = L^2/2Mr^2 - \vec{\mu} \cdot \vec{B}$,
where M = reduced mass of monopole + nucleus, r = separation, and μ
= magnetic moment of nucleus, he gets a binding energy

$$B.E. = - \frac{(Z-A\mu/\mu_N)\nu h^2}{2Mr^2}$$

where μ_N = nuclear magneton and ν = eg/hc = integer/2 due to Dirac
quantization of the magnetic charge. Table 3 lists the nuclear
magnetic moments and average abundances of nuclei with $\mu > 2\mu_N$. the
most abundant of these nuclei, ^{27}Al, will have a binding energy of
~ 1 MeV at a distance of ~ 5 F, corresponding to the nuclear radius.
Goebel has argued that a binding energy this large is
uncontroversial, depending as it does on the electromagnetic
interaction. What happens at subnuclear distances depends on the
strong interaction. It is conceivable that at very small
separations a catastrophe could occur: the monopole might catalyze
nucleon decay. The upper limit that we will derive is based on the
assumption that this does not happen. Goebel estimates a cross
section for radiative capture of ^{27}Al to be $\sigma \approx 30$ μb$(\beta/10^{-3})^{-2}$ and
a mean free path in earth of $\lambda \approx 7$ km$(\beta/10^{-3})^2$. The nucleus will
not be stripped off, provided $\beta \lesssim \sqrt{2B.E./Mc^2}$. For ^{27}Al, $\beta_{max} \approx 10^{-2}$
$\sqrt{B.E.(MeV)}$. The range of the monopole + ^{27}Al will exceed the radius
of the earth for $\beta \gtrsim 10^{-3}(10^{16}$ amu/M$_{monopole})$. The binding

Table 3. Nuclei Capable of Binding to Monopoles

Nucleus	μ/μ_N	Average abundance (Si$\equiv 10^6$)
^{23}Na	2.22	6.0×10^4
^{27}Al	3.64	8.5×10^4
^{45}Sc	4.76	35
^{51}V	5.15	261
^{55}Mn	3.47	9300
^{59}Co	4.65	2210
$20 \leq Z \leq 29$	–	1.2×10^4
$30 \leq Z \leq 39$	–	1.72
$40 \leq Z \leq 49$	–	1.59
$50 \leq Z \leq 59$	–	2.4
$60 \leq Z \leq 69$	–	0.08
$70 \leq Z \leq 79$	–	0.11
^{209}Bi	4.08	0.15

energies, capture mean free paths, stripping velocities, and ranges of monopoles bound to other nuclei in Table 3 are qualitatively similar to those for ^{27}Al.

Although mica is far less sensitive than CR-39 or Lexan as a track-recording solid, it has the advantage of an enormous area × time factor, many orders of magnitude greater than could be attained with any modern detector. Experiments by Walker,[18] Maurette,[19] and their colleagues[20-22] have shown that in the velocity regime of interest to us, corresponding to energies of ~0.1 to ~3 keV/amu, nuclei with Z ≳ 8 produce etchable tracks in mica with a diameter that increases with the time of etching in hydrofluoric acid and with the nuclear stopping power. The track depth is extremely shallow, <0.1 μm, corresponding to the short range of the ion. Figure 12 shows examples of etch pits due to 1 keV/amu Xe ions and 0.4 keV/amu Th ions in mica observed with interference constrast microscopy.

Figure 13 shows calculations[15] of stopping power of mica for various ions and for monopoles bound to ^{27}Al, ^{55}Mn, and ^{209}Bi. The family of curves that peak at $\beta \approx 10^{-3}$ give the nuclear stopping power, obtained by using the analytic expressions of Wilson et al.,[23] which fit experimental measurements better than does the classic Lindhard-Scharff-Schiott nuclear-stopping theory. The solid points on these curves indicate values of Z and β for which etch pits like those in Fig. 12 have been seen in accelerator bombardments of mica.[19-22] Note that the effect of a massive monopole bound to a nucleus is to increase the stopping power of the nucleus.

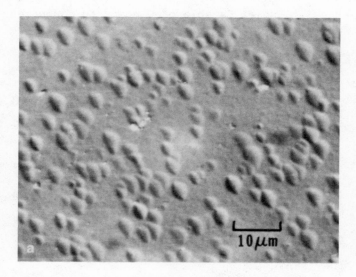

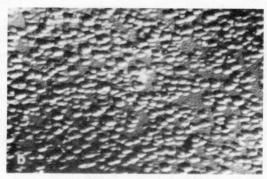

Fig. 12. Shallow etch pits of (a) 1 keV/amu ^{132}Xe ions
         (β = 0.0015) and (b) 0.4 keV/amu ^{228}Th ions (β = 0.001)
         seen in mica with interference contrast microscopy.

     The family of curves in Fig. 13 that peak at $\beta \approx 0.05$ give the
electronic contribution to stopping power. Track-etching has been
observed throughout the solid regions along these curves.[24-26] Only
one experiment, with 30 keV/amu Ni ions, has been done in the
region well between the maxima in the two families of curves.[22] The
data at $\beta \gtrsim 10^{-2}$ are not particularly relevant because a nucleus
would be stripped off a monopole at such velocities.

     Based on observations in the nuclear stopping regime, we
assume that a monopole-nucleus system with a stopping power greater
than 1.75 MeV/mg/cm^2, indicated by the horizontal band in Fig. 13,

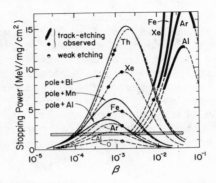

Fig. 13.   Calculated dependence of nuclear and electronic stopping
power on velocity for ions (dashed lines) and for
monopoles bound to nuclei (solid lines) in mica.[15] Etched
tracks have been seen in regions indicated by shading and
by solid circles.

will produce a continuously etchable track throughout the slab of
mica it traverses. This leads to a lower and upper limit on
velocity for which a particular monopole-nucleus system would be
detectable. Several effects must be taken into account before one
can determine an upper limit on flux as a function of velocity.
The maximum detectable velocity is either that read off the
appropriate curve in Fig. 13 or the stripping velocity, whichever
is smaller. If the range of the monopole-nucleus system with the
minimum detectable velocity is less than the diameter of the earth,
a correction must be made for the solid-angle factor as a function
of velocity.

The area × time factor depends on the area of the mica
crystals studied and the time during which they were storing
radiation-damaged tracks capable of being etched. One can
determine this time interval experimentally by measuring the
"fission-track age" of the mica,[11] which is based on a count of the
number of etchable tracks due to spontaneous fission of ^{233}U
impurities in the mica and on a determination of the concentration
of these uranium impurity successive alpha decays of U and Th and
their daughters (typical recoil energy ~0.4 keV/amu) showed that
the ratio of density of recoil tracks to density of spontaneous
fission tracks was 3500 ± 1000 in samples of widely different ages,
implying comparable resistance of the two types of tracks to
thermal fading. Because the track produced by a monopole-nucleus
system would have a damage distribution similar to that of a recoil
track from alpha decay, we use the fission-track age of the mica
samples as the measure of collection time.

We are re-examining a large muscovite mica crystal from North Carolina previously studied by Fleischer, Woods and me[27] in a search for relativistic monopoles with charge $g \gtrsim 2g_0$. We etched this mica for 184 hours in concentrated HF in order to produce tracks large enough to detect readily in a low-power stereomicroscope. The background of ~40 fission tracks per cm^2 are ~400 μm in diameter and have a depth up to ~15 μm. The area scanned is 310 cm^2 and the track-retention age is $(248 \pm 27) \times 10^6$ years. We are looking for etched tracks that either extend through the entire ~2 mm-thick mica crystal or have corresponding pits on opposite surfaces.

Figure 11 gives the upper limits for several monopole-nucleus systems assuming that we obtain a negative result in our search for long tracks in mica. Because of its large concentration in the earth, ^{27}Al provides the most stringent limit. The heavier nuclei enable one to probe somewhat lower velocities--down to ~8×10^{-5} c in the case of ^{209}Bi and other nuclei with $Z > 60$--but the flux limit is much worse than for ^{27}Al because of the low concentration of these nuclei in the earth.

It should be emphasized that the mica limit that we hope to obtain--approximately nine orders of magnitude lower than the Cabrera limit--applies in just the velocity regime where Dimopoulos et al.[28] have proposed an escape clause to keep the Cabrera event from being inconsistent with the Parker limit. They have argued that monopoles could be trapped in orbits about the sun at a much higher concentration than the average throughout the Galaxy and with a velocity distribution similar to that of meteoroids--in the range $(0.3 - 2) \times 10^{-4}$ c--which would make them undetectable with devices based on ionization loss. The mica limit would close that loophole, provided GUT monopoles do pick up nuclei. Other methods of GUT monopole detection that depend on nuclear attachment, such as trapping in ferromagnetic deposits,[29] will have to be capable of setting a limit lower than 10^{-19} cm^{-2} sr^{-1} s^{-1} to do better than the mica method can do.

ACKNOWLEDGEMENTS

I am indebted to S. P. Ahlen, S. W. Barwick, T. E. Coan, R. L. Fleischer, J. D. Garnett, C. Goebel, Shi-lun Guo, K. Kinoshita, M. L. Tincknell and T. Tombrello for useful discussion and assistance. This work was supported in part by NSF Grant PHY-8024128.

REFERENCES

1. R. R. Ross, in "New Pathways in High-Energy Physics I.
    Magnetic Charge and Other Fundamental Approaches," A.
    Perlmutter, ed., Plenum, New York (1976), p. 151.
2. G. Giacomelli, in "Proc. Conf. on Monopoles in Quantum Field
    Theory," Trieste (December, 1981).
3. J. K. Yoh et al., Phys. Rev. Lett. $\underline{41}$, 684 (1978).
4. K. Kinoshita, Ph.D. Thesis, University of California, Berkeley
    (1981).
5. K. Kinoshita, P. B. Price, and D. Fryberger, Phys. Rev. Lett.
    $\underline{48}$, 77 (1982).
6. B. Aubert, P. Musset, M. Price, and J. P. Vialle, these
    proceedings.
7. P. B. Price and K. Kinoshita, Fermilab Experiment No. P-713.
8. M. H. Salamon and S. P. Ahlen, Nucl. Instr. Meth. $\underline{195}$, 557
    (1982).
9. S. P. Ahlen and G. Tarlé, submitted to Phys. Rev. Lett.
    (1982).
10. S. P. Ahlen, these proceedings.
11. R. L. Fleischer, P. B. Price, and R. M. Walker, "Nuclear
    Tracks in Solids: Principles and Applications," University
    of California Press, Berkeley (1975).
12. W. T. Diamond, unpublished manuscript.
13. S. P. Ahlen and K. Kinoshita, Phys. Rev. D (1982), in press.
14. K. Kinoshita and P. B. Price, Phys. Rev. D $\underline{24}$, 1707 (1981).
15. P. B. Price, Shi-lun Guo, S. P. Ahlen, and R. L. Fleischer, to
    be submitted to Phys. Rev. Lett.
16. D. Sivers, Phys. Rev. D $\underline{2}$, 2048 (1970).
17. C. Goebel, Proc. Monopole Seminars, University of Wisconsin
    (1981).
18. W. H. Huang and R. M. Walker, Science $\underline{155}$, 1103 (1967).
19. J. Borg, J. C. Dran, Y. Langevin, M. Maurette, and J. C.
    Petit, Rad. Effects $\underline{65}$, 133 (1982).
20. E. Zinner, R. M. Walker, J. Borg, and M. Maurett, "Proc. Fifth
    Lunar Conf. 3," Pergamon, New York (1974), p. 2975.
21. J. Bibring, J. Borg, J. Dran, M. Maurette, R. Meunier, J.
    Peters, and R. Walker, "Proc. 8th Inter. Conf. Nucl.
    Photography and Solid State Track Detectors," Bucharest $\underline{1}$,
    485 (1972).
22. R. M. Walker, E. Zinner, and M. Maurette, "Apollo 17
    Preliminary Science Report," NASA SP-330, pp. 19-2 to 19-15
    (1973).
23. W. D. Wilson, L. G. Haggmark, and J. P. Biersack, Phys. Rev. B
    $\underline{15}$, 2458 (1977).
24. P. B. Price, R. L. Fleischer, and C. D. Moak, Phys. Rev. $\underline{167}$,
    277 (1968).
25. J. B. Natowitz, A. Khodai-Joopari, J. Alexander, and T. D.
    Thomas, Phys. Rev. $\underline{169}$, 993 (1968).

26. H. Blok, F. M. Kiely, and B. D. Pate, Nucl. Instr. Meth. 100,
     403 (1972).
27. R. L. Fleischer, P. B. Price, and R. T. Woods, Phys. Rev. 184,
     1398 (1969).
28. S. Dimopoulos, S. L. Glashow, E. M. Purcell, and F. Wilczek,
     Nature 298, 824 (1982).
29. D. B. Cline, these proceedings.
30. D. F. Bartlett, D. Soo, R. L. Fleischer, H. R. Hart, and A.
     Mogro-Campero, Phys. Rev. D 24, 612 (1981).
31. R. L. Fleischer, H. R. Hart, G. E. Nichols, and P. B. Price,
     Phys. Rev. D 4, 24 (1971).

SEARCH FOR MAGNETIC MONOPOLES IN PROTON-ANTIPROTON

INTERACTIONS AT 540 GEV C.M. ENERGY

P. Musset[2], B. Aubert[1], M. Price[2] and J. P. Vialle[2]

[1]LAPP, Annecy, France
[2]CERN, Geneva, Switzerland

The concept of magnetic monopole arose very long ago from experimental observation of the forces exerted by magnets on each other. In addition, the existence of magnetic sources and currents would symmetrize the Maxwell equations. However, the strongest argument in favor of the magnetic monopole is provided by quantum mechanics, from which P.A.M. Dirac[1] was able to predict the strength g_D of the magnetic pole g_D = hc/2e. All magnetic charges g would be multiples of g_D : g = Ng_D. If the electron is elementary, the least magnetic charge is g_D. If fractional charged quarks are elementary, the least magnetic charge is 3 g_D.

So far, experimental searches for monopoles have failed.[2] Nevertheless, a recent publication[3] has reported a possible candidate from cosmic radiation observed in an experiment using a superconducting coil.

The masses of the monopoles are not predicted by theory, except in some of the Grand Unified Theories. For this reason, the new energy domain of the proton-antiproton collider, with 540 GeV energy in the center of mass, was explored in the present experiment.[4]

The experiment aims at identifying the monopoles by their high ionization rate.[5]

$$(dE/dx)/(dE/dx_m) = N^2 \ g_D^2/e^2 \ f(\beta)$$

where dE/dx_m is the ionization rate for a minimum ionizing particle, β is the monopole velocity. Plastic detectors,[6] in which tracks left by heavily ionizing particles are developed by chemical etching after exposure, are well adapted to this search. They have the advantage of being insensitive to minimum ionizing particles, which are copiously produced at accelerators. They are thin and have a moderate stopping power. Finally, by choosing various materials, it is possible to cover a large range of ionization.

As the ionization loss of monopoles is expected to be very large, of the order of $N^2 \times 10$ GeV/g/cm^{-2} for relativistic monopoles, their range may be short. Moreover, possible binding of slow monopoles with nuclei or electrons would diminish the detection efficiency. Consequently, the presence of any material between the region of production and the detector has to be avoided as much as possible. This is particularly the case if monopoles have large magnetic charges. In the present experiment, part of the detector is placed inside the vacuum pipe of the collider around the collision region. After a series of tests, the plastic Kapton was chosen for this purpose. This plastic resists the high temperatures encountered during baking of the intersection region and has a very low degassing rate. The portion of the detector which is located inside the vacuum pipe consists of 3 cylindrical sections. These sections are each 1 m long with a diameter of 12 cm and are separated from each other by 2 m. Background due to short range spallation nuclear fragments is reduced by employing a double layer of Kapton in this region. In order to eliminate this background, it is sufficient to choose the thickness of the detector, 2×75 μ in the present case, to be much larger than the range of the spallation products (of the order of one micron). Monopoles, on the other hand, would be much more penetrating.

Around the beam pipe, which is a stainless steel corrugated cylinder of 0.2 mm thickness, a third layer is wrapped over a 6 m length. Large plastic sheets are also placed around the central detector of the UA1 colliding experiment in order to possibly track the monopoles through the magnetic field. The results presented here were obtained using one of the three interior sections and the complete detector placed outside the pipe during the first period of operation of the proton-antiproton collider from August to December 1981.[4] The total integrated luminosity during this period is (2.5 ± 1) 10^{32}/cm^2, the uncertainty on the luminosity coming from the lack of knowledge of the beam shapes and from uncertainties in the intensity measurement.

The Kapton foils were developed by immersion in a 15% NaC10 solution for 4 hours. The action of the developer is to etch the bulk of the material at the rate of about 5 μm per hour while etching zones which have been damaged by the passage of highly ionizing particles at several times this rate. The net result of this differential attack is to leave a hole along the path of the particle. To search for such holes the developed plastic sheet is placed between two sponges wetted with an electrolyte; these sponges are in turn connected to a battery in series with a microammeter. In the case of a traversing hole, the electrolyte connects the two sponges and a current of some microamperes flows. During the scanning process the apparatus is continually calibrated using holes of known diameter. No hole was found which could be attributed to the passage of a highly ionizing and penetrating particle.

The main difficulty which arises in extracting limits for the production cross section of monopoles is the complete absence of a good production model, because perturbation methods are inapplicable. In the following, the hypothesis is made that monopole production is not due directly to a proton-antiproton interaction, but is rather similar to a Drell-Yan process, i.e., $q\bar{q} \rightarrow g\bar{g}$. The x_1 and x_2 distribution of quarks and antiquarks in the proton and antiproton are taken to be $(1-x)^3$, as indicated by lepton production experiments. The production can only occur for $x_1 x_2$ S $\geq$ $(2 M_g)^2$. The contribution of the quarks from the sea is neglected. No angular dependence of the cross section is explicitly assumed. Note that the effect due to the x-distribution of quarks is generally omitted in the discussion of previous experiments, whereas it is important in hadron reactions.

The external detector only was used to set a limit on the monopole production cross section, because the developed internal detector covers a small solid angle and then adds little information. Furthermore, the corresponding result would have been very model-dependent.

The ionization threshold for Kapton has been taken into account. It has a significant effect for $g = g_D$, but is almost ineffective for $g = 3 g_D$. A small energy loss, of the order of $N^2 \times 1.5$ GeV, is due to the traversal of the vacuum pipe by monopoles and is also taken into account. The results are given in Fig. 1. Typical values of the limiting cross sections lie between 10^{-32} and 10^{-31} cm^2, depending on monopole mass and charge values, assuming that the above production mechanism is a valid hypothesis.

An order of magnitude increase in sensitivity is expected to be obtained in the next year of operation of the collider.

We would like to thank all our colleagues who were involved in the collider construction, and in particular B. Angerth, G. Engelmann and M. Genet for their active contributions in the preparation of the detector. We also wish to thank the ISR vacuum group for earlier technical developments, O. Piolatto for his help in the development and scanning of the foils and our colleagues of the UA1 experiment for the opportunity they gave us to place our detector inside their apparatus.

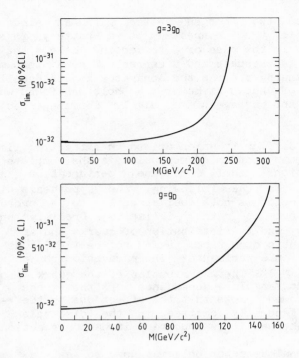

Fig. 1. Cross sections for monopole production at the SPSC as a
         function of monopole mass.   Top figure is for $g = 3g_D$,
         lower figure for $g = g_D$.

REFERENCES

1. P. A. M. Dirac, Proc. Royal Soc. of London A 133, 60 (1931).
2. For a recent compilation see for example, Review of Particle
        Properties, Phys. Lett. 111B, 111 (1982).
3. B. Cabrera, Phys. Rev. Lett. 48, 1378 (1982).
4. The staff of the CERN proton-antiproton project, Phys. Lett.
        107B, 306 (1981).
5. S. P. Ahlen, Phys. Rev. D 17, 229 (1978).
6. For a review, see for example: R. L. Fleisher, P. B. Price,
        R. M. Waler, "Nuclear Tracks in Solids," University of
        California Press, Berkeley (1975); M. Monnin, University of
        Clermont-Ferrand (1977).

PARTICIPANTS

Steven Ahlen, Physics Department, University of California, Berkeley CA 94720, USA.

Carl Akerlof, SLAC, PO Box 4349, Stanford CA 94305, USA.

M. Arik, Physics Department, Bogazici Universitesi, PK 2 Bebek, Istanbul, TURKEY.

Barry C. Barish, Lauritsen Laboratory, California Institute of Technology, Pasadena CA 91125, USA.

Arie Bodek, Physics Department, University of Rochester, Rochester NY 14627, USA.

Bruno Borgia, Istituto di Fisica, Universita di Roma, Piazzale A. Moro 2, I-00815 Roma, ITALY.

Blas Cabrera, Physics Department, Stanford University, Stanford CA 94305, USA.

Curtis G. Callan, Physics Department, Princeton University, Princeton NJ 08540, USA.

Richard A. Carrigan, Jr., Fermilab MS 208, PO Box 500, Batavia IL 60510, USA.

Norman H. Christ, Pupin Labs Box 124, Columbia University, New York NY 10027, USA.

David Cline, Physics Department, University of Wisconsin, Madison WI 53706, USA.

Sidney Coleman, Physics Department, Harvard University, Cambridge MA 02138, USA.

George B. Collins, Physics Department, Virginia Tech, Blacksburg VA 24061, USA.

329

Neil Craigie, ICTP, PO Box 568, Miramare-Strada Costiera 11, I-34100 Trieste, ITALY.

Timir Datta, Physics Department, University of South Carolina, Columbia SC 29208, USA.

Bascom Deaver, Jr., Physics Department, University of Virginia, Charlottesville VA 22903, USA.

A.K. Drukier, Nuclear Medicine Department, Klinikum Rechts der Isar, Ismaningerstrasse 22, 8000 Munchen 80, WEST GERMANY.

Philippe Eberhard, Lawrence Berkeley Laboratory, Berkeley CA 94720, USA.

John Ellis, Theory Division, SLAC, Stanford CA 94305, USA.

Steven Errede, Physics Department, University of Michigan, Ann Arbor MI 48109, USA.

William M. Fairbank, Physics Department, Stanford University, Stanford CA 94305, USA.

John R. Ficenec, Physics Department, Virginia Tech, Blacksburg VA 24061, USA.

Robert Fleischer, General Electric R&D Center, PO Box 8, Schenectady NY 12301, USA.

Henry Frisch, Physics Department, HEP 320, University of Chicago, Chicago IL 60637, USA.

Miguel Furman, Physics Department, Columbia University, New York NY 10027, USA.

Giorgio Giacomelli, Istituto di Fisica, Via Irnerio 46, Univ di Bologna, I-40126 Bologna, ITALY.

Charles Goebel, Physics Department, University of Wisconsin, Madison WI 53706, USA.

Alfred S. Goldhaber, Physics Department, SUNY, Stony Brook NY 11794, USA.

Maury Goodman, Fermilab MS 220, PO Box 500, Batavia IL 60510, USA.

Richard Gustafson, Fermilab MS 103, PO Box 500, Batavia IL 60510, USA.

A.H. Guth, Center for Theoretical Physics, MIT 6-318, 77 Mass Ave, Cambridge MA 02138, USA.

Carl Hagen, Physics Department, University of Rochester, Rochester NY 14627, USA.

Ray Hagstrom, High Energy Physics Division, Argonne National Laboratory, 9700 South Cass Avenue, Argonne IL 60439, USA.

K. Hayashi, Physics Department, KISO, Kinki University, Higashi-Osaka, JAPAN.

Chris Hill, Fermilab MS 106, PO Box 500, Batavia IL 60510, USA.

Roberto Iengo, ICTP, P O Box 586, Miramare-Strada Costiera 11, 34100 Trieste, ITALY.

Keith Jones, Editorial Office of Nuclear Physics, Blegdamsvej 17, DK-2100 Copenhagen, DENMARK.

Yi-Han Kao, Physics Department, SUNY, Stony Brook NY 11794, USA.

Kay Kinoshita, Wilson Synchrotron Laboratory, Cornell University, Ithaca NY 14850, USA.

Wiley P. Kirk, Physics Department, Texas A&M University, College Station TX 77843, USA.

Moyses Kuchnir, Fermilab MS 330, PO Box 500, Batavia IL 60510, USA.

Joseph Lach, Fermilab, PO Box 500, Batavia IL 60510, USA.

Kenneth Lande, Physics Department, University of Pennsylvania, Philadelphia PA 19174, USA.

George Lazarides, Physics Department, Rockefeller University, 1230 York Avenue, New York NY 10021, USA.

P. Lehmann, Departement de Physique des Particules Elementaires, CEN-SACLAY, 91191 Gif-Sur-Yvette Cedex, FRANCE.

Eugene Loh, Physics Department, 201 JFB, University of Utah, Salt Lake City UT 84112, USA.

Michael J. Longo, Physics Department, University of Michigan, Ann Arbor MI 48109, USA.

E. Menichetti, Istituto di Fisica, Universita di Torino, Corso M. d'Azeglio 46, I-10125 Torino, ITALY.

R. Mignani, Istituto di Fisica, Universita digli Studi, Piazzale Aldo Moro, I-00815 Rome, ITALY.

N.K. Mondal, Tata Institute of Fundamental Research, Homi Bhabha Road, Bombay 400 005, INDIA.

Don Morris, Lawrence Berkeley Laboratory, Berkeley CA 94720, USA.

Paul Musset, CERN-EP, CH-1211 Geneva 23, SWITZERLAND.

Frank Nezrick, Fermilab MS 344, PO Box 500, Batavia IL 60510, USA.

Antti Niemi, Room 6-415 CTP, M.I.T., Cambridge MA 02139, USA.

David Olive, Physics Department, University of Virginia, Charlottesville VA 22903, USA.

N.K. Pak, ICTP, PO Box 586, Miramare-Strada Costiera 11, I-34100 Trieste, ITALY.

C. Panagiotakopoulos, ICTP, PO Box 586, Miramare-Strada Costiera 11, I-34100 Trieste, ITALY.

J.G. Park, Blackett Laboratory, Imperial College, Prince Consort Road, London SW7 2BZ, ENGLAND.

Stephen Parke, SLAC, P O Box 4349, Stanford CA 94305, USA.

Jogesh Pati, Physics Department, University of Maryland, College Park MD 20742, USA.

Murray Peshkin, Argonne National Laboratory, 9700 South Cass Avenue, Argonne IL 60439, USA.

Earl A. Peterson, Physics Department, 116 Church Street SE, University of Minnesota, Minneapolis MN 55455, USA.

Andrew Pickering, Science Studies Unit, University of Edinburgh, Edinburgh EH9 3J4, SCOTLAND.

Tazio Pinelli, INFN, University of Pavia, Via Bassi 6, I-27100 Pavia, ITALY.

J.P. Preskill, Physics Department, Harvard University, Cambridge MA 02138, USA.

Buford Price, Physics Department, University of California, Berkeley CA 94720, USA.

Edward Purcell, Physics Department, Harvard University, Cambridge MA 02138, USA.

David Ritson, Physics Department, Stanford University, Stanford CA 94305, USA.

Ronald R. Ross, Physics Department, University of California, Berkeley CA 94720, USA.

Tetsuo Sawada, Physics Department, University of Tsukuba, Ibaraki-ken 305, JAPAN.

Kenneth Schatten, Planetary Aeronomy Branch, NASA, Goddard Space Flight Center, Greenbelt MD 20771, USA.

David Schramm, Physics Department, University of Chicago, Chicago IL 60637, USA.

Qaisar Shafi, ICTP, P O Box 586, Miramare-Strada Costiera 11, I-34100 Trieste, ITALY.

James Stone, Randall Physics Laboratory, University of Michigan, Ann Arbor, MI 48109, USA.

Lawrence Sulak, Randall Physics Laboratory, University of Michigan, Ann Arbor MI 48109, USA.

L.J. Tassie, Department of Theoretical Physics, Australian National University, PO Box 4. Canberra ACT 2600, AUSTRALIA.

Robert Thews, HEP Division, Department of Energy, Washington DC 20545, USA.

Dietrick Thomsen, Science News, 1719 N Street NW, Washington DC 20036, USA.

W. Peter Trower, Physics Department, Virginia Tech, Blacksburg VA 24061, USA.

Chang C. Tsuei, IBM Watson Research Center, PO Box 218, Yorktown Heights NY 10598, USA.

Michael Turner, Astronomy & Astrophysics Center, University of Chicago, Chicago IL 60637, USA.

Jack Ullman, Physics Department, Herbert Lehman College, Bronx NY 10464, USA.

Ira Wasserman, Physics Department, Cornell University, Ithaca NY 14853, USA.

Robert C. Webb, Physics Department, Texas A&M University, College Station TX 77843, USA.

Erik J. Weinberg, Physics Department, Columbia University, New York NY 10027, USA.

C.N. Yang, Institute for Theoretical Physics, SUNY, Stony Brook NY 11794, USA.

Anthony Zee, Physics Department, University of Washington, Seattle WA 98195, USA.

Daniel Zwanziger, Physics Department, New York University, 4 Washington Place, New York NY 10012, USA.

INDEX

Abundance, 81
Accelerator searches, 47, <u>309</u>, <u>325</u>
Acoustic detection, <u>219</u>, 283
Annihilation, 37, 53, 74, 82
Astrophysical constraints, 35, 133, 220
Astrophysics, <u>127</u>

Bar mode deformations, 151
Bardeen-Cooper-Schreiffer theory, 179
Baryon decay, 20, 24, 27, 36, 89, 107
Baryonsynthesis, 24
Binding, <u>245</u>, 283, 318
Black hole (monohole), 8
Bogomol'nyi bound, 92
Bosonization, 103
Bubbles, 77, 85

Cabibbo mixing, 22, 30, 31
Cabrera flux, 34, 111, 221
Catalysis (monopole), 11, 23, 31, 62, <u>97</u>, 139
  decay modes, 12, 33
Causality bound, 74
Charge quantization, 2, 28, 41, 112
Coleman—Callan mechanism, 85
Coleman-E. Weinberg, 77, 85, 129
Collider experiments, 310, <u>325</u>
Color-magnetic, 13, 98, 113, 160
Cooper pairs, 179

Correlation length, 81
Cosmic-ray searches, 48, <u>291</u>, 309
Cosmological constraints, 35, 74, <u>81</u>
Cosmology, <u>71</u>, <u>81</u>, <u>127</u>
CP violation, 4, 100, 117
Cross section (GUM-Baryon), 31

Debye length (magnetic), 153
Density fluxuations in universe, 78, 85, 93
Density in universe, 76, 83
Desert hypothesis, 21, 160
Detection, 37, 219, 279
Dirac monopoles, 28, 41, 112
Dirac quantization, 2, 22, 28, 112
Dumand detector, 241
Dynamo (galactic), 155
Dyons, 3, 30, 102, 117

Earth, 135, 249, 278
Eddy current losses, 135
Electronic searches, 58, <u>291</u>
Energy loss, 44, 222, <u>259</u>, 292, 312
  stars, 270
Escape velocities, 54, 145
Extraordinary electromagnetism, 118

Fairbank, 111
Faraday induction, 2, 202
Faraday rotation, 141, 155